Graduate Texts in Mathematics **75**

Gerhard P. Hochschild

Basic Theory of Algebraic Groups and Lie Algebras

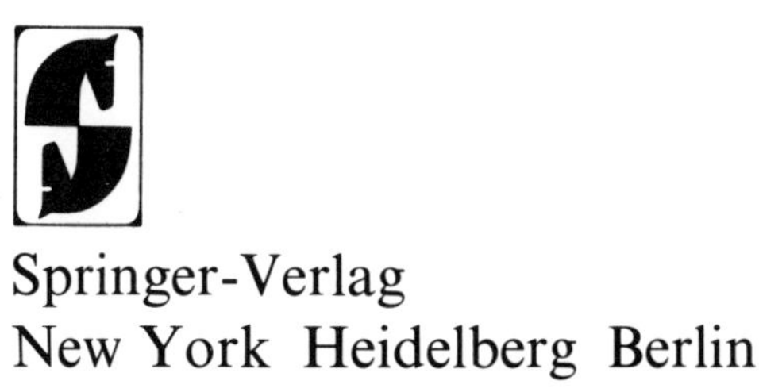

Springer-Verlag
New York Heidelberg Berlin

Gerhard P. Hochschild
Department of Mathematics
University of California
Berkeley, CA 94720
USA

AMS Subject Classification (1981): 14–01, 20–01, 20GXX

Library of Congress Cataloging in Publication Data

Hochschild, Gerhard Paul, 1915–
  Basic theory of algebraic groups and Lie
algebras.
  (Graduate texts in mathematics; 75)
  Bibliography: p.
  Includes index.
  1. Lie algebras. 2. Linear algebraic groups.
I. Title. II. Series.
QA252.3.H62   512′.55   80–27983

ISBN 0–387–90541–3 Springer-Verlag New York
ISBN 3–540–90541–3 Springer-Verlag Berlin Heidelberg

Preface

The theory of algebraic groups results from the interaction of various basic techniques from field theory, multilinear algebra, commutative ring theory, algebraic geometry and general algebraic representation theory of groups and Lie algebras. It is thus an ideally suitable framework for exhibiting basic algebra in action. To do that is the principal concern of this text. Accordingly, its emphasis is on developing the major general mathematical tools used for gaining control over algebraic groups, rather than on securing the final definitive results, such as the classification of the simple groups and their irreducible representations. In the same spirit, this exposition has been made entirely self-contained; no detailed knowledge beyond the usual standard material of the first one or two years of graduate study in algebra is presupposed.

The chapter headings should be sufficient indication of the content and organisation of this book. Each chapter begins with a brief announcement of its results and ends with a few notes ranging from supplementary results, amplifications of proofs, examples and counter-examples through exercises to references. The references are intended to be merely suggestions for supplementary reading or indications of original sources, especially in cases where these might not be the expected ones.

Algebraic group theory has reached a state of maturity and perfection where it may no longer be necessary to re-iterate an account of its genesis. Of the material to be presented here, including much of the basic support, the major portion is due to Claude Chevalley. Although Chevalley's decisive classification results, contained in [6], have not been included here, a glimpse of their main ingredients can be had from Chapters XVII and XIII. The subject of Chapter XIII is Armand Borel's fundamental theory of maximal solvable subgroups and maximal toroids, which has made it possible tc recreate the combinatorial features of the Cartan–Weyl theory of semisimple Lie algebras, dealt with in Chapter XVII, in terms of subgroups of semisimple algebraic groups. In particular, this has freed the theory from the classical restriction to base fields of characteristic 0.

I was encouraged to write this exposition chiefly by the appearance of James Humphreys's *Linear Algebraic Groups*, where the required algebraic geometry has been cut down to a manageable size. In fact, the algebraic-geometric developments given here have resulted from Humphreys's treatment simply by adding proofs of the underlying facts from commutative algebra. Moreover, much of the general structure theory in arbitrary charac-teristic has been adapted from Borel's lecture notes [1] and Humphreys's book.

I have made use of valuable advice from my friends, given in the course of several years on various occasions and in various forms, including print. It is a pleasure to express my thanks for their help to Walter Ferrer-Santos, Oscar Goldman, Bertram Kostant, Andy Magid, Calvin Moore, Brian Peterson, Alex Rosenberg, Maxwell Rosenlicht, John B. Sullivan, Moss Sweedler and David Wigner. However, it must be emphasized that no one but me has had an opportunity to remedy any of the defects of my actual manuscript.

Gerhard P. Hochschild

Contents

Chapter I

Representative Functions
and Hopf Algebras

This chapter introduces the basic algebraic machinery arising in the study of group representations. The principal notion of a Hopf algebra is developed here as an abstraction from the systems of functions associated with the representations of a group by automorphisms of finite-dimensional vector spaces. This leads to an initializing discussion of our main objects of study, affine algebraic groups.

1. Given a non-empty set S and a field F, we denote the F-algebra of all F-valued functions on S by F^S. In the statement of the following lemma, and frequently in the sequel, we use the symbol δ_{ij}, which stands for 1 if $i = j$, and for 0 if $i \neq j$.

Lemma 1.1. *Let V be a non-zero finite-dimensional sub F-space of F^S. There is a basis $(v_1, \ldots, v_n)$ of V and a corresponding subset $(s_1, \ldots, s_n)$ of S such that $v_i(s_j) = \delta_{ij}$ for all indices i and j.*

PROOF. Suppose that we have already found elements $s_1, \ldots, s_k$ of S and a basis $(v_{1,k}, \ldots, v_{n,k})$ of V such that the $v_{i,k}$'s and the s_j's satisfy the requirements of the lemma for each i from $(1, \ldots, n)$ and each j from $(1, \ldots, k)$. If $k < n$, there is an element s_{k+1} in S such that $v_{k+1,k}(s_{k+1}) \neq 0$. We set

$$v_{k+1,k+1} = v_{k+1,k}(s_{k+1})^{-1} v_{k+1,k}.$$

For the indices i other than $k + 1$, we set

$$v_{i,k+1} = v_{i,k} - v_{i,k}(s_{k+1})v_{k+1,k+1}.$$

Now the sets $(s_1, \ldots, s_{k+1})$ and $(v_{1,k+1}, \ldots, v_{n,k+1})$ satisfy our requirements at level $k + 1$. The lemma is obtained by induction, starting with an arbitrary basis of V at level $k = 0$. $\qquad\square$

For non-empty sets S and T, we examine the canonical morphism of F-algebras, π, from the tensor product $F^S \otimes F^T$ to $F^{S \times T}$, where

$$\pi(\textstyle\sum f \otimes g)(s, t) = \sum f(s)g(t).$$

Proposition 1.2. *The canonical morphism* $\pi: F^S \otimes F^T \to F^{S \times T}$ *is injective, and its image consists of all functions h with the property that the F-space spanned by the partial functions h_t, where t ranges over T and $h_t(s) = h(s, t)$, is finite-dimensional.*

PROOF. Let $\sum_{j=1}^m f_j \otimes g_j$ be an element of the kernel of π, and let V be the sub F-space of F^S spanned by $f_1, \ldots, f_m$. If $V = (0)$ then our element is 0. Otherwise choose $(v_1, \ldots, v_n)$ and $(s_1, \ldots, s_n)$ as in Lemma 1.1, and write our element in the form $\sum_{i=1}^n v_i \otimes h_i$. Applying π and evaluating at (s_k, t) yields $h_k(t) = 0$. This shows that each h_i is 0, and we conclude that π is injective.

It is clear that if h is an element of the image of π then it has the property stated in the proposition. Conversely, suppose that h is an element of $F^{S \times T}$ having this property. This means that there are elements $f_1, \ldots, f_n$ in F^S such that each h_t is an F-linear combination of the f_i's. Choosing coefficients from F for each t in T, we obtain elements $g_1, \ldots, g_n$ of F^T such that, for each t,

$$h_t = \sum_{i=1}^n g_i(t) f_i.$$

This means that

$$h = \pi\left(\sum_{i=1}^n f_i \otimes g_i \right). \qquad \square$$

Let us consider the above in the case where both S and T coincide with the underlying set of a monoid G, with composition $m: G \times G \to G$. This composition transposes in the natural fashion to a morphism of F-algebras $m^*: F^G \to F^{G \times G}$, where $m^*(f)$ is the composite $f \circ m$. We abbreviate $m(x, y)$ by xy, so that $m^*(f)(x, y) = f(xy)$. By transposing the right and left translation actions of G on itself, we obtain a two-sided G-module structure on F^G, which we indicate as follows

$$(x \cdot f)(y) = f(yx), \qquad (f \cdot x)(y) = f(xy).$$

Now we see from Proposition 1.2 that $m^*(f)$ belongs to the image of $\pi: F^G \otimes F^G \to F^{G \times G}$ if and only if the F-space spanned by the functions $x \cdot f$, with x ranging over G, is finite-dimensional. If this is so, we say that f is a *representative function*. We denote the F-algebra of all F-valued representative functions on G by $\mathcal{R}_F(G)$, but we shall permit ourselves to suppress the subscript F when there is no danger of confusion. Clearly, $\mathcal{R}_F(G)$ is a two-sided sub G-module of F^G, as well as a sub F-algebra.

Proposition 1.3. *The image of the morphism of F-algebras*

$$\pi^{-1} \circ m^* : \mathscr{R}_F(G) \to F^G \otimes F^G$$

actually lies in $\mathscr{R}_F(G) \otimes \mathscr{R}_F(G)$.

PROOF. Let f be an element of $\mathscr{R}_F(G)$. Proceeding as in the proof of Proposition 1.2, we find elements $s_1, \ldots, s_n$ in G and elements $v_1, \ldots, v_n$ in F^G as in Lemma 1.1 such that we may write

$$\pi^{-1}(m^*(f)) = \sum_{i=1}^{n} v_i \otimes h_i .$$

Evaluating this at (s_j, t), we find $h_j(t) = f(s_j t)$, whence $h_j = f \cdot s_j$. This shows that h_j belongs to $\mathscr{R}_F(G)$. The conclusion is that the image of $\pi^{-1} \circ m^*$ lies in $F^G \otimes \mathscr{R}_F(G)$. Changing sides throughout, we find that this image also lies in $\mathscr{R}_F(G) \otimes F^G$. Clearly, the last two conclusions imply the assertion of Proposition 1.3. $\qquad\square$

The morphism of F-algebras $\mathscr{R}(G) \to \mathscr{R}(G) \otimes \mathscr{R}(G)$ defined by Proposition 1.3 is called the *comultiplication* of $\mathscr{R}(G)$, and we shall denote it by δ. For an element s of G, let $s^* : \mathscr{R}(G) \to F$ denote the evaluation at s, so that $s^*(f) = f(s)$. Then δ is characterized by the formula

$$(x^* \otimes y^*)(\delta(f)) = f(xy).$$

More explicity, if

$$\delta(f) = \sum_i f_i' \otimes f_i'' ,$$

then

$$f(xy) = \sum_i f_i'(x) f_i''(y).$$

We adopt some general terminology, as follows. The structure of an F-algebra A is understood to consist of

(1)   the structure of A as an F-space;
(2)   the *multiplication* of A, viewed as an F-linear map $\mu : A \otimes A \to A$;
(3)   the *unit* of A, viewed as an F-linear map $u : F \to A$ sending each element α of F onto the α-multiple $\alpha 1_A$ of the identity element of A.

In writing the axioms, it is convenient to name the canonical identification maps coming from the F-space structure of A. These are

$$p_1 : F \otimes A \to A \quad \text{and} \quad p_2 : A \otimes F \to A.$$

Generally, we use i_S to denote the identity map on a set S. The axioms of an F-algebra structure may now be written as follows

$$\mu \circ (u \otimes i_A) = p_1, \qquad \mu \circ (i_A \otimes u) = p_2,$$

$$\mu \circ (\mu \otimes i_A) = \mu \circ (i_A \otimes \mu).$$

Purely formal dualization of this system yields the notion of an *F-coalgebra*. Thus, the structure of an F-coalgebra C consists of the following:

(1)   the structure of C as an F-space;
(2)   the *comultiplication* of C, an F-linear map $\delta: C \to C \otimes C$;
(3)   the *counit* of C, an F-linear map $\varepsilon: C \to F$.

Let $q_1: C \to F \otimes C$ and $q_2: C \to C \otimes F$ denote the canonical identifications coming from the F-space structure of C. Then the axioms of an F-coalgebra structure are the following

$$(\varepsilon \otimes i_C) \circ \delta = q_1, \qquad (i_C \otimes \varepsilon) \circ \delta = q_2,$$

$$(\delta \otimes i_C) \circ \delta = (i_C \otimes \delta) \circ \delta.$$

It is worth noting that, in the case where the given F-space C is F, the maps q_1 and q_2 coincide with a comultiplication δ which, together with the identity map $\varepsilon = i_F$, makes F into an F-coalgebra. When F is regarded as a coalgebra in this way, then the counit ε of any coalgebra C is a morphism of coalgebras. This is the formal dual of the familiar fact that the unit u of an F-algebra A is a morphism of F-algebras when F is viewed as an F-algebra in the canonical fashion.

Now let us return to the F-algebra $\mathcal{R}_F(G)$ and its comultiplication δ. If we define $\varepsilon: \mathcal{R}_F(G) \to F$ as the evaluation at the neutral element 1_G of the monoid G then the F-space structure of $\mathcal{R}_F(G)$, together with δ and ε, makes $\mathcal{R}_F(G)$ into an F-coalgebra. A notable feature here is that δ and ε are morphisms of F-algebras.

In general terms, let us recall that, if (A, μ, u) and (A', μ', u') are F-algebras, then a morphism of F-algebras $h: A \to A'$ is an F-linear map satisfying

$$h \circ \mu = \mu' \circ (h \otimes h) \quad \text{and} \quad h \circ u = u'.$$

Dually, if (C, δ, ε) and $(C', \delta', \varepsilon')$ are F-coalgebras, then a morphism of F-coalgebras $h: C \to C'$ is an F-linear map satisfying

$$\delta' \circ h = (h \otimes h) \circ \delta \quad \text{and} \quad \varepsilon' \circ h = \varepsilon.$$

Now suppose that B is an F-space carrying both, an algebra structure (μ, u), and a coalgebra structure (δ, ε). Suppose in addition that δ and ε are morphisms of F-algebras. Then $(B, \mu, u, \delta, \varepsilon)$ is called an *F-bialgebra*. Our above discussion of $\mathcal{R}_F(G)$ amounts to the definition of a bialgebra structure.

The usual definition of the tensor product of two F-algebras (A, μ, u) and (A', μ', u') yields $(A \otimes A', \mu \boxtimes \mu', u \otimes u')$, where

$$\mu \boxtimes \mu' = (\mu \otimes \mu') \circ (i_A \otimes s \otimes i_{A'}).$$

Here, s stands for the canonical switch of tensor factors $A' \otimes A \to A \otimes A'$, and in writing $u \otimes u'$ we have identified $F \otimes F$ with F.

Similarly, if (C, δ, ε) and $(C', \delta', \varepsilon')$ are F-coalgebras, one defines the tensor product F-coalgebra $(C \otimes C', \delta \boxtimes \delta', \varepsilon \otimes \varepsilon')$ by making

$$\delta \boxtimes \delta' = (i_C \otimes s \otimes i_{C'}) \circ (\delta \otimes \delta'),$$

where now s is the switch $C \otimes C' \to C' \otimes C$.

We remark that, in the notation used in defining a bialgebra, *the condition that δ and ε be morphisms of F-algebras is equivalent to the condition that μ and u be morphisms of F-coalgebras.*

Returning to $\mathscr{R}(G)$, let us now suppose that G is a group. Then the inversion of G transposes into a map $\eta: \mathscr{R}(G) \to \mathscr{R}(G)$, defined by $\eta(f)(x) = f(x^{-1})$. This is called the *antipode* of $\mathscr{R}(G)$.

In order to give the proper setting for the antipode, we need to develop more of the machinery of algebras and coalgebras. Let (C, δ, ε) be an F-coalgebra and let (A, μ, u) be an F-algebra. Then we obtain an F-algebra structure on the F-space $\mathrm{Hom}_F(C, A)$ of all F-linear maps from C to A as follows. The product of two elements h and k of $\mathrm{Hom}_F(C, A)$ is defined as the composite $\mu \circ (h \otimes k) \circ \delta$. It is verified directly that this does in fact define an associative multiplication, for which $u \circ \varepsilon$ is the neutral element.

Now let $(H, \mu, u, \delta, \varepsilon)$ be a bialgebra. In the definition just made, put $C = H$ and $A = H$, so that we obtain an F-algebra structure on

$$\mathrm{End}_F(H) = \mathrm{Hom}_F(H, H).$$

The multiplication of this structure is called the *convolution*. One calls H a *Hopf algebra* if i_H has an inverse with respect to the convolution. The inverse of i_H is called the *antipode* of the Hopf algebra H. Denoting it by η, the defining property is

$$\mu \circ (\eta \otimes i_H) \circ \delta = u \circ \varepsilon = \mu \circ (i_H \otimes \eta) \circ \delta.$$

An immediate verification shows that the map η defined above for $\mathscr{R}(G)$ is indeed an *antipode* in the general sense. It can be shown that, for every Hopf algebra, the antipode is an antimorphism of algebras, as well as an antimorphism of coalgebras, where "anti" signifies the intervention of the usual switching of tensor factors. Also, if one of δ or μ is commutative, then $\eta \circ \eta = i_H$. In our case, $H = \mathscr{R}(G)$, these properties of η are evident. For the general situation, see the notes at the end of this chapter.

2. Let (C, δ, ε) be an F-coalgebra, and write C° for $\mathrm{Hom}_F(C, F)$. Viewing F as an F-algebra, we have the structure of an F-algebra on C°. Let δ° denote the multiplication and ε° the unit of this algebra. Note that the neutral element for δ° is simply ε, so that $\varepsilon^\circ(a) = a\varepsilon$ for every element a of F. If we identify $F \otimes F$ with F then δ° may be written simply as composition with δ, i.e., $\delta^\circ(z) = z \circ \delta$ for every element z of $C^\circ \otimes C^\circ$.

With every element τ of C°, we associate the F-linear endomorphisms $\tau_{[}$ and $\tau_{]}$ of C, as defined by the following formulas, where we identify $C \otimes F$ and $F \otimes C$ with C.

$$\tau_{[} = (i_C \otimes \tau) \circ \delta,$$

$$\tau_{]} = (\tau \otimes i_C) \circ \delta.$$

Proposition 2.1. *The map $\tau \mapsto \tau_{[}$ is an injective morphism of F-algebras from C° to $\mathrm{End}\,_F(C)$, and the map $\tau \mapsto \tau_{]}$ is an injective antimorphism of F-algebras from C° to $\mathrm{End}_F(C)$. Every $\tau_{[}$ commutes with every $\sigma_{]}$, so that these maps define the structure of a two-sided C°-module on C. This module is locally finite, in the sense that C coincides with the sum of the family of its finite-dimensional two-sidedly C°-stable subspaces.*

PROOF. Throughout the computations below, we identify $F \otimes F$ with F, $C \otimes F$ with C, $F \otimes C$ with C, etc. Let σ and τ be elements of C°. We have

$$\delta \circ \sigma_{[} = \delta \circ (i_C \otimes \sigma) \circ \delta = (\delta \otimes \sigma) \circ \delta$$

$$= (i_C \otimes i_C \otimes \sigma) \circ (\delta \otimes i_C) \circ \delta.$$

Now we use the associativity of δ, replacing $(\delta \otimes i_C) \circ \delta$ with $(i_C \otimes \delta) \circ \delta$. This gives

$$\delta \circ \sigma_{[} = (i_C \otimes (i_C \otimes \sigma) \circ \delta) \circ \delta = (i_C \otimes \sigma_{[}) \circ \delta.$$

On composition with $i_C \otimes \tau$, this yields

$$\tau_{[} \circ \sigma_{[} = (i_C \otimes \tau \circ \sigma_{[}) \circ \delta.$$

It is clear from the definitions that $\tau \circ \sigma_{[}$ coincides with the product $\tau\sigma$ in C°. Therefore, the last equality above means that $\tau_{[} \circ \sigma_{[} = (\tau\sigma)_{[}$. Since the identity element of C° is ε and since $\varepsilon_{[} = i_C$, we may now conclude that the map $\tau \mapsto \tau_{[}$ is the structure of a C°-module on C. This map is injective, because $\varepsilon \circ \tau_{[} = \tau$.

In the exactly analogous way, one verifies that the map $\tau \mapsto \tau_{]}$ is injective and a right C°-module structure on C. Note that $\sigma \circ \tau_{]}$ coincides with the product $\tau\sigma$ in C°. In particular, $\varepsilon \circ \tau_{]} = \tau$.

Next, it follows directly from the definitions that

$$\sigma_{]} \circ \tau_{[} = (\sigma \otimes i_C \otimes \tau) \circ (\delta \otimes i_C) \circ \delta,$$

while

$$\tau_{[} \circ \sigma_{]} = (\sigma \otimes i_C \otimes \tau) \circ (i_C \otimes \delta) \circ \delta.$$

By the associativity of δ, this shows that $\sigma_{]}$ and $\tau_{[}$ commute with each other, so that we have indeed the structure of a two-sided C°-module on C.

It remains to be proved that C is locally finite. Let c be any element of C, and write $\delta(c) = \sum_i c_i' \otimes c_i''$. Then it is clear from the definitions that the left C°-orbit $C_{[}^\circ(c)$ lies in the F-space spanned by the c_i''s, while the right C°-

orbit $C_1^\circ(c)$ lies in the F-space spanned by the c_i'''s. Thus, each of these two orbits is finite-dimensional. Using the second fact for the elements c_i', we see that the two-sided C°-orbit $C_1^\circ(C_1^\circ(c))$ is finite-dimensional. $\square$

Proposition 2.2. *Each of C_1° and C_1° is the commuting algebra of the other, in $\operatorname{End}_F(C)$. An element e of $\operatorname{End}_F(C)$ belongs to C_1° if and only if $\delta \circ e = (i_C \otimes e) \circ \delta$. If this holds, then $e = (\varepsilon \circ e)_1$.*

PROOF. We have already shown that the elements of C_1° and C_1° commute with each other. Now suppose that e is any element of $\operatorname{End}_F(C)$ that commutes with every element of C_1°. This means that, for every element τ of C°, we have

$$e \circ (\tau \otimes i_C) \circ \delta = (\tau \otimes i_C) \circ \delta \circ e.$$

We may write this in the form

$$(\tau \otimes i_C) \circ ((i_C \otimes e) \circ \delta - \delta \circ e) = 0.$$

Since this holds for every τ in C°, it follows that $\delta \circ e = (i_C \otimes e) \circ \delta$. Composing this with $i_C \otimes \varepsilon$, we obtain $e = (\varepsilon \circ e)_1$. The part of Proposition 2.2 not yet proved, namely that C_1° is the commuting algebra of C_1° in $\operatorname{End}_F(C)$, is proved by changing sides throughout the above. $\square$

Recall that, in the case where C is the coalgebra $\mathscr{R}(G)$ of a monoid G, the evaluation at an element x of G was denoted by x^*, so that x^* is an element of C°. Now x_1^* is the automorphism $f \mapsto x \cdot f$, corresponding to the action of x on G from the right, while x_1^* is the automorphism $f \mapsto f \cdot x$ corresponding to the action of x on G from the left.

Now let us consider a locally finite G-module over a field F. This is an F-space V, together with a morphism of monoids ρ from G to $\operatorname{End}_F(V)$, such that every element of V is contained in a finite-dimensional G-stable sub F-space of V. Let V° denote the dual space $\operatorname{Hom}_F(V, F)$ of V. For each element γ of V° and each element v of V, we define an F-valued function γ/v on G by putting

$$(\gamma/v)(x) = \gamma(x \cdot v).$$

where $x \cdot v$ is the customary abbreviation for $\rho(x)(v)$. Referring to the G-module structure of F^G, we have, as a direct consequence of the definitions, $x \cdot (\gamma/v) = \gamma/(x \cdot v)$. Since V is locally finite as a G-module, this shows that γ/v belongs to $\mathscr{R}(G)$. Hence, we have an F-linear map

$$\rho' : V \to \operatorname{Hom}_F(V^\circ, \mathscr{R}(G)),$$

where $\rho'(v)(\gamma) = \gamma/v$. On the other hand, consider the canonical F-linear map

$$\tau : V \otimes \mathscr{R}(G) \to \operatorname{Hom}_F(V^\circ, \mathscr{R}(G)).$$

Clearly, τ is injective. We claim that the image of ρ' lies in the image of τ. In order to see this, choose an F-basis $(v_1, \ldots, v_n)$ for the space spanned by the transforms $x \cdot v$ of the fixed given element v of V by the elements x of G. Next, choose elements $\gamma_1, \ldots, \gamma_n$ of V° such that $\gamma_i(v_j) = \delta_{ij}$. One verifies directly that

$$\tau\left(\sum_{i=1}^n v_i \otimes (\gamma_i/v)\right)(\gamma) = \sum_{i=1}^n \gamma(v_i)\gamma_i/v = \gamma/v,$$

which shows that

$$\rho'(v) = \tau\left(\sum_{i=1}^n v_i \otimes (\gamma_i/v)\right).$$

Since τ is injective, it follows that there is one and only one linear map

$$\rho^* : V \to V \otimes \mathscr{R}(G)$$

such that $\rho' = \tau \circ \rho^*$. Viewing the elements of $V \otimes \mathscr{R}(G)$ as maps from G to V in the evident way, we may write

$$\rho^*(v)(x) = x \cdot v,$$

which shows that

$$\rho(x) = (i_V \otimes x^*) \circ \rho^*.$$

The fact that the neutral element of G acts as the identity map on V is expressed by the formula

$$(i_V \otimes \varepsilon) \circ \rho^* = i_V,$$

while the fact that $\rho(xy) = \rho(x)\rho(y)$ for all elements x and y of G is expressed by the formula

$$(i_V \otimes \delta) \circ \rho^* = (\rho^* \otimes i_{\mathscr{R}(G)}) \circ \rho^*.$$

Generally, if (C, δ, ε) is any F-coalgebra, then a *C-comodule* is an F-space V, together with an F-linear map

$$\sigma : V \to V \otimes C$$

satisfying

$$(i_V \otimes \varepsilon) \circ \sigma = i_V \quad \text{and} \quad (i_V \otimes \delta) \circ \sigma = (\sigma \otimes i_C) \circ \sigma.$$

The above connections between ρ and ρ^* show that *the category of locally finite G-modules over the field F is naturally equivalent to the category of $\mathscr{R}_F(G)$-comodules.*

In the general situation, we may let C° take the place of G in the above discussion to show that *the category of C-comodules is naturally equivalent to the category of those locally finite C°-modules which are of type C*, in the sense that the associated representative functions belong to the canonical image of C in $C^{\circ\circ}$. Note that δ is a C-comodule structure on C, and that the corresponding C°-module structure is given by the map $\gamma \mapsto \gamma_{\Gamma}$. *If V is any locally finite C°-module of type C, then the corresponding C-comodule structure $\sigma: V \to V \otimes C$ is a morphism of C°-modules when C° acts on $V \otimes C$ via the factor C alone.*

Let $(B, \mu, u, \delta, \varepsilon)$ be an F-bialgebra. The multiplication μ of B comes into play with the construction of the *tensor product of comodules*, which is as follows. Suppose that

$$\sigma: S \to S \otimes B \quad \text{and} \quad \tau: T \to T \otimes B$$

are B-comodules. Then we define the map

$$\sigma \boxtimes \tau: S \otimes T \to S \otimes T \otimes B$$

as the composite of the ordinary tensor product of $\sigma \otimes \tau$ from $S \otimes T$ to $S \otimes B \otimes T \otimes B$ with the switch $s_{2,3}$ of the middle two tensor factors and the map $i_S \otimes i_T \otimes \mu$ from $S \otimes T \otimes B \otimes B$ to $S \otimes T \otimes B$. One verifies directly that this is indeed the structure of a B-comodule on $S \otimes T$.

In the case where $B = \mathscr{R}(G)$, if σ and τ are the comodule structures α^* and β^* corresponding to G-module structures α and β, we have $\alpha^* \boxtimes \beta^* = \gamma^*$, where $\gamma(x) = \alpha(x) \otimes \beta(x)$ for every element x of G.

Finally, let $(H, \mu, u, \delta, \varepsilon, \eta)$ be a Hopf algebra, where η is the antipode. The representation-theoretical significance of η is that it yields the construction of *dual comodules*. This is as follows. Let

$$\sigma: S \to S \otimes H$$

be an H-comodule, and define the linear map

$$\sigma': S^\circ \to \operatorname{Hom}_F(S, H)$$

by

$$\sigma'(\gamma) = (\gamma \otimes \eta) \circ \sigma.$$

Now *assume that S is finite-dimensional.* Then the canonical map

$$\alpha: S^\circ \otimes H \to \operatorname{Hom}_F(S, H)$$

is an isomorphism, so that we can form $\sigma^\circ = \alpha^{-1} \circ \sigma'$, One verifies directly that

$$\sigma^\circ: S^\circ \to S^\circ \otimes H$$

is the structure of an H-comodule on S°, called the *dual* of σ.

In the case where $H = \mathscr{R}(G)$, with G a group, if $\sigma = \rho^*$, we have $\sigma^\circ = \gamma^*$, where γ is the familiar dual of ρ, given by $\gamma(x)(f) = f \circ \rho(x^{-1})$ for every f in S°.

In the general case, with S finite-dimensional, the tensor product $\sigma^\circ \boxtimes \tau$ gives a comodule structure on $\operatorname{Hom}_F(S, T)$, because this F-space may be identified with $S^\circ \otimes T$.

3. We introduce some terminology from algebraic geometry, as follows. An *affine algebraic F-set* is a non-empty set S, together with a *finitely generated* sub F-algebra $\mathscr{P}(S)$ of F^S such that the following requirements are satisfied

(1) $\mathscr{P}(S)$ separates the points of S; i.e., for every pair (s_1, s_2) of distinct points of S, there is an f in $\mathscr{P}(S)$ with $f(s_1) \neq f(s_2)$;
(2)   every F-algebra homomorphism from $\mathscr{P}(S)$ to F is the evaluation s^* at an element s of S.

Frequently, we shall abbreviate "affine algebraic F-set" to "algebraic set". Such a set S is viewed as a topological space, the topology being the *Zariski topology*, which is defined by declaring that the closed sets be the annihilators in S of subsets of $\mathscr{P}(S)$. If T is a non-empty closed subset of S then the annihilator, J_T say, of T in $\mathscr{P}(S)$ is a proper ideal, and T inherits the structure of an algebraic set, with $\mathscr{P}(T)$ the F-algebra of restrictions to T of the elements of $\mathscr{P}(S)$, so that $\mathscr{P}(T)$ is isomorphic with $\mathscr{P}(S)/J_T$.

If A and B are affine algebraic F-sets then a *morphism of affine algebraic F-sets* from A to B is a set map α from A to B such that $\mathscr{P}(B) \circ \alpha \subset \mathscr{P}(A)$. Note that this implies that α is continuous.

The elements of $\mathscr{P}(S)$ are called the *polynomial functions* on S. A morphism of affine algebraic sets is also called a *polynomial map*.

Let S and T be algebraic sets. We known from Proposition 1.2 that the canonical map from $\mathscr{P}(S) \otimes \mathscr{P}(T)$ to $F^{S \times T}$ is injective. By considering elements of the form $f \otimes 1 + 1 \otimes g$, we see that the image of $\mathscr{P}(S) \otimes \mathscr{P}(T)$ in $F^{S \times T}$ separates the points of $S \times T$. Now it is clear that $S \times T$ is made into an algebraic set if $\mathscr{P}(S \times T)$ is defined as the image of $\mathscr{P}(S) \otimes \mathscr{P}(T)$ in $F^{S \times T}$. Evidently, the projection maps from $S \times T$ to S and T are morphisms of algebraic sets. Moreover, if $\sigma : A \to S$ and $\tau : A \to T$ are morphisms of algebraic sets, then their direct product $\sigma \times \tau$ as set maps is clearly a morphism of algebraic sets from A to $S \times T$. This shows that our definition of $\mathscr{P}(S \times T)$ satisfies the categorical requirements of a direct product in the category of affine algebraic F-sets.

An *affine algebraic F-group* is a group G, equipped with the structure $\mathscr{P}(G)$ of an affine algebraic F-set, such that the composition map $G \times G \to G$ and the inversion map $G \to G$ are morphisms of affine algebraic F-sets. Note that the last two requirements are equivalent to the requirement that $\mathscr{P}(G)$ *be a sub Hopf algebra of* $\mathscr{R}_F(G)$.

Now let $(A, \mu, u, \delta, \varepsilon, \eta)$ be a Hopf algebra over F. The F-algebra homomorphisms from A to F constitute a group, with the composition

$$ab = (a \otimes b) \circ \delta$$

whose neutral element is ε, and where the inverse of an element a is $a \circ \eta$. We denote this group by $\mathscr{G}(A)$. We know from Proposition 2.1 that A is locally finite as an A°-module. In particular, this implies that the evident map from A to $F^{\mathcal{G}(A)}$ sends A into $\mathcal{R}_F(\mathcal{G}(A))$. Moreover, it is seen directly from the definitions that this map is a morphism of Hopf algebras. Let $G = \mathcal{G}(A)$, and let $\mathcal{P}(G)$ be the image of A in $\mathcal{R}_F(G)$. If $\mathcal{P}(G)$ is finitely generated as an F-algebra, it makes G into an affine algebraic F-group. Conversely, if we are given an affine algebraic F-group G and make the above construction with $A = \mathcal{P}(G)$, then we recover the given affine algebraic F-group. Accordingly, we shall permit ourselves to identify the elements x of G with the corresponding F-algebra homomorphisms x^* from $\mathcal{P}(G)$ to F.

As with algebraic sets, we shall frequently abbreviate "affine algebraic F-group" to "algebraic group."

There are three basic examples of algebraic groups, which we present in detail. The first is the additive group of F. For this, $\mathcal{P}(F)$ consists of all those F-valued functions on F which can be written as polynomials in the identity map $x : F \to F$. The comultiplication δ, the counit ε and the antipode η are the unique morphisms of F-algebras satisfying $\delta(x) = x \otimes 1 + 1 \otimes x$ and $\varepsilon(x) = 0$ and $\eta(x) = -x$.

The second basic example is the multiplicative group of F, which we denote by F^*. For this group, $\mathcal{P}(F^*)$ consists of all those F-valued functions on F^* which can be written as polynomial in u and its reciprocal u^{-1}, where u denotes the identity map on F^*. The formulas characterizing δ, ε and η as morphisms of F-algebras are

$$\delta(u) = u \otimes u \quad \text{and} \quad \varepsilon(u) = 1 \quad \text{and} \quad \eta(u) = u^{-1}.$$

The third basic example is of a more general nature. In fact, it contains the last example as a very special case. Let E be any finite-dimensional F-algebra, and let E^* denote the group of units of E, whose composition comes from the multiplication of E. We define $\mathcal{P}(E^*)$ as the smallest sub Hopf algebra of $\mathcal{R}_F(E^*)$ containing the restrictions to E^* of the elements of E°. It is easy to check that this makes E^* into an affine algebraic F-group. We proceed to obtain an explicit description of $\mathcal{P}(E^*)$.

Let ρ be an injective representation of E by linear endomorphisms of a finite-dimensional F-space V, and let σ denote the restriction of ρ to E^*. Let $S(\sigma)$ be the F-space of *representative functions associated with σ*, i.e., the functions $\gamma \circ \sigma$ with γ in $\mathrm{End}_F(V)^\circ$. Let d_σ be the function on E^* that maps every element e onto the determinant $d_\sigma(e)$ of $\sigma(e)$. Clearly, $S(\sigma)$ is contained in the image of E° in $\mathcal{P}(E^*)$. Since d_σ is a polynomial in elements of $S(\sigma)$, it follows that d_σ belongs to $\mathcal{P}(E^*)$. The reciprocal d_σ^{-1} of d_σ coincides with $\eta(d_\sigma)$, so that it also belongs to $\mathcal{P}(E^*)$. The explicit formula for the inverse of a matrix shows that, if f is an element of $S(\sigma)$, then $\eta(f)$ is of the form $d_\sigma^{-1}g$, where g is a polynomial in elements of $S(\sigma)$. Therefore, if A is the subalgebra of $\mathcal{P}(E^*)$ that is generated by d_σ^{-1} and the elements of $S(\sigma)$, then A is stable under the antipode η, as well as under the right and left translation actions of E^*, which evidently stabilize $S(\sigma)$. Therefore, A coincides with $\mathcal{P}(E^*)$.

Finally, let $(v_1, \ldots, v_n)$ be an F-basis of V. For e in E, write

$$\rho(e)(v_j) = \sum_{i=1}^{n} f_{ij}(e)v_i.$$

This defines elements f_{ij} of E°. Let g_{ij} be the restriction of f_{ij} to E^*. Then $\mathscr{P}(E^*)$ is generated as an F-algebra by the g_{ij}'s and d_σ^{-1}. We have

$$\varepsilon(g_{ij}) = \delta_{ij} \quad \text{and} \quad \delta(g_{ij}) = \sum_{k=1}^{n} g_{ik} \otimes g_{kj}.$$

4. Let (C, δ, ε) be a coalgebra. A subspace J of C is called a *coideal* if

$$\varepsilon(J) = (0) \quad \text{and} \quad \delta(J) \subset C \otimes J + J \otimes C.$$

These are precisely the conditions needed to ensure that C/J inherit a co-algebra structure. By a *biideal* of a bialgebra one means a coideal that is also an ideal. Note that, since a biideal is annihilated by the counit, it is always a proper ideal. Finally, a *Hopf ideal* of a Hopf algebra is a biideal that is stable under the antipode, so that the factor space with respect to a Hopf ideal inherits the structure of a Hopf algebra.

Now let G be an affine algebraic group, and let K be a subgroup of G. Let J_K denote the annihilator of K in $\mathscr{P}(G)$. Evidently, J_K is a Hopf ideal of $\mathscr{P}(G)$. Conversely, if J is any Hopf ideal of $\mathscr{P}(G)$, then the annihilator of J in G is a subgroup of G.

A subgroup K of G is called an *algebraic subgroup* if it is closed in G. The restrictions of the elements of $\mathscr{P}(G)$ to K make up the algebra $\mathscr{P}(K)$ of polynomial functions of an algebraic group structure on the closed subgroup K. Clearly, $\mathscr{P}(K)$ is isomorphic, as a Hopf algebra, with $\mathscr{P}(G)/J_K$, via the restriction morphism.

Proposition 4.1. *Let G be an algebraic group, K a submonoid of G. Then the closure of K in G is an algebraic subgroup of G.*

PROOF. From the fact that K is a submonoid of G, it follows immediately that J_K is a biideal of $\mathscr{P}(G)$. This implies that the annihilator, K' say, of J_K in G is a submonoid of G.

Now let x be an element of K', and consider the translation operator x_l^* on $\mathscr{P}(G)$. Evidently, this stabilizes J_K, and so induces an injective linear endomorphism on J_K. Since $\mathscr{P}(G)$ is locally finite as a G-module, it follows that this endomorphism of J_K is also surjective. Therefore, the inverse $(x^{-1})_l^*$ of x_l^* also stabilizes J_K. This implies that x^{-1} belongs to K'. Thus, we conclude that K' is a subgroup of G. Evidently, K' is the closure of K in G. $\square$

Let G and H be algebraic groups. It is clear that the direct product structure of $G \times H$ as an algebraic set, together with its structure of an abstract group, is the structure of an algebraic group. Note that the coalgebra structure of $\mathscr{P}(G \times H)$, via the canonical identification, becomes the tensor product coalgebra structure of $\mathscr{P}(G) \otimes \mathscr{P}(H)$.

By a *morphism of affine algebraic F-groups* one means a group homomorphism that is also a morphism of affine algebraic F-sets. Evidently, the direct product $G \times H$, as defined above, satisfies the categorical requirements of a direct product in the category of affine algebraic groups.

Since we have been casual about categorical concerns, a warning in the form of a classical example may be appropriate at this point. The exponential map from the additive group of complex numbers to the multiplicative group is *not* a morphism of affine algebraic groups!

Let V be a finite-dimensional F-space, and let E stand for the finite-dimensional F-algebra $\mathrm{End}_F(V)$. Then the affine algebraic F-group E^*, as defined in Section 3, is called the *full linear group on* V. If G is an affine algebraic F-group, then a *polynomial representation* of G on V is a morphism of affine algebraic F-groups from G to E^*. It is equivalent to say that V has the structure of a G-module such that the associated representative functions on G belong to $\mathscr{P}(G)$. One then refers to V as a *polynomial G-module*.

Notes

1. Let $(H, \mu, u, \delta, \varepsilon, \eta)$ be a Hopf algebra over the field F. At the end of Section 1, we mentioned some formal properties of the antipode η without proof. We sketch a procedure by which the reader can establish these properties. In order to show that η is an antimorphism of coalgebras, verify first that $\varepsilon \circ \eta = \varepsilon$ by writing

$$\varepsilon = \varepsilon \circ (u \circ \varepsilon) = \varepsilon \circ \mu \circ (\eta \otimes i_H) \circ \delta = \cdots.$$

In order to show that

$$\delta \circ \eta = (\eta \otimes \eta) \circ s \circ \delta,$$

consider the F-algebras $\mathrm{Hom}_F(H, H)$ and $\mathrm{Hom}_F(H, H \otimes H)$ constructed from the coalgebra structure of H and the algebra structures of H and $H \otimes H$, and note that the map $\alpha \mapsto \delta \circ \alpha$ is a morphism of F-algebras from the first to the second. Since η is the inverse of i_H in $\mathrm{Hom}_F(H, H)$, this shows that $\delta \circ \eta$ is the inverse of δ in $\mathrm{Hom}_F(H, H \otimes H)$. Hence, it suffices to show that $(\eta \otimes \eta) \circ s \circ \delta$ is the inverse of δ. This can be done by a direct calculation.

Similarly, by considering the morphism of F-algebras from $\mathrm{Hom}_F(H, H)$ to $\mathrm{Hom}_F(H \otimes H, H)$ sending each α onto $\alpha \circ \mu$, one can show that η is an antimorphism of F-algebras.

Using this, one obtains

$$\mu \circ ((\eta \circ \eta) \otimes \eta) \circ \delta = \mu \circ (\eta \otimes \eta) \circ (\eta \otimes i_H) \circ \delta$$

$$= \eta \circ \mu \circ s \circ (\eta \otimes i_H) \circ \delta$$

Now, if one of μ or δ is commutative, one finds that the above reduces to $u \circ \varepsilon$, whence one obtains $\eta \circ \eta = i_H$.

2. With the help of Proposition 2.1 and the above results concerning the antipode, one shows that every Hopf algebra with commutative multiplication is the union of the family of its finitely algebra generated sub Hopf algebras.

3. Let C be an F-space, and suppose that C° is endowed with the structure of an F-algebra, and C is endowed with the structure of a locally finite two-sided C°-module such that $\alpha(\beta \cdot c) = \beta(c \cdot \alpha)$ for all elements c of C and all elements α and β of C°. Define a coalgebra structure δ, ε on C such that the given two-sided C°-module structure of C is that of Proposition 2.1.

4. Basic references for the machinery of Hopf algebras and the associated module theory are [9] and [17].

Chapter II

Affine Algebraic Sets and Groups

We begin with the basic facts concerning the irreducible components of algebraic sets in general, and the irreducible component of the neutral element in an algebraic group. The main result of Section 2 is the fact that algebraic subgroups are determined by their semi-invariants in the algebra of polynomial functions of the containing group. Section 3 contains the fundamental results on homomorphisms from commutative algebras to the base field, culminating in Hilbert's Nullstellensatz. Section 4 applies this to yield an important tool theorem about polynomial maps between algebraic groups, and then establishes the principal general result concerning factor groups of algebraic groups.

1. A topological space is said to be *irreducible* if it is non-empty and not the union of two non-empty closed proper subsets. Equivalently, a topological space is irreducible if it is non-empty and every pair of non-empty open subsets has a non-empty intersection. This notion is of importance for us in the case where the space is an affine algebraic set, with its Zariski topology.

Let S be such a set, and let T be a non-empty subset of S. We show that *T is irreducible if and only if its annihilator J_T in $\mathscr{P}(S)$ is a prime ideal.*

First, suppose that T is irreducible, and let a and b be elements of $\mathscr{P}(S)$ such that ab belongs to J_T. If A and B are the sets of zeros in S of a and b, respectively, then $T = (A \cap T) \cup (B \cap T)$. Since T is irreducible, it follows that one of $A \cap T$ or $B \cap T$ coincides with T, whence one of a or b belongs to J_T. Thus, J_T is a prime ideal.

Now suppose that T is not irreducible, so that $T = (A \cap T) \cup (B \cap T)$, where A and B are closed subsets of S neither of which contains T. Now $J_A \cap J_B \subset J_T$, while neither J_A nor J_B is contained in J_T. This shows that J_T is not a prime ideal.

From the fact that $\mathscr{P}(S)$ is a Noetherian ring, it follows immediately that S is *Noetherian as a topological space*, in the sense that it satisfies the maximal condition for open sets. We use this in showing that S *is the union of a finite family* $(S_1, \ldots, S_n)$ *of maximal irreducible subsets, the uniquely determined irreducible components of S.*

Consider the family $\mathscr{B}$ of non-empty closed subsets of S that are *not* finite unions of irreducible closed sets. Since S is Noetherian, it satisfies the minimal condition for closed sets. Therefore, if $\mathscr{B}$ is not empty, it has a minimal member, T say. Now T is not irreducible, so that $T = T_1 \cup T_2$, where T_1 and T_2 are non-empty closed proper subsets of S. By the minimality of T, each T_i is a finite union of closed irreducible subsets. But this implies that T is such a union, so that we have a contradiction. Thus, $\mathscr{B}$ is empty.

In particular, S is therefore the union of a finite family of closed irreducible subsets. Let $(S_1, \ldots, S_n)$ be a family of closed irreducible subsets obtained by discarding redundant members from any such family whose union is S. Then it is easy to see that every irreducible subset of S is contained in one of the S_i's, and hence that the family $(S_1, \ldots, S_n)$ satisfies all of our requirements. Note that each irreducible component is closed.

It is easily seen from the definition of irreducibility that *every non-empty open subset of an irreducible space S is irreducible, and dense in S. Also, the closure of an irreducible subset is irreducible.*

Proposition 1.1. *Let $\alpha : S \to T$ be a morphism of algebraic sets. If A is an irreducible subset of S, then $\alpha(A)$ is an irreducible subset of T.*

PROOF. This follows direcly from the definition of irreducibility; using only the continuity of α. $\qquad\square$

Lemma 1.2. *Let F be a field, S an irreducible affine algebraic F-set, B an integral domain F-algebra. Then $\mathscr{P}(S) \otimes B$ is an integral domain.*

PROOF. Let u and v be elements of $\mathscr{P}(S) \otimes B$ such that $uv = 0$. Write

$$u = \sum_{i=1}^{n} u_i \otimes b_i \quad \text{and} \quad v = \sum_{i=1}^{n} v_i \otimes b_i,$$

where the b_i's are F-linearly independent elements of B, and the u_i's and v_i's belong to $\mathscr{P}(S)$. Every element s of S, via evaluation at s, defines a B-algebra homomorphism s^B from $\mathscr{P}(S) \otimes B$ to B. If there is an index i such that $v_i(s) \neq 0$, then $s^B(v) \neq 0$, whence $s^B(u) = 0$, so that $u_j(s) = 0$ for each j. Therefore, in any case, we have $u_j(s)v_i(s) = 0$ for all indices i and j and all elements s of S. Thus, $u_j v_i = 0$ for all indices i and j. If $v \neq 0$ then one of the v_i's must be different from 0, and it follows that $u = 0$. $\qquad\square$

Proposition 1.3. *If S and T are irreducible algebraic sets, so is $S \times T$.*

PROOF. We know from the beginning of this Section that $\mathscr{P}(S)$ and $\mathscr{P}(T)$ are integral domains. By Lemma 1.2, it follows that $\mathscr{P}(S) \otimes \mathscr{P}(T)$ is an integral domain. Since this is $\mathscr{P}(S \times T)$, we conclude that $S \times T$ is irreducible. $\square$

Theorem 1.4. *Let G be an affine algebraic group. The irreducible components of G are mutually disjoint. The component G_1 containing the neutral element of G is a closed normal subgroup of G, and the irreducible components of G are the cosets of G_1 in G. Moreover, G_1 is the only irreducible closed subgroup of finite index in G.*

PROOF. Suppose that U and V are irreducible components of G and that each contains the neutral element of G. The product set UV in G is the image of $U \times V$ under the composition map of G. By Propositions 1.3 and 1.1, it is therefore irreducible. Since it contains U and V, we conclude that

$$U = UV = V.$$

Thus, only one of the irreducible components of G contains the neutral element. This defines G_1.

We have just seen that $G_1 G_1 = G_1$. Since the inversion of G is a homeomorphism, G_1^{-1} is an irreducible component of G. Since it contains the neutral element, it therefore coincides with G_1. Thus, G_1 is a subgroup of G. By considering the translation actions of G on itself, we see immediately that the set of left cosets of G_1 in G, as well as the set of right cosets, coincides with the set of irreducible components of G. This evidently implies that G_1 is a normal subgroup of finite index in G.

Finally, let K be any closed irreducible subgroup of finite index in G. Clearly, the set of left (or right) cosets of K in G coincides with the set of irreducible components of G, i.e., with the set of cosets of G_1. Hence, $K = G_1$. $\qquad\square$

2. Let F be a field, V an F-space. The *exterior algebra* $\bigwedge(V)$ built over V is defined as the factor algebra of the tensor algebra $\otimes(V)$ mod the ideal generated by the squares of the elements of V. If G is a group and V is a G-module, then $\bigwedge(V)$ inherits the structure of a G-module via the tensor product construction, and G acts on V by F-algebra automorphisms respecting the grading of V by its homogeneous components $\bigwedge^k(V)$ $(k = 0, 1, \ldots)$. A module of this type plays the decisive role in the proof of the following theorem.

Theorem 2.1. *Let G be an affine algebraic F-group, H an algebraic subgroup of G. There is a finite subset E of $\mathcal{P}(G)$, and an element f of $\mathcal{P}(G)$ whose restriction to H is a group homomorphism from H to F^*, such that*

(1) $x \cdot e = f(x)e$ *for every x in H and every e in E*
(2) *if x is an element of G such that $x \cdot e$ belongs to Fe for every element e of E then x belongs to H.*

PROOF. Let I denote the annihilator of H in $\mathcal{P}(G)$. There is a finite-dimensional left G-stable sub F-space V of $\mathcal{P}(G)$ such that $V \cap I$ generates I as an ideal. Let d denote the dimension of $V \cap I$, and consider the action of G on

$\bigwedge^d(V)$. Let S denote the canonical image of $\bigwedge^d(V \cap I)$ in $\bigwedge^d(V)$. Clearly, S is 1-dimensional, and we write $S = Fs$, fixing any non-zero element s of S.

Since $H \cdot I \subset I$, it is clear that S is an H-stable subspace of $\bigwedge^d(V)$. Thus, for x in H, we have $x \cdot s = g(x)s$, where g is a group homomorphism from H to F^*. Now choose σ from $(\bigwedge^d(V))^\circ$ such that $\sigma(s) = 1$, and let f be the representative function σ/s on G. It is clear from the comodule form of the construction of tensor products of G-modules that f is an element of $\mathscr{P}(G)$. Evidently, the restriction of f to H coincides with g.

Now let $(\sigma_1, \ldots, \sigma_p)$ be an F-basis of the annihilator of S in $(\bigwedge^d(V))^\circ$, and consider the elements σ_i/s of $\mathscr{P}(G)$. For every element x of H, we have

$$x \cdot (\sigma_i/s) = \sigma_i/(x \cdot s) = g(x)\sigma_i/s.$$

Conversely, suppose that x is an element of G such that $x \cdot (\sigma_i/s)$ is an F-multiple of σ_i/s for each i. Evaluating at the neutral element of G, we obtain $\sigma_i(x \cdot s) = 0$ for each i. This shows that $x \cdot s$ belongs to S, so that $x \cdot S = S$. Let t be an element of $V \cap I$. Then, in $\bigwedge^{d+1}(V)$, we have $tS = (0)$, whence also $x \cdot (tS) = (0)$. But

$$x \cdot (tS) = (x \cdot t)(x \cdot S) = (x \cdot t)S.$$

Thus, we have $(x \cdot t)S = (0)$, which means that $x \cdot t$ belongs to $V \cap I$. Since $V \cap I$ generates I as an ideal, our result shows that $x \cdot I \subset I$, whence x belongs to H. This proves the theorem, with $E = (\sigma_1/s, \ldots, \sigma_p/s)$. $\qquad\square$

If e is an element of $\mathscr{P}(G)$, and g is a group homomorphism from H to F^* such that $x \cdot e = g(x)e$ for every element x of H, then e is called a *semi-invariant of H*, and g is called the *weight* of e.

Theorem 2.2. *Let H be a normal algebraic subgroup of the algebraic group G. There is a finite subset Q of $\mathscr{P}(G)$ such that the left element-wise fixer of Q in G is precisely H.*

PROOF. Let E be a finite set of semi-invariants of H, such as given by Theorem 2.1, and let g denote the common weight of the elements of E. Let J be the smallest left G-stable subspace of $\mathscr{P}(G)$ that contains E. Since $\mathscr{P}(G)$ is locally finite as a G-module, J is of finite dimension over the base field F. Those elements of J which are H-semi-invariants of weight g evidently constitute a sub H-module, J_1 say, of J.

If x is an element of G then $x \cdot J_1$ is clearly the sub H-module of J consisting of those elements which are semi-invariants of weight g_x, where $g_x(y) = g(x^{-1}yx)$ for every element y of H. Since J is finite-dimensional, it is therefore a finite direct H-module sum $J_1 + \cdots + J_k$, where the J_i's are all the distinct $x \cdot J_1$'s.

Let U denote the sub F-algebra of $\mathrm{End}_F(J)$ consisting of the endomorphisms stabilizing each J_i, and let $\rho: G \to \mathrm{End}_F(J)$ be the representation of G on J coming from the action of G on $\mathscr{P}(G)$ from the left. It is clear from the definitions that, if u is an element of U and x is an element of G, then $\rho(x)u\rho(x)^{-1}$ belongs to U. Thus, we have a representation σ of G on U, where

$$\sigma(x)(u) = \rho(x)u\rho(x)^{-1}.$$

It is easy to see that the representative functions associated with σ belong to $\mathscr{P}(G)$. In fact, $S(\rho)$ is the smallest left and right G-stable subspace of $\mathscr{P}(G)$ containing J, and $S(\sigma) \subset S(\rho)\eta(S(\rho))$, where η is the antipode of $\mathscr{P}(G)$. Since every element of H acts as a scalar multiplication on each J_i, the kernel of σ contains H, so that the representative functions associated with σ must actually belong to the left H-fixed part of $\mathscr{P}(G)$, which we denote by $\mathscr{P}(G)^H$ (the normality of H implies that this coincides with the right H-fixed part ${}^H\mathscr{P}(G)$).

Now let Q be any finite subset of $\mathscr{P}(G)^H$ spanning $S(\sigma)$. Let x be an element of G such that $x \cdot q = q$ for every q in Q. Then x belongs to the kernel of σ, which means that $\rho(x)$ commutes with every element of U. It follows from this that $\rho(x)$ stabilizes each J_i, and that the restriction of $\rho(x)$ to J_i is a scalar multiplication. Since $E \subset J_1$, the element x therefore satisfies condition (2) of Theorem 2.1, so that x belongs to H. $\square$

We make an immediate simple application of Theorem 2.2 to the situation of Theorem 1.4. It is clear from Theorem 2.2 that $\mathscr{P}(G)^{G_1}$ separates the elements of G/G_1. If S is a finite set, and A is a sub F-algebra of F^S separating the elements of S, then A must coincide with F^S. Hence, viewed as an F-algebra of F-valued functions on G/G_1, the algebra $\mathscr{P}(G)^{G_1}$ coincides with F^{G/G_1}. In particular, the characteristic functions of the irreducible components of G are elements of $\mathscr{P}(G)$, whence we have the following result.

Theorem 2.3. *Let G be an affine algebraic F-group. As an F-algebra, $\mathscr{P}(G)$ is isomorphic, via the restriction maps, with the direct F-algebra sum of the algebras of polynomial funcions on the irreducible components of G.*

3. Lemma 3.1. *Let R be a subring of a field K, and suppose that J is a proper ideal of R. For every element u of K, if $R[u]J = R[u]$ then*

$$R[u^{-1}]J \neq R[u^{-1}].$$

PROOF. Suppose this is false. Then there is an element u in K for which we have relations

$$\sum_{i=1}^{m} a_i u^i = 1 = \sum_{j=0}^{n} b_j u^{-j},$$

where the a_i's and b_j's are elements of J. We assume that the relations have been so chosen that $m + n$ is as small as possible. Replacing u with u^{-1}, if necessary, we arrange to have $n \leq m$. Since $J \neq R$, these indices m and n must be greater than 0. From the second relation, we obtain

$$(1 - b_0)u^m = \sum_{j=1}^{n} b_j u^{m-j}.$$

Multiplying the first relation by $1 - b_0$ and then substituting for $(1 - b_0)u^m$, we obtain

$$1 - b_0 = \sum_{i=0}^{m-1} (1 - b_0)a_i u^i + a_m \sum_{j=1}^{n} b_j u^{m-j}.$$

This may be written in the form

$$\sum_{i=0}^{m-1} c_i u^i = 1,$$

where each c_i is an element of J, so that we have a contradiction to the minimality of $m + n$. $\qquad\square$

A subring S of a field K is called a *valuation subring* if, for every element u of K not belonging to S, the reciprocal u^{-1} does belong to S.

Proposition 3.2. *Let R be a subring of a field K, and suppose that ρ is a ring homomorphism from R to an algebraically closed field F. Then ρ can be extended to a ring homomorphism from a valuation subring S of K to F, where $R \subset S$.*

PROOF. An evident application of Zorn's lemma shows that, among the subrings T of K containing R and such that ρ can be extended to a ring homomorphism $T \to F$, there is a maximal one. Therefore, we assume without loss of generality that R is already maximal, and we show that R is a valuation subring of K.

Let u be an element of K. We must show that one of u or u^{-1} belongs to R. Let J denote the kernel of ρ. By virtue of Lemma 3.1, we may suppose that $R[u]J \neq R[u]$, and it suffices to show that then u belongs to R. The last assumption implies that J is contained in some maximal ideal, M say, of $R[u]$. Evidently, ρ can be extended to a ring homomorphism from the ring of fractions $R[(R \setminus J)^{-1}]$ to F, where $R \setminus J$ denotes the complement of J in R. Therefore, the maximality of R implies that this ring of fractions coincides with R. This means that $\rho(R)$ is a sub*field* of F, so that J is a maximal ideal of R. Therefore, $M \cap R = J$.

Now consider the canonical homomorphism

$$\pi: R[u] \to R[u]/M.$$

Since the kernel of π in R is J, there is an isomorphism

$$\sigma: \pi(R) \to \rho(R)$$

such that the restriction to R of $\sigma \circ \pi$ coincides with ρ. Next, we observe that u must be algebraic over R, because otherwise the evident extension of ρ to a ring homomorphism $R[u] \to F$ sending u onto 0 would contradict the maximality of R. Hence, $\pi(u)$ is algebraic over the subfield $\pi(R)$ of

$R[u]/M$. Since F is algebraically closed, the isomorphism σ can therefore be extended to a homomorphism

$$\tau: R[u]/M \to F.$$

Now the homomorphism $\tau \circ \pi$ from $R[u]$ to F is an extension of ρ, and the maximality of R implies that u belongs to R. $\qquad\square$

Theorem 3.3. *Let R be a subring of a field K, and let P be a finite subset of K. For every non-zero element u of $R[P]$, there is a non-zero element u' in R such that every homomorphism from R to an algebraically closed field F not annihilating u' extends to a homomorphism from $R[P]$ to F not annihilating u.*

PROOF. Evidently, the statement of the theorem is adapted to an induction on the cardinality of P. Therefore, we suppose without loss of generality that P consists of a single element p. First, we deal with the case where p is not algebraic over the field of fractions of R, which we denote by $[R]$. Write

$$u = r_0 + \cdots + r_n p^n,$$

with each r_i in R and $r_n \neq 0$. Let ρ be a ring homomorphism from R to F not annihilating r_n. There is an element t in F such that

$$\rho(r_0) + \cdots + \rho(r_n)t^n \neq 0.$$

Evidently, ρ can be extended to a ring homomorphism σ from $R[p]$ to F such that $\sigma(p) = t$, and our choice of t ensures that $\sigma(u) \neq 0$, so that we have the desired conclusion, with $u' = r_n$.

Now suppose that p is algebraic over $[R]$. Then we can find a non-zero element u' in R such that $u'p$ and u'/u are integral over R, which implies that p and u^{-1} are integral over $R[u'^{-1}]$. Suppose that ρ is a homomorphism from R to F not annihilating u'. Evidently, we can extend ρ to a homomorphism σ from $R[u'^{-1}]$ to F. By Proposition 3.2, there is a valuation subring S of K containing $R[u'^{-1}]$ and a homomorphism τ from S to F extending σ. Since S is a valuation subring of K, it is integrally closed in K, so that p and u^{-1} belong to S. The restriction of τ to $R[p]$ is an extension of ρ, and we have $\tau(u) \neq 0$, because u^{-1} belongs to the domain of τ and $\tau(u)\tau(u^{-1}) = 1$. $\qquad\square$

Lemma 3.4. *Let B be a commutative ring. The intersection of the family of all prime ideals of B coincides with the set of all nilpotent elements.*

PROOF. Evidently, every nilpotent element of B belongs to every prime ideal. Conversely, suppose that b is an element of B that belongs to every prime ideal of B. Consider the polynomial ring $B[x]$, where x is an auxiliary variable. The assumption on b clearly implies that b, and hence bx, belongs to every maximal ideal of $B[x]$. Therefore, $1 - bx$ is a unit of $B[x]$, so that there are elements b_i in B such that

$$(1 + \cdots + b_n x^n)(1 - bx) = 1.$$

One reads off from this that $b_1 = b, \ldots, b_n = b_{n-1}b$, and $b_n b = 0$. This gives $b^{n+1} = 0$. $\qquad\square$

The following theorem is a version of the Hilbert *Nullstellensatz*.

Theorem 3.5. *Let L be a field, B a finitely generated L-algebra having no nilpotent elements other than 0. Let F be an algebraically closed field containing L. Then the L-algebra homomorphisms from B to F separate the elements of B.*

PROOF. By Lemma 3.4, the assumption on B means that the intersection of the family of all prime ideals of B is (0). Hence, if b is any non-zero element of B, there is a prime ideal J in B not containing b. We identify L with its canonical image in the integral domain B/J, and we regard B/J as an L-algebra. Evidently, it is finitely generated as such. Now we apply Theorem 3.3, with L in the place of R, and B/J in the place of $R[P]$. This shows that there is an L-algebra homomorphism from B/J to F not annihilating the canonical image of b. The composite of this with the canonical homomorphism from B to B/J is an L-algebra homomorphism from B to F not annihilating b. $\qquad\square$

Proposition 3.6. *Let $F \subset L \subset K$ be a tower of fields. Suppose that K is finitely field-generated over F. Then the same is true for L.*

PROOF. There is a transcendence basis $(s_1, \ldots, s_m, t_1, \ldots, t_n)$ for K over F such that $(s_1, \ldots, s_m)$ is a transcendence basis for L over F and $(t_1, \ldots, t_n)$ is one for K over L. Now K is finite algebraic over $F(s_1, \ldots, s_m, t_1, \ldots, t_n)$, whence the same is true for the subextension $L(t_1, \ldots, t_n)$. Write P for $F(s_1, \ldots, s_m)$. We have just seen that $L(t_1, \ldots, t_n)$ is of finite dimension over $P(t_1, \ldots, t_n)$. Since the t_i's are algebraically independent over L, this implies that L is of finite dimension over P. In particular, L is therefore finitely field-generated over F. $\qquad\square$

Proposition 3.7 (Artin–Tate). *Let R, B, A be commutative rings, with $R \subset B \subset A$. Suppose that R is Noetherian, that A is finitely generated as an R-algebra and also that A is finitely generated as a B-module. Then B is finitely generated as an R-algebra.*

PROOF. Exhibiting the assumptions on the generation of A, we write

$$R[a_1, \ldots, a_n] = A = Bu_1 + \cdots + Bu_m$$

choosing $u_1 = 1$. Then we have

$$a_i = \sum_{j=1}^{m} b_{ij} u_j \quad \text{and} \quad u_i u_j = \sum_{k=1}^{m} b_{ijk} u_k,$$

where the b_{ij}'s and the b_{ijk}'s are elements of B. Let C denote the sub R-algebra of B that is generated by these elements. Since R is Noetherian and C is finitely generated as an R-algebra, C is Noetherian. Clearly, $Cu_1 + \cdots + Cu_m$ is a subring of A containing R as well as each a_i. Therefore, this subring coincides with A, showing that A is finitely generated as a C-module. Since C is Noetherian, the sub C-module B of A is also finitely generated as a C-module. Now, if $(b_1, \ldots, b_q)$ is a set of C-module generators of B, then B is generated as an R-algebra by the b_p's, the b_{ij}'s and the b_{ijk}'s. $\qquad\square$

Lemma 3.8. *Let A be a commutative algebra over the field F that can be generated as such by a finite set of cardinality n. Then every chain of prime ideals of A has length at most n. If P and Q are prime ideals of A such that P is properly contained in Q then the degree of transcendence of $[A/Q]$ over F is strictly smaller than that of $[A/P]$.*

PROOF. It is clear that the transcendence degree of $[A/P]$ cannot exceed n. Hence, it suffices to prove the second assertion of the lemma.

There is a transcendence basis $(y_1, \ldots, y_t)$ of $[A/Q]$ relative to F consisting of elements of A/Q. For each y_i, we choose an element x_i from A/P whose canonical image in A/Q is y_i. Let x_0 be any non-zero element of Q/P. It suffices to show that the x_i's are algebraically independent over F.

Suppose that this is not the case, and choose a non-zero polynomial f with coefficients in F of the smallest possible total degree such that $f(x_0, \ldots, x_t) = 0$. We may write this in the form

$$g(x_1, \ldots, x_t) + h(x_0, \ldots, x_t)x_0 = 0$$

where g and h are polynomials with coefficients in F. The canonical image in A/Q of the element on the left is $g(y_1, \ldots, y_t)$. Since the y_i's are algebraically independent over F, it follows that g must be the zero polynomial, whence $h(x_0, \ldots, x_t) = 0$. This contradicts the minimality of the degree of f. $\qquad\square$

4. Theorem 4.1. *Let F be an algebraically closed field, and let G and H be affine algebraic F-groups, G being irreducible. Suppose that ρ is a polynomial map from G to H sending the neutral element of G onto that of H. Then the products of finite sequences of elements of $\rho(G)$ constitute an irreducible algebraic subgroup P of H, and there is a natural number n such that every element of P is the product of n elements of $\rho(G)$.*

PROOF. For every positive natural number m, let G_m denote the direct product of m copies of G. Let ρ_m be the map from G_m to H defined by

$$\rho_m(x_1, \ldots, x_m) = \rho(x_1) \cdots \rho(x_m).$$

Clearly, ρ_m is a polynomial map. Let J_m denote the annihilator of $\rho_m(G_m)$ in $\mathscr{P}(H)$. Since G is irreducible, we have from Proposition 1.3 that G_m is irreducible. By Proposition 1.1, this implies that $\rho_m(G_m)$ is irreducible, whence J_m is a prime ideal of $\mathscr{P}(H)$.

Since $\rho_m(G_m) \subset \rho_{m+1}(G_{m+1})$, we have $J_{m+1} \subset J_m$, and it follows from Lemma 3.8 that there is a natural number q such that $J_m = J_q$ for every $m \geq q$. Clearly, J_q is the annihilator of P in $\mathcal{P}(H)$, where P is the union of the family of $\rho_m(G_m)$'s. Let Q denote the closure of P in H. Since P is a submonoid of H, we have from Proposition I.4.1 that Q is an algebraic subgroup of H. Since J_q is a prime ideal, Q is irreducible.

Since F is algebraically closed, we can apply Theorem 3.3 to find that there is a non-zero element f in $\mathcal{P}(H) \circ \rho_q$ such that every F-algebra homomorphism from $\mathcal{P}(H) \circ \rho_q$ to F not annihilating f extends to an F-algebra homomorphism from $\mathcal{P}(G_q)$ to F, i.e., is the restriction of an evaluation y^* with y in G_q.

Write $f = g \circ \rho_q$, with g in $\mathcal{P}(H)$. Since $f \neq 0$, we have $g \notin J_q$. Let x be any element of Q. Noting that J_q is stable under the action of Q on $\mathcal{P}(H)$, as well as under the antipode, η say, we conclude that $x \cdot \eta(g) \notin J_q$. This means that there is an element u in G_q such that

$$(x \cdot \eta(g))(\rho_q(u)) \neq 0, \quad \text{i.e.} \quad g(x^{-1}\rho_q(u)^{-1}) \neq 0.$$

Now $x^{-1}\rho_q(u)^{-1}$ is an element of Q. Therefore, it annihilates J_q, so that it defines an F-algebra homomorphism

$$\sigma \colon \mathcal{P}(H) \circ \rho_q \to F,$$

where $\sigma(h \circ \rho_q) = h(x^{-1}\rho_q(u)^{-1})$ for every h in $\mathcal{P}(H)$. In particular,

$$\sigma(f) = g(x^{-1}\rho_q(u)^{-1}) \neq 0.$$

By the choice of f, the homomorphism σ is therefore the restriction of a y^* with y in G_q.

Thus, for every h in $\mathcal{P}(H)$, we have

$$h(x^{-1}\rho_q(u)^{-1}) = h(\rho_q(y)),$$

so that $x^{-1}\rho_q(u)^{-1} = \rho_q(y)$, or

$$x^{-1} = \rho_q(y)\rho_q(u)$$

showing that every element of Q is the product of $2q$ elements of $\rho(G)$, whence also $Q = P$. $\qquad\square$

Let F be an arbitrary field, G an affine algebraic F-group, K a field containing F as a subfield. We can construct the K-Hopf algebra $\mathcal{P}(G) \otimes K$ and the associated affine algebraic K-group $\mathcal{G}(\mathcal{P}(G) \otimes K)$, which we denote by G^K. It is easy to see that G^K separates the elements of $\mathcal{P}(G) \otimes K$. In fact, the canonical extension of F-algebra homomorphisms $\mathcal{P}(G) \to F$ to K-algebra homomorphisms $\mathcal{P}(G) \otimes K \to K$ defines an injective group homomorphism from G to G^K, and the image of G in G^K already separates the elements of $\mathcal{P}(G) \otimes K$. Hence, we may identify $\mathcal{P}(G^K)$ with $\mathcal{P}(G) \otimes K$, and G with a dense subgroup of G^K. If G is irreducible, we see from Lemma 1.2 that $\mathcal{P}(G) \otimes K$ is an integral domain, which means that G^K is irreducible.

Lemma 4.2. *Let F be a field, G an irreducible affine algebraic F-group, B a sub Hopf algebra of $\mathscr{P}(G)$. Then $[B] \cap \mathscr{P}(G) = B$.*

PROOF. By Note I.2, B is the union of a family of sub Hopf algebras that are finitely generated as F-algebras. Therefore, we assume, without loss of generality, that B is finitely generated as an F-algebra.

Let L be an algebraically closed field containing F, and consider the extended irreducible affine algebraic L-group G^L whose algebra of polynomial functions is $\mathscr{P}(G) \otimes L$. If we prove that the intersection of $[B \otimes L]$ with $\mathscr{P}(G) \otimes L$ coincides with $B \otimes L$, then it clearly follows that

$$[B] \cap \mathscr{P}(G) = B.$$

Therefore, we assume, without loss of generality, that F is algebraically closed.

Consider an element f of $[B] \cap \mathscr{P}(G)$. Let J be the ideal of B consisting of all elements b with the property that $(x \cdot f)b$ belongs to B for every element x of G. Since all the transforms $x \cdot f$ lie in a finite-dimensional subspace of $[B]$, we have $J \neq (0)$. Evidently, J is stable under the action of G. Let j be a non-zero element of J, and let y be an element of $\mathscr{G}(B)$. Then $j \cdot y \neq 0$. Since G separates the elements of B, there is an element x in G such that $(j \cdot y)(x) \neq 0$. But $(j \cdot y)(x) = (x \cdot j)(y)$. Thus, we have $(x \cdot j)(y) \neq 0$ and $x \cdot j \in J$. This shows that J has no zero in $\mathscr{G}(B)$.

Since F is algebraically closed, this last fact implies that $J = B$, as is seen by applying Theorem 3.5 as follows. Let J' denote the radical of J. If $J' \neq B$, we obtain a contradiction by applying Theorem 3.5 to B/J'. Hence $J' = B$, which evidently implies that $J = B$. From this, it is clear that f belongs to B. $\qquad\square$

Theorem 4.3. *Let G be an affine algebraic F-group, B a sub Hopf algebra of $\mathscr{P}(G)$. Then B is finitely generated as an F-algebra. If F is algebraically closed then the restriction map $G \to \mathscr{G}(B)$ is surjective.*

PROOF. First, we deal with the case where G is irreducible. In that case, we see from Proposition 3.6 that $[B]$ is finitely field-generated over F. Let $(u_1 v_1^{-1}, \ldots, u_n v_n^{-1})$ be a finite system of field generators for $[B]$ over F, where each u_i and each v_i belongs to B. Let B_1 denote the smallest sub Hopf algebra of B containing all these u_i's and v_i's. By Note I.2, B_1 is finitely generated as an F-algebra, while $[B_1] = [B]$. Using Lemma 4.2, we obtain

$$B = [B] \cap \mathscr{P}(G) = [B_1] \cap \mathscr{P}(G) = B_1,$$

so that B is finitely generated as an F-algebra.

In the case where G is not irreducible, we use Theorem 2.3, as follows. Let f_1 denote the characteristic function of the irreducible component G_1 of the neutral element in G, and let $f_2, \ldots, f_m$ be the characteristic functions of the other irreducible components of G. We know from Theorem 2.3 that these are elements of $\mathscr{P}(G)$. Now $\mathscr{P}(G)f_1$ may be identified with $\mathscr{P}(G_1)$, and this identifies Bf_1 with a sub Hopf algebra of $\mathscr{P}(G_1)$. By what we have already proved, Bf_1 is therefore finitely generated as an F-algebra. For each i, there is an element x_i in G such that $f_i = x_i \cdot f_1$, and then $Bf_i = x_i \cdot (Bf_1)$. It follows that the sub F-algebra $Bf_1 + \cdots + Bf_m$ of $\mathscr{P}(G)$ is finitely generated. Clearly, this sub F-algebra contains B. Now it follows immediately from Proposition 3.7 that B is finitely generated as an F-algebra.

Finally, suppose that F is algebraically closed, and consider the restriction morphism $\rho\colon G \to \mathscr{G}(B)$. By Theorem 4.1, $\rho(G_1)$ is closed in $\mathscr{G}(B)$. Since $\rho(G)$ is the union of a finite family of translates of $\rho(G_1)$, it follows that $\rho(G)$ is closed in $\mathscr{G}(B)$. Since G separates the points of B, it is clear that $\rho(G)$ is dense in $\mathscr{G}(B)$. Therefore, we have $\rho(G) = \mathscr{G}(B)$. $\qquad\square$

Our results combine to yield the following main theorem concerning factor groups.

Theorem 4.4. *Let F be an algebraically closed field, G an affine algebraic F-group, H a normal algebraic subgroup of G. Then G/H has the structure of an affine algebraic F-group such that, via the transpose of the canonical morphism $\pi\colon G \to G/H$, the Hopf algebra $\mathscr{P}(G/H)$ is isomorphic with $\mathscr{P}(G)^H$. If $\gamma\colon G \to K$ is a morphism of affine algebraic F-groups whose kernel contains H, then the induced group homomorphism $\gamma^H\colon G/H \to K$, satisfying $\gamma^H \circ \pi = \gamma$, is a morphism of affine algebraic F-groups.*

PROOF. Clearly, $\mathscr{P}(G)^H$ is stable under the left and right actions of G on $\mathscr{P}(G)$, as well as under the antipode. Hence, $\mathscr{P}(G)^H$ is a sub Hopf algebra of $\mathscr{P}(G)$. It follows from Theorem 2.2 that the kernel of the restriction morphism from G to $\mathscr{G}(\mathscr{P}(G)^H)$ coincides with H. By Theorem 4.3, this morphism is surjective, and $\mathscr{P}(G)^H$ is finitely generated as an F-algebra. Thus, $\mathscr{G}(\mathscr{P}(G)^H)$ is canonically isomorphic with G/H, and its algebra of polynomial functions may be identified with $\mathscr{P}(G)^H$ as indicated in the theorem.

Now let $\gamma\colon G \to K$ be as described in the theorem. Then we have $\mathscr{P}(K) \circ \gamma \subset \mathscr{P}(G)^H$, showing that γ^H is a morphism of affine algebraic groups, because $\mathscr{P}(K) \circ \gamma^H$ coincides with $\mathscr{P}(K) \circ \gamma$ when $\mathscr{P}(G/H)$ has been identified with $\mathscr{P}(G)^H$. $\qquad\square$

Notes

1. It will become evident later on that, in Theorem 2.2, the condition that H be normal is not superfluous, so that the much weaker Theorem 2.1 cannot be strengthened. The simplest example illustrating the difficulty with non-normal subgroups is as follows. Let G be the multiplicative group of all matrices

$$x = \begin{pmatrix} \alpha(x) & \beta(x) \\ \gamma(x) & \delta(x) \end{pmatrix}$$

of determinant 1, with entries in a field F. This has the structure of an affine algebraic F-group with $\mathscr{P}(G) = F[\alpha, \beta, \gamma, \delta]$ (where $\alpha\delta - \beta\gamma = 1$). Let H be the algebraic subgroup consisting of the elements x such that $\gamma(x) = 0$. It is not difficult to exhibit a set of H-semi-invariants characterizing H as in Theorem 2.1. On the other hand, taking F to be an infinite field, one can verify directly, though somewhat painfully, that $\mathscr{P}(G)^H = F$.

2. Over non algebraically closed base fields, the theory of factor groups is deficient, because Theorem 4.3 fails easily. For example, let F be the field of real numbers, and let G be the multiplicative group of matrices

$$
x = \begin{pmatrix} \alpha(x) & \beta(x) \\ -\beta(x) & \alpha(x) \end{pmatrix}
$$

with non-zero determinant $d(x) = \alpha(x)^2 + \beta(x)^2$. Regard G as an affine algebraic F-group, with $\mathscr{P}(G) = F[\alpha, \beta, d^{-1}]$. Let H be the normal algebraic subgroup consisting of the elements of determinant 1 (i.e., the group of rotations of the real plane). First, one shows that $\mathscr{P}(G)^H = F[d, d^{-1}]$, and then one sees that the restriction map $G \to \mathscr{G}(\mathscr{P}(G)^H)$ is not surjective.

3. The surprisingly elementary proof of the finite generation of B in Theorem 4.3 is due to J. B. Sullivan.

Chapter III

Derivations and Lie Algebras

Here, we introduce some concepts and techniques that could be described as the differential calculus of algebra and group theory. As in analysis, this is a tool for linearizing problems.

Sections 1 and 2 deal with the basic separability and transcendence questions in field theory from the point of view of derivations. The results will be needed later on in connection with dimension-theoretical problems. Sections 3 and 4 begin the Lie algebra theory for algebraic groups.

1. If R is a commutative ring, and S is an R-module, then a *derivation* from R to S is a homomorphism τ from the additive group of R to that of S such that

$$\tau(xy) = x \cdot \tau(y) + y \cdot \tau(x)$$

for all elements x and y of R. This notion is the basis for the following definition of separability of a field extension, which combines the case of a separable *algebraic* extension, in the usual sense, with the case of a purely transcendental extension in an appropriate way.

Definition 1.1. *Let K be an extension field of a field F. We say that K is separable over F if, for every K-space S, every derivation from F to S extends to one from K to S.*

The natural heredity pattern is described in the following proposition.

Proposition 1.2. *Let $F \subset K \subset L$ be a tower of fields. If K is separable over F and L is separable over K then L is separable over F. If L is separable over F, so is K.*

PROOF. The first part is clear from the definition. In order to prove the second part, let τ be a derivation from F to a K-space S. We form the L-space $L \otimes S$, and write it as a direct K-space sum $S + T$. Now we may view τ as a derivation from F to $L \otimes S$. By assumption on L, this derivation extends to a derivation, σ say, from L to $L \otimes S$. If π is the K-space projection from $L \otimes S$ to S corresponding to our above decomposition, then the restriction of $\pi \circ \sigma$ to K is evidently a derivation from K to S extending τ. $\square$

In the following proposition, "separably algebraic" has the usual meaning.

Proposition 1.3. *Let K be a field, F a subfield of K, and u an element of K. Let τ be a derivation from F to an $F(u)$-space S. If u is not algebraic over F then, for every element s of S, there is one and only one extension of τ to a derivation from $F(u)$ to S sending u onto s. If u is separably algebraic over F then τ has precisely one extension to a derivation from $F(u)$ to S.*

PROOF. First, consider the case where u is not algebraic over F. Clearly, there is one and only one derivation σ from $F[u]$ to S sending u onto s and coinciding with τ on F. In fact, σ is given by

$$\sigma\left(\sum_i c_i u^i\right) = \sum_i (u^i \cdot \tau(c_i) + i c_i u^{i-1} \cdot s).$$

Now σ extends in one and only one way to a derivation from $F(u)$ to S by the usual formula for the derivative of a fraction:

$$\sigma(ab^{-1}) = b^{-2} \cdot (b \cdot \sigma(a) - a \cdot \sigma(b)).$$

Next, suppose that u is separably algebraic over F. Let f denote the monic minimum polynomial for u relative to F, and let f' denote the formal derivative of f. The assumption on u means that $f'(u) \neq 0$. Let us denote the coefficients of f by c_i ($i = 0, \ldots, n$), with $c_n = 1$. Let x be an auxiliary variable, and let us regard S as an $F[x]$-module via the F-algebra homomorphism from $F[x]$ to $F[u]$ sending x onto u. Let ρ be the derivation from $F[x]$ to S that is determined by the conditions that ρ be an extension of τ and that

$$\rho(x) = -f'(u)^{-1} \cdot \sum_{i=0}^{n} u^i \cdot \tau(c_i).$$

Then ρ annihilates the ideal $F[x]f(x)$, and therefore induces a derivation from $F[u]$ to S extending τ. If σ is any such extension of τ, we must have

$$0 = \sigma(f(u)) = f'(u) \cdot \sigma(u) + \sum_{i=0}^{n} u^i \cdot \tau(c_i),$$

which shows that σ must coincide with the derivation induced by ρ. $\square$

Via an evident application of Zorn's Lemma, Proposition 1.3 shows that, *in characteristic 0, every field extension is separable,* and also that *every purely transcendental field extension is separable.*

Lemma 1.4. *Let F be a field of non-zero characteristic p, let S be an F-space, s an element of S, and u an element of F that is not the p-th power of an element of F. There is a derivation τ from F to S such that $\tau(u) = s$.*

PROOF. Let $F^{[p]}$ denote the subfield of F consisting of the p-th powers of the elements of F, let L be a subfield of F containing $F^{[p]}$, and let v be an element of F not belonging to L. Let $f(x)$ denote the minimum polynomial for v relative to L. Then $f(x)$ divides $x^p - v^p = (x - v)^p$ in $F[x]$, so that we must have $f(x) = (x - v)^q$, with $0 < q \leq p$. Now v^q and v^p lie in L. If $q \neq p$, there are integers r and s such that $rp + sq = 1$, so that $v = (v^p)^r(v^q)^s \in L$, contrary to assumption. Therefore, we have $q = p$, so that $f(x) = x^p - v^p$.

Let ρ be any derivation from L to S. Clearly, ρ can be extended to a derivation from $L[x]$ to S sending x to s. This extension sends $f(x)$ to 0 and hence induces an extension of ρ to a derivation from $L[v]$ to S sending v to s. Using this result in an evident application of Zorn's Lemma, we obtain the required derivation τ. $\qquad\square$

Proposition 1.5. *Let F be a field of non-zero characteristic p, and let K be a field extension of F. Then K is separable over F if and only if, for every F-linearly independent subset U of K, the set $U^{[p]}$ of p-th powers of the elements of U is F-linearly independent.*

PROOF. First, suppose that the condition is satisfied. Then the multiplication map from $F \otimes_{F^{[p]}} K^{[p]}$ to K is injective, so that the subfield $F[K^{[p]}]$ of K is F-algebra isomorphic with $F \otimes_{F^{[p]}} K^{[p]}$. Let τ be a derivation from F to a K-space S. Evidently, τ annihilates $F^{[p]}$, so that it is an $F^{[p]}$-linear map. As such, it extends naturally to a $K^{[p]}$-linear map from $F \otimes_{F^{[p]}} K^{[p]}$ to S, which is clearly a derivation. Because of the isomorphism noted above, this means that τ extends to a derivation from $F[K^{[p]}]$ to S. It is clear from Lemma 1.4 that we can apply Zorn's Lemma in the usual way in order to extend this further to a derivation from K to S. Thus, K is separable over F.

Now suppose that the condition of the proposition is not satisfied, and choose an F-linearly independent subset $(u_1, \ldots, u_n)$ of K such that the u_i^p's are not linearly independent over F, with n as small as possible. Then there are elements $c_2, \ldots, c_n$ in F such that

$$u_1^p + \sum_{i=2}^{n} c_i u_i^p = 0.$$

Suppose that, contrary to what we must prove, K is separable over F. Then every derivation τ from F to F extends to a derivation σ from K to K. Applying σ to our above relation, we obtain

$$\sum_{i=2}^{n} \tau(c_i) u_i^p = 0.$$

By the minimality of n, this gives $\tau(c_i) = 0$ for each i from 2 to n. Thus, each c_i is annihilated by every derivation from F to F. By Lemma 1.4, this implies that each c_i is the p-th power d_i^p of an element d_i of F. Our original relation may now be written

$$\left(u_1 + \sum_{i=2}^{n} d_i u_i\right)^p = 0.$$

This contradicts the F-linear independence of the u_i's. The conclusion is that if the condition of the proposition is not satisfied then K is not separable over F. $\square$

If R is any commutative ring, let us agree to call a derivation from R to R simply a derivation of R. The last part of the proof of Proposition 1.5 has shown that if every derivation of F extends to a derivation of K then the condition of Proposition 1.5 is satisfied. Hence, *if K is an extension field of the field F such that every derivation of F extends to one of K, then K is separable over F.*

It follows from Zorn's Lemma and Proposition 1.3 that an extension that is separably algebraic in the usual sense is separable also in the sense of Definition 1.1. Conversely, if K is an algebraic field extension of F that is separable in the sense of Definition 1.1, it follows from Proposition 1.5 that K is separably algebraic over F in the usual sense. Thus, *for algebraic field extensions, separability in the sense of Definition 1.1 is equivalent to separability in the usual sense.*

A field F is called *perfect* if either F is of characteristic 0, or F is of non-zero characteristic p and coincides with $F^{[p]}$. Since every extension in characteristic 0 is separable, it follows from Proposition 1.5 that *every field extension of a perfect field is separable.*

2. Theorem 2.1. *Suppose that K is a separable finitely generated field extension $F(u_1, \ldots, u_n)$ of a field F. Then some subset X of $(u_1, \ldots, u_n)$ is a transcendence basis for K over F such that K is separably algebraic over $F(X)$. The degree of transcendence of K over F is equal to the dimension of the K-space of all F-linear derivations of K.*

PROOF. Let S denote the K-space of all F-linear derivations of K. Since every element of S is determined by its values at the u_i's, it is identifiable with a subspace of the space of all maps from the set $(u_1, \ldots, u_n)$ to K. Hence, we can apply Lemma I.1.1 to conclude that there is a K-basis $(\sigma_1, \ldots, \sigma_r)$ of S and corresponding elements $v_1, \ldots, v_r$, chosen from $(u_1, \ldots, u_n)$, such that $\sigma_i(v_j) = \delta_{ij}$. Relabelling, we arrange to have $v_i = u_i$ for each $i \leq r$, and we take X to be the set $(u_1, \ldots, u_r)$.

First, we show that K is separably algebraic over $F(X)$. Let t be the smallest index $\geq r$ such that K is separably algebraic over $F(u_1, \ldots, u_t)$. We shall obtain a contradiction from the assumption that t is strictly greater than r. By the choice of t, the field K is not separably algebraic over

$F(u_1, \ldots, u_{t-1})$. By Proposition 1.2, it follows that $F(u_1, \ldots, u_t)$ is not separably algebraic over $F(u_1, \ldots, u_{t-1})$. By Proposition 1.3, the element u_t is therefore not separably algebraic over $F(u_1, \ldots, u_{t-1})$.

Let us first deal with the case where u_t is algebraic over $F(u_1, \ldots, u_{t-1})$. In this case, F must be of non-zero characteristic p, and the monic minimum polynomial, f say, for u_t relative to $F(u_1, \ldots, u_{t-1})$ must satisfy $f(x) = g(x^p)$, where x is an auxiliary variable, and g is a polynomial with coefficients in $F(u_1, \ldots, u_{t-1})$. This shows that every $F(u_1, \ldots, u_{t-1})$-linear derivation of $F(u_1, \ldots, u_{t-1})[x]$ annihilates $f(x)$, so that it yields an $F(u_1, \ldots, u_{t-1})$-linear derivation of $F(u_1, \ldots, u_t)$ in the evident way. In particular, it follows that there exists a non-zero $F(u_1, \ldots, u_{t-1})$-linear derivation of $F(u_1, \ldots, u_t)$. Since K is separable over $F(u_1, \ldots, u_t)$, this extends to a derivation of K. On the other hand, each of $u_1, \ldots, u_r$ belongs to $F(u_1, \ldots, u_{t-1})$ and is therefore annihilated by this derivation. We have the contradiction that our derivation is the 0-map, because its coefficients with respect to $\sigma_1, \ldots, \sigma_r$ are its values at $u_1, \ldots, u_r$.

Now consider the case where u_t is not algebraic over $F(u_1, \ldots, u_{t-1})$. In this case, Proposition 1.3 gives us the existence of a non-zero $F(u_1, \ldots, u_{t-1})$-linear derivation of $F(u_1, \ldots, u_t)$, whence we have the same contradiction as in the first case.

Our conclusion so far is that K is separably algebraic over $F(u_1, \ldots, u_r)$, and it remains only to show that the set $(u_1, \ldots, u_r)$ is algebraically free over F. Suppose that this is not the case, and let f be a non-zero polynomial with coefficients in F of the smallest possible total degree such that $f(u_1, \ldots, u_r) = 0$. Let f_i denote the formal derivative of f with respect to the i-th variable. Applying the derivation σ_i to our relation, we obtain $f_i(u_1, \ldots, u_r) = 0$. By the minimality of the degree of f, this implies that $f_i = 0$. Therefore, F must be of non-zero characteristic p, and there must be a polynomial g with coefficients in F such that

$$f(x_1, \ldots, x_r) = g(x_1^p, \ldots, x_r^p),$$

where the x_i's are independent auxiliary variables. Writing g as an F-linear combination of monomials, we see from this that there is a non-empty F-linearly independent set $(w_1, \ldots, w_m)$ of monomials formed from $u_1, \ldots, u_r$ such that the set $(w_1^p, \ldots, w_m^p)$ is not F-linearly independent. This contradicts Proposition 1.5, because K is separable over F. $\square$

Theorem 2.2. *A field extension K of a field F is separable if and only if, for every field L containing F, the tensor product $K \otimes_F L$ has no non-zero nilpotent element.*

PROOF. First, suppose that K is separable over F, and let u be a nilpotent element of $K \otimes_F L$. We shall prove that $u = 0$. Clearly, there is a field K_1 between F and K that is finitely field-generated over F and such that u belongs to the canonical image of $K_1 \otimes_F L$ in $K \otimes_F L$, which we may identify with $K_1 \otimes_F L$. Therefore, we assume without loss of generality that K is finitely field-generated over F. Then, by Theorem 2.1, there is a transcendence basis X for K over F such that K is separably algebraic over $F(X)$. In fact, since K is also a *finite* algebraic extension of $F(X)$, there is an element s in K that is separably algebraic over $F(X)$ and such that $K = F(X)[s]$.

Let X' be a set of independent variables over L that is in bijective correspondence with X. Then $F(X) \otimes_F L$ may evidently be identified with an $F(X')$-subalgebra of the purely transcendental field extension $L(X')$ of L. Accordingly, we may write

$$K \otimes_F L = K \otimes_{F(X)} (F(X) \otimes_F L) \subset K \otimes_{F(X)} L(X'),$$

where $L(X')$ is viewed as an $F(X)$-algebra via an F-algebra isomorphism from $F(X)$ to $F(X')$ extending a bijection from X to X'.

Let f denote the monic minimum polynomial for s relative to $F(X)$, but view f as a polynomial with coefficients in $F(X')$ via the isomorphism just mentioned. Since s is separable over $F(X)$, the polynomial f is the product of a set of *mutually distinct* monic irreducible factors with coefficients in $L(X')$. It follows from this that $K \otimes_{F(X)} L(X')$ is a direct sum of fields, one for each irreducible factor of f. Therefore, our nilpotent element u must be 0.

Now suppose that the condition of the theorem is satisfied. In showing that K is separable over F, we may assume that F is of non-zero characteristic p. We show that then the condition of Proposition 1.5 is satisfied, so that K is separable over F. Let $(u_1, \ldots, u_n)$ be an F-linearly independent subset of K, and suppose that $c_1, \ldots, c_n$ are elements of F such that

$$\sum_{i=1}^{n} c_i u_i^p = 0.$$

We construct a field extension $L = F[t_1, \ldots, t_n]$ of F such that $t_i^p = c_i$ for each i. Then the p-th power of the element $\sum_{i=1}^{n} u_i \otimes t_i$ of $K \otimes_F L$ is equal to 0, so that, because of the present assumption on K, the element itself must be 0. This gives $t_i = 0$, and hence $c_i = 0$ for each i. Our conclusion is that the set $(u_1^p, \ldots, u_n^p)$ is F-linearly independent. $\square$

Theorem 2.3. *Let K be a field, and let A be a group of field automorphisms of K. Then K is separable over its A-fixed part K^A.*

PROOF. By Theorem 2.2, it suffices to prove that $K \otimes_{K^A} L$ has no nilpotent element other than 0, for every field L containing K^A. Since there is nothing to prove if K is of characteristic 0, we assume that K is of non-zero characteristic p. Suppose that the result is false for some L. Then there is a non-zero element x in $K \otimes_{K^A} L$ such that $x^p = 0$. Write

$$x = \sum_{i=1}^{n} k_i \otimes t_i,$$

with each k_i in K and each t_i in L. We suppose that x has been so chosen that n is as small as possible. Clearly, we must have $n > 1$, and we can arrange to have $k_1 = 1$. Then

$$x = t_1 + \sum_{i=2}^{n} k_i \otimes t_i.$$

Now we let A act by L-algebra automorphisms on $K \otimes_{K^A} L$, via the factor K. For α in A, we have

$$\alpha(x) = t_1 + \sum_{i=2}^{n} \alpha(k_i) \otimes t_i,$$

so that

$$x - \alpha(x) = \sum_{i=2}^{n} (k_i - \alpha(k_i)) \otimes t_i.$$

Since $(x - \alpha(x))^p = 0$, it follows from the minimality of n that $x = \alpha(x)$. Also, it is clear from the minimality of n that the t_i's are linearly independent over K^A. Therefore, it follows from the fact that $x = \alpha(x)$ for every α in A that each k_i belongs to K^A. This makes $n = 1$, and we have a contradiction.
$\square$

Proposition 2.4. *Let L be a field, A a finitely generated integral domain L-algebra, B a sub L-algebra of A. Let F be an algebraically closed field containing L. Suppose that x is an element of A with the property that L-algebra homomorphisms from A to F whose restrictions to B coincide take the same value at x. Then x is purely inseparably algebraic over $[B]$.*

PROOF. First, we obtain a contradiction from the assumption that x is not algebraic over $[B]$. By Theorem II.3.3, there is a non-zero element y in $B[x]$ such that every L-algebra homomorphism from $B[x]$ to F not annihilating y extends to an L-algebra homomorphism from A to F. Write

$$y = b_0 + b_1 x + \cdots + b_n x^n,$$

with each b_i in B and $b_n \neq 0$. By Theorem II.3.5, b_n is not annihilated by every L-algebra homomorphism from A to F. A fortiori, there is an L-algebra homomorphism σ from B to F such that $\sigma(b_n) \neq 0$. Since x is not algebraic over $[B]$, σ has infinitely many extensions to L-algebra homomorphisms from $B[x]$ to F not annihilating y and therefore extending further to L-algebra homomorphisms from A to F. This contradicts the assumption on x. Thus, we conclude that x must be algebraic over $[B]$.

Let p denote the characteristic of L if that is not 0; otherwise, let $p = 1$. Suppose that x is not purely inseparable over $[B]$. Then there is a non-negative exponent e such that x^{p^e} is separably algebraic over $[B]$, but does not belong to $[B]$. We can find a non-zero element b in B such that the monic minimum polynomial, f say, for bx^{p^e} relative to $[B]$ has all its coefficients in B. Write z for bx^{p^e}.

As before, there is a non-zero element y in $B[z]$ such that every L-algebra homomorphism from $B[z]$ to F not annihilating y extends to an L-algebra homomorphism from A to F. We write

$$y = b_0 + b_1 z + \cdots + b_n z^n,$$

with each b_i in B and $b_n \neq 0$, and n strictly smaller than the degree of f. Let g denote the polynomial whose coefficients are these b_i's, so that $y = g(z)$. There are polynomials u and v with coefficients in B such that $uf + vg$ is a non-zero element, s say, of B. On the other hand, since f is a separable polynomial, there are polynomials q and r with coefficients in B such that $qf + rf'$ is a non-zero element, t say, of B.

Appealing to Theorem II.3.5 as in the first part of this proof, we find that there is an L-algebra homomorphism ρ from B to F such that $\rho(st) \neq 0$. Let $\rho(f)$ denote the polynomial with coefficients in F obtained by applying ρ to the coefficients of f. Define $\rho(g)$ and $\rho(f')$ in the same way. Then, since $\rho(s)$ and $\rho(t)$ are non-zero elements of F, the polynomials $\rho(f)$ and $\rho(g)$ are relatively prime, and the same is true for $\rho(f)$ and $\rho(f')$. Moreover, since f is monic, the degree of $\rho(f)$ equals that of f. Call this degree m, and note that $m > 1$.

Now it is clear that $\rho(f)$ has m distinct roots in F, and that none of them is a root of $\rho(g)$. Using these roots as values for z, we obtain m different extensions of ρ to L-algebra homomorphisms from $B[z]$ to F, none of which annihilates y, so that each extends further to an L-algebra homomorphism from A to F. The images of x under these homomorphisms are mutually distinct, so that again we have a contradiction. $\quad\square$

3. A *Lie algebra* over a field F is an F-space L that is equipped with a bilinear composition $L \times L \to L$, indicated by $(x, y) \mapsto [x, y]$, satisfying

(1) $[x, x] = 0$,
(2) $[x, [y, z]] + [y, [z, x]] + [z, [x, y]] = 0$.

The identity (2) is called the *Jacobi identity*. It is better to think of this structure as follows. For every element x of L, let D_x denote the linear endomorphism of L given by $D_x(y) = [x, y]$. Then, in the presence of (1), the identity (2) means that D_x is a *derivation* with respect to the composition of L, i.e., that

$$D_x([u, v]) = [D_x(u), v] + [u, D_x(v)].$$

If A is an associative F-algebra, not necessarily having a unit, we obtain a Lie algebra structure on the F-space A by setting $[x, y] = xy - yx$. We denote this Lie algebra by $\mathscr{L}(A)$.

By a *derivation* of A we mean an F-linear endomorphism τ of A such that

$$\tau(xy) = \tau(x)y + x\tau(y).$$

The derivations of A constitute a sub Lie algebra of $\mathscr{L}(\mathrm{End}_F(A))$, which we denote by $\mathscr{D}(A)$, or more fully by $\mathscr{D}_F(A)$.

Now let us consider a bialgebra $(B, \mu, u, \delta, \varepsilon)$. Recall that the dual B° of B has an algebra structure coming from the comultiplication δ, where $\sigma\tau = (\sigma \otimes \tau) \circ \delta$. We are interested in a certain sub Lie algebra of the Lie algebra $\mathcal{L}(B^\circ)$. The elements of this sub Lie algebra are the *differentiations* of B, by which we mean those elements τ of B° which satisfy

$$\tau(ab) = \varepsilon(a)\tau(b) + \tau(a)\varepsilon(b)$$

for all elements a and b of B. One can verify directly that the differentiations do indeed constitute a sub Lie algebra of $\mathcal{L}(B^\circ)$. However, we shall see that this is the case by the following connection between the differentiations of B and the F-linear derivations of B.

Consider the two-sided B°-module structure of B, as dealt with in Proposition I.2.1. Recall that the element $\tau_{\mathfrak{l}}$ of $\mathrm{End}_F(B)$ corresponding to an element τ of B° in the left B°-module structure of B is defined by

$$\tau_{\mathfrak{l}} = (i_B \otimes \tau) \circ \delta.$$

One verifies directly that τ *is a differentiation if and only if* $\tau_{\mathfrak{l}}$ *is a derivation.* By Proposition I.2.2, the left B°-module structure of B induces an isomorphism of F-Lie algebras from $\mathcal{L}(B^\circ)$ to the sub Lie algebra of $\mathcal{L}(\mathrm{End}_F(B))$ consisting of the endomorphisms that commute with the action of B° on B from the right. From our last statement about the differentiations, it is now clear that *the differentiations of B constitute a sub Lie algebra of $\mathcal{L}(B^\circ)$ which, via the B°-module structure of B, is isomorphic with the sub Lie algebra of $\mathcal{D}(B)$ consisting of those derivations of B which are also right B°-module endomorphisms.*

Now let G be an affine algebraic F-group. *We define the Lie algebra of G as the Lie algebra of differentiations of $\mathcal{P}(G)$, and we denote it by $\mathcal{L}(G)$.*

A morphism $\rho \colon G \to K$ of affine algebraic F-groups naturally induces a morphism of F-algebras from $\mathcal{P}(G)^\circ$ to $\mathcal{P}(K)^\circ$; namely, the dual of the transpose of ρ. This restricts to a morphism of Lie algebras $\rho^{\cdot} \colon \mathcal{L}(G) \to \mathcal{L}(K)$ where

$$\rho^{\cdot}(\tau)(g) = \tau(g \circ \rho)$$

for every element τ of $\mathcal{L}(G)$ and every element g of $\mathcal{P}(K)$. We call $\rho^{\cdot}$ the *differential* of ρ, and we note that the association of $\rho^{\cdot}$ with ρ makes $\mathcal{L}$ a functor from the category of affine algebraic F-groups to the category of F-Lie algebras, i.e., the composite of differentials coincides with the differential of the composite.

In particular, consider the case where ρ is the injection $G_1 \to G$. Let Q denote the annihilator of G_1 in $\mathcal{P}(G)$. It is clear from Theorem II.2.3 that $QQ = Q$. Since ε annihilates Q, it follows from this that every element of $\mathcal{L}(G)$ annihilates Q, whence $\rho^{\cdot}$ is injective, in the present case. Looking at Theorem II.2.3 once more, we see immediately that $\rho^{\cdot}$ is also surjective. Thus, *via the injection morphism from G_1 to G, $\mathcal{L}(G_1)$ is isomorphic with $\mathcal{L}(G)$.*

Note that $\mathscr{L}(G)$ is finite-dimensional, because a differentiation of $\mathscr{P}(G)$ is already determined by its values on a finite system of F-algebra generators.

Theorem 3.1. *Let G be an affine algebraic F-group. The $\mathscr{P}(G)$-module homomorphism*

$$\mathscr{P}(G) \otimes \mathscr{L}(G) \to \mathscr{D}_F(\mathscr{P}(G))$$

determined by the condition that it send each element τ of $\mathscr{L}(G)$ onto $\tau_{\mathfrak{l}}$ is an isomorphism.

PROOF. Let $\sum_{i=1}^{n} a_i \otimes \tau_i$ be an element of $\mathscr{P}(G) \otimes \mathscr{L}(G)$ whose image in $\mathscr{D}(\mathscr{P}(G))$ is 0, where the τ_i's are F-linearly independent elements of $\mathscr{L}(G)$. Let a be an element of $\mathscr{P}(G)$, let x be an element of G, and apply the formal image of our element in $\mathscr{D}(\mathscr{P}(G))$ to the element $a \cdot x^{-1}$ of $\mathscr{P}(G)$. This yields the following relation in $\mathscr{P}(G)$:

$$\sum_{i=1}^{n} a_i(\tau_i)_{\mathfrak{l}}(a \cdot x^{-1}) = 0.$$

Now operate from the right with x, and use the fact that each $(\tau_i)_{\mathfrak{l}}$ commutes with this operation. This gives

$$\sum_{i=1}^{n} (a_i \cdot x)(\tau_i)_{\mathfrak{l}}(a) = 0.$$

Now apply ε, obtaining

$$\sum_{i=1}^{n} a_i(x)\tau_i(a) = 0.$$

Since this holds for every element a of $\mathscr{P}(G)$, and since the τ_i's are F-linearly independent, we must therefore have $a_i(x) = 0$ for each i. Since this holds for every element x of G, it follows that $a_i = 0$. Our conclusion is that the homomorphism of the theorem is injective.

Now let e be an element of $\mathscr{D}(\mathscr{P}(G))$. For every element x of G, define the endomorphism $e \cdot x$ of $\mathscr{P}(G)$ by

$$(e \cdot x)(a) = e(a \cdot x^{-1}) \cdot x.$$

One verifies directly that $e \cdot x$ is in fact an element of $\mathscr{D}(\mathscr{P}(G))$, so that the composite $\varepsilon \circ (e \cdot x)$ is an element of $\mathscr{L}(G)$. Let $(\tau_1, \ldots, \tau_n)$ be an F-basis of $\mathscr{L}(G)$, and write

$$\varepsilon \circ (e \cdot x) = \sum_{i=1}^{n} a_i(x)\tau_i$$

By applying this element of $\mathscr{L}(G)$ to suitable elements of $\mathscr{P}(G)$ one sees that the functions a_i defined by the above relation belong to $\mathscr{P}(G)$ (cf. Lemma I.1.1). It is verified directly that

$$e = \sum_{i=1}^{n} a_i(\tau_i)_{\mathfrak{l}}$$

showing that the homomorphism of the theorem is also surjective. $\qquad\square$

Assuming that G is irreducible, we show that *tensoring the isomorphism of Theorem 3.1 with* $[\mathscr{P}(G)]$ *yields an isomorphism of* $[\mathscr{P}(G)]$-*spaces from* $[\mathscr{P}(G)] \otimes \mathscr{L}(G)$ *to* $\mathscr{D}_F([\mathscr{P}(G)])$.

Let us write A for $\mathscr{P}(G)$. It suffices to show that $[A] \otimes_A \mathscr{D}(A)$ may be identified with $\mathscr{D}([A])$. Let $\mathscr{D}(A, [A])$ denote the $[A]$-space of all F-linear derivations from A to $[A]$. Clearly, the canonical $[A]$-linear map

$$[A] \otimes_A \mathscr{D}(A) \to \mathscr{D}(A, [A])$$

is injective. From the fact that A is finitely generated as an F-algebra, we see that, for every element σ of $\mathscr{D}(A, [A])$, there is a non-zero element s in A such that $s\sigma$ sends A into A. This shows that the canonical map is also surjective. Finally, since every element of $\mathscr{D}(A, [A])$ extends in one and only one way to an element of $\mathscr{D}([A])$, we may identify $\mathscr{D}(A, [A])$ with $\mathscr{D}([A])$.

Theorem 3.2. *Let G be an irreducible algebraic group. The dimension of $\mathscr{L}(G)$ is equal to the degree of transcendence of $[P(G)]$ over the base field.*

PROOF. By the isomorphism established just above, the dimension of $\mathscr{L}(G)$ as a vector space over the base field F equals the dimension of $\mathscr{D}_F([\mathscr{P}(G)])$ as a vector space over $[\mathscr{P}(G)]$. By Lemma II.1.2 and Theorem 2.2, $[\mathscr{P}(G)]$ is separable over F. Therefore, the theorem is an immediate consequence of Theorem 2.1. $\square$

If S is an irreducible affine algebraic F-set, one defines the *dimension* of S as the degree of transcendence of $[\mathscr{P}(S)]$ over F. Thus, Theorem 3.2 says that the dimension of G as an affine algebraic set is equal to the dimension of $\mathscr{L}(G)$.

Theorem 3.3. *Let $\rho: G \to H$ be a morphism of algebraic groups, where G and H are irreducible. Suppose that $\rho(G)$ is dense in H, and that $[\mathscr{P}(G)]$ is separable over $[\mathscr{P}(H) \circ \rho]$. Then the differential of ρ is surjective from $\mathscr{L}(G)$ to $\mathscr{L}(H)$.*

PROOF. Since $\rho(G)$ is dense in H, the transpose of ρ is injective from $\mathscr{P}(H)$ to $\mathscr{P}(G)$. Accordingly, we identify the elements g of $\mathscr{P}(H)$ with their images $g \circ \rho$ in $\mathscr{P}(G)$, so that we have $\mathscr{P}(H) \subset \mathscr{P}(G)$. Now let τ be an element of $\mathscr{L}(H)$, and consider the corresponding element τ_1 of $\mathscr{D}(\mathscr{P}(H))$. This has one and only one extension to an element of $\mathscr{D}([\mathscr{P}(H)])$. Since $[\mathscr{P}(G)]$ is separable over $[\mathscr{P}(H)]$, our element of $\mathscr{D}([\mathscr{P}(H)])$ extends further to yield an element, σ of $\mathscr{D}([\mathscr{P}(G)])$. Since $\mathscr{P}(G)$ is finitely generated as an algebra, there is a non-zero element a in $\mathscr{P}(G)$ such that $a\sigma$ stabilizes $\mathscr{P}(G)$. Choose an element x from G such that $a(x) \neq 0$, and consider the transform $(a\sigma) \cdot x$, defined as in the proof of Theorem 3.1. This is equal to $(a \cdot x)(\sigma \cdot x)$. We have

$$\varepsilon(a \cdot x) = a(x) \neq 0.$$

On the other hand, the restriction of $\sigma \cdot x$ to $\mathscr{P}(H)$ coincides with

$$\tau_{[} \cdot \rho(x) = \tau_{[}.$$

Thus, replacing σ with $\sigma \cdot x$, if necessary, we arrange to have $\varepsilon(a) \neq 0$. This ensures that the composite $\varepsilon \circ \sigma$ is defined on $\mathscr{P}(G)$. Clearly, it is a differentiation of $\mathscr{P}(G)$, i.e., an element of $\mathscr{L}(G)$. Its restriction to $\mathscr{P}(H)$ coincides with $\varepsilon \circ \tau_{[} = \tau$, and this means that $\tau = \rho'(\varepsilon \circ \sigma)$. $\square$

Let G be an algebraic group, H a closed subgroup of G. If J_H is the annihilator of H in $\mathscr{P}(G)$ then we may identify $\mathscr{P}(H)$ with $\mathscr{P}(G)/J_H$. Accordingly, we identify $\mathscr{L}(H)$ with the sub Lie algebra of $\mathscr{L}(G)$ consisting of the differentiations that annihilate J_H. Moreover, as we have seen earlier in this section, this sub Lie algebra remains the same if we replace H with H_1, i.e., the annihilator of J_H in $\mathscr{L}(G)$ annihilates also the Hopf ideal J_{H_1}, which contains J_H.

Now let K be another closed subgroup of G. If $H_1 \subset K_1$ (in particular, if $H \subset K$) then it is clear from the above that $\mathscr{L}(H) \subset \mathscr{L}(K)$. Moreover, *if $H_1 \subset K_1$ and $\mathscr{L}(H) = \mathscr{L}(K)$, then it follows that $H_1 = K_1$.* In order to see this, write P for J_{K_1} and Q for J_{H_1}, so that $P \subset Q$. By Theorem 3.2, our assumption that $\mathscr{L}(H_1) = \mathscr{L}(K_1)$ implies that $[\mathscr{P}(G)/P]$ and $[\mathscr{P}(G)/Q]$ have the same degree of transcendence over the base field. By Lemma II.3.8, this implies that $P = Q$, whence $H_1 = K_1$.

In general, these results concerning the relations between algebraic subgroups and sub Lie algebras cannot be strengthened. In particular, if F has non-zero characteristic, there are cases where two distinct irreducible algebraic subgroups have the same Lie algebra. Also, in any characteristic, there are sub Lie algebras of an $\mathscr{L}(G)$ that do not belong to any algebraic subgroup of G.

4. We determine the Lie algebra of the main general example of Chapter I. Here, we have a finite-dimensional algebra E over a field F, and we consider the group E^* of units of E, made into an affine algebraic F-group whose algebra of polynomial functions is generated by the restrictions to E^* of the elements of E° and their antipodes. In the case where F is a finite field, E^* is a finite group, $\mathscr{P}(E^*)$ is the algebra of all F-valued functions on E^*, and $(E^*)_1$ is the trivial group, so that $\mathscr{L}(E^*) = (0)$. Therefore, *we assume that F is an infinite field.* In this case, it is easy to see that the restriction images in $\mathscr{P}(E^*)$ of the elements of an F-basis of E° are algebraically independent over F. Moreover, the sub F-algebra of $\mathscr{P}(E^*)$ generated by these and the reciprocal of a certain polynomial in them (the restriction to E^* of the determinant of any injective finite-dimensional representation of E) is a sub Hopf algebra of $\mathscr{P}(E^*)$, and so coincides with $\mathscr{P}(E^*)$. In particular, $\mathscr{P}(E^*)$ is therefore an integral domain. Thus, *if the base field is infinite then E^* is irreducible.*

Let τ be an element of $\mathscr{L}(E^*)$. Via the restriction map from E° to $\mathscr{P}(E^*)$, τ yields an F-linear map from E° to F. Therefore, there is one and only one element τ' in E such that this linear map is the evaluation at τ'. Clearly, the map $\tau \mapsto \tau'$ is an injective F-linear map from $\mathscr{L}(E^*)$ to E. Conversely, given an element e of E, it is clear from our above description of $\mathscr{P}(E^*)$ that there is a differentiation τ of $\mathscr{P}(E^*)$ such that $\tau' = e$. Thus, the map $\tau \mapsto \tau'$ is an isomorphism of F-spaces from $\mathscr{L}(E^*)$ to E.

Since the image of E° in $\mathscr{P}(E^*)$ is stable under the right and left translation actions of E^*, the comultiplication δ of $\mathscr{P}(E^*)$ sends the image of E° into its tensor square. Hence, if f is an element of E°, there are elements f_i' and f_i'' of E° such that

$$\delta([f]) = \sum_{i=1}^{n} [f_i'] \otimes [f_i''],$$

where we have used $[\ \ ]$ to indicate restriction of a function on E to E^*. Thus, if x and y are elements of E^*, we have

$$f(xy) = \sum_{i=1}^{n} f_i'(x) f_i''(y).$$

This says that a certain polynomial function on $E \times E$ vanishes on the subset $E^* \times E^*$, which is easily seen to imply that that polynomial function is identically 0. Therefore, the above equality holds for all elements x and y of E.

Now let σ and τ be elements of $\mathscr{L}(E^*)$. Then we have

$$f([\sigma, \tau]') = [\sigma, \tau]([f]) = (\sigma \otimes \tau - \tau \otimes \sigma)(\delta([f]))$$

$$= \sum_{i=1}^{n} \sigma([f_i'])\tau([f_i'']) - \sum_{i=1}^{n} \tau([f_i'])\sigma([f_i''])$$

$$= \sum_{i=1}^{n} (f_i'(\sigma')f_i''(\tau') - f_i'(\tau')f_i''(\sigma'))$$

$$= f(\sigma'\tau' - \tau'\sigma').$$

Since this holds for every element f of E°, it follows that

$$[\sigma, \tau]' = \sigma'\tau' - \tau'\sigma'.$$

Our conclusion is that *the map $\tau \mapsto \tau'$ is an isomorphism of F-Lie algebras from $\mathscr{L}(E^*)$ to $\mathscr{L}(E)$.*

In particular, let us consider the case where E is the F-algebra $\mathrm{End}_F(V)$ of endomorphisms of a finite-dimensional F-space V. Let G be an affine algebraic F-group, and suppose we are given a morphism of affine algebraic F-groups $\rho\colon G \to E^*$. Let ρ' denote the morphism of Lie algebras from $\mathscr{L}(G)$ to $\mathscr{L}(\mathrm{End}_F(V))$ that is obtained by following up the differential $\rho^\cdot$ of ρ with the above isomorphism from $\mathscr{L}(E^*)$ to $\mathscr{L}(E)$. Thus, for τ in $\mathscr{L}(G)$, $\rho'(\tau)$ denotes the F-linear endomorphism $\rho^\cdot(\tau)'$ of V.

The definition of ρ' evidently generalizes to the situation where V is a locally finite G-module whose associated representative functions belong to $\mathscr{P}(G)$. We call such a G-module a *polynomial G-module*, and we let ρ denote the group homomorphism from G to the group $\mathrm{Aut}_F(V)$ of all F-linear automorphisms of V that defines the G-module structure. In the evident way, ρ is the direct limit of the system of finite-dimensional polynomial representations of G on the finite-dimensional G-stable subspaces of V. It is easy to see that the differentials of these finite-dimensional representations, extended as above, fit together to yield a morphism of F-Lie algebras $\rho': \mathscr{L}(G) \to \mathscr{L}(\mathrm{End}_F(V))$, which we call the *extended differential* of ρ.

Recall from Section I.2 that ρ defines the structure

$$\rho^*: V \to V \otimes \mathscr{P}(G)$$

of a $\mathscr{P}(G)$-comodule on V, from which ρ is recovered by the formula

$$\rho(x) = (i_V \otimes x^*) \circ \rho^*$$

for every x in G. The following lemma records the description of ρ' in the context of $\mathscr{P}(G)$-comodules.

Lemma 4.1. *Let G be an affine algebraic F-group, and suppose that*

$$\rho: G \to \mathrm{Aut}_F(V)$$

is the structure of a polynomial G-module. If ρ^* denotes the corresponding comodule structure, then the following equations hold for every element τ of $\mathscr{L}(G)$:

(1) $\rho'(\tau) = (i_V \otimes \tau) \circ \rho^*$;

(2) $\rho^* \circ \rho'(\tau) = (i_V \otimes \tau_{\mathfrak{l}}) \circ \rho^*$

PROOF. Consider an element γ/v of $\mathrm{End}_F(V)^\circ$, where γ is an element of V° and v is an element of V. The composite function $(\gamma/v) \circ \rho$ is an element of $\mathscr{P}(G)$, and the definition of ρ' gives, for every τ in $\mathscr{L}(G)$,

$$\tau((\gamma/v) \circ \rho) = (\gamma/v)(\rho'(\tau)).$$

Using the above formula connecting ρ and ρ^*, we write the value of $(\gamma/v) \circ \rho$ at an element x of G in the form

$$\gamma((i_V \otimes x^*)(\rho^*(v))) = (\gamma \otimes x^*)(\rho^*(v)).$$

This shows that

$$(\gamma/v) \circ \rho = (\gamma \otimes i_{\mathscr{P}(G)}) \circ \rho^*(v).$$

Substituting this in the above equation involving τ, we obtain

$$\gamma(\rho'(\tau)(v)) = (\gamma \otimes \tau)(\rho^*(v)).$$

Since this holds for every element γ of V° and every element v of V, formula (1) of the lemma is established.

Using this, we may write

$$\rho^* \circ \rho'(\tau) = (i_V \otimes i_{\mathscr{P}(G)} \otimes \tau) \circ (\rho^* \otimes i_{\mathscr{P}(G)}) \circ \rho^*.$$

Now (2) is obtained from this by replacing $(\rho^* \otimes i_{\mathscr{P}(G)}) \circ \rho^*$ with $(i_V \otimes \delta) \circ \rho^*$, where δ is the comultiplication of $\mathscr{P}(G)$. $\qquad\square$

If L is any Lie algebra over F, if V is an F-space and ρ a morphism of Lie algebras from L to $\mathscr{L}(\mathrm{End}_F(V))$, then we call ρ a *representation of L on V*, and we refer to V as an *L-module*. Thus, if V is a polynomial module for an algebraic group G, then the extended differential of the representation of G on V makes V into an $\mathscr{L}(G)$-module. It is seen directly that, *with this, a morphism of polynomial G-modules is also a morphism of $\mathscr{L}(G)$-modules.*

Proposition 4.2. *Let A and B be polynomial modules for an algebraic group G. Let $\alpha, \beta, \alpha \otimes \beta$ denote the representations of G on A, B, $A \otimes B$, respectively. For every element τ of $\mathscr{L}(G)$, one has*

$$(\alpha \otimes \beta)'(\tau) = \alpha'(\tau) \otimes i_B + i_A \otimes \beta'(\tau).$$

PROOF. By (1) of Lemma 4.1, we have

$$(\alpha \otimes \beta)'(\tau) = (i_A \otimes i_B \otimes \tau) \circ (\alpha \otimes \beta)^*.$$

In Section I.2, we saw that

$$(\alpha \otimes \beta)^* = (i_A \otimes i_B \otimes \mu) \circ s_{2,3} \circ (\alpha^* \otimes \beta^*),$$

where μ denotes the multiplication of $\mathscr{P}(G)$, and $s_{2,3}$ is the switching of the 2nd and 3rd tensor factors. We substitute the expression on the right for $(\alpha \otimes \beta)^*$ in the above equation involving τ, and we note that

$$\tau \circ \mu = \varepsilon \otimes \tau + \tau \otimes \varepsilon.$$

We obtain

$$
\begin{aligned}
(\alpha \otimes \beta)'(\tau) =& (i_A \otimes i_B \otimes \varepsilon \otimes \tau) \circ s_{2,3} \circ (\alpha^* \otimes \beta^*) \\
&+ (i_A \otimes i_B \otimes \tau \otimes \varepsilon) \circ s_{2,3} \circ (\alpha^* \otimes \beta^*) \\
=& (i_A \otimes \varepsilon \otimes i_B \otimes \tau) \circ (\alpha^* \otimes \beta^*) \\
&+ (i_A \otimes \tau \otimes i_B \otimes \varepsilon) \circ (\alpha^* \otimes \beta^*).
\end{aligned}
$$

The formula of the proposition follows upon noting that one has

$$(i_A \otimes \varepsilon) \circ \alpha^* = i_A \quad \text{and} \quad (i_A \otimes \tau) \circ \alpha^* = \alpha'(\tau)$$

as well as the analogous relations for β^*. $\qquad\square$

Proposition 4.3. *Let V be a finite-dimensional polynomial module for an affine algebraic F-group G, and let ρ denote the representation of G on V. For every element τ of $\mathscr{L}(G)$, the endomorphism $\rho'(\tau)$ of V is an F-linear combination of endomorphisms of the form $\rho(x) - i_V$, with x in G.*

PROOF. Suppose that α is an element of $\mathrm{End}_F(V)^\circ$ that annihilates every endomorphism of the form $\rho(x) - i_V$. This means that the representative function $\alpha \circ \rho$ is a constant, whence $\tau(\alpha \circ \rho) = 0$. Hence,

$$\alpha(\rho'(\tau)) = \rho'(\tau)(\alpha) = \tau(\alpha \circ \rho) = 0.$$

Thus, $\rho'(\tau)$ is annihilated by every element of $\mathrm{End}_F(V)^\circ$ annihilating each $\rho(x) - i_V$. $\qquad\square$

As we shall see in the next chapter, if F is of characteristic 0, Proposition 4.3 has a converse, saying that every endomorphism of the form $\rho(x) - i_V$ is an F-linear combination of products of elements of $\rho'(\mathscr{L}(G))$. With this, the Lie algebra becomes an extremely powerful tool for the structure and representation theory of algebraic groups *over fields of characteristic* 0.

Notes

1. At the end of Section 1, we gave an intrinsic definition of a perfect field, and we saw that every extension of a perfect field is separable. One should complete the picture by showing that every non-perfect field has an inseparable finite algebraic extension (whose degree is equal to the characteristic).

2. In connection with the discussion at the end of Section 3, concerning the correspondence between closed subgroups and sub Lie algebras, consider the following example. Let F be an infinite field of non-zero characteristic p. Let G be the direct product of two copies of the multiplicative group F^* of F. Let H be the subgroup of G consisting of the elements of the form (a, a^{p-1}), and let K be the subgroup consisting of the elements of the form (a, a^{-1}). It is easy to see that H and K are irreducible algebraic subgroups of G, and that $\mathscr{L}(H) = \mathscr{L}(K)$. However, if $p \neq 2$, we have $H \neq K$.

3. Let A be a finite-dimensional general F-algebra, i.e., a finite-dimensional F-space that is equipped with a bilinear composition. Let G be the group of all algebra automorphisms of A, regarded as an algebraic subgroup of the group of all F-linear automorphisms of A. By viewing the composition of A as a linear map from $A \otimes A$ to A and using Proposition 4.2, one shows that $\mathscr{L}(G)$, when identified with a sub Lie algebra of $\mathscr{L}(\mathrm{End}_F(A))$ in the canonical fashion, becomes a sub Lie algebra of the Lie algebra of derivations of A.

Chapter IV

Lie Algebras and Algebraic Subgroups

This chapter establishes the Lie algebra technique for the structure and representation theory of algebraic groups. Section 1 contains only special field-theoretical preparations. Section 2 develops the connections between the algebraic subgroups of an algebraic group G and the sub Lie algebras of $\mathscr{L}(G)$ fully, under the assumption that the base field be of characteristic 0. This assumption is retained in Section 3, which is devoted to reducing, as far as is possible in general, the representation theory of an algebraic group to that of its Lie algebra.

Section 4 determines the structurally basic adjoint representation of an algebraic group, as well as its differential. Again, the principal benefits can be had only in characteristic 0.

Section 5 returns to the characterization of algebraic subgroups by invariants, as given in Chapter II, providing concomitant characterizations of the Lie algebras of algebraic subgroups. These results play an important technical role in the algebraic-geometric theory of coset spaces given in Chapter XII. As a first application, this section establishes the expected relation between the dimensions of G, H and G/H, where H is a normal algebraic subgroup of the algebraic group G.

1. Lemma 1.1. *Let K be a field, and let $(u_1, \ldots, u_r)$ be a set of independent variables over K. Let S be a finite group of field automorphisms of $K(u_1, \ldots, u_r)$ that stabilizes K as well as the multiplicative group generated by the u_i's. Suppose that the representation of S on K is injective. Then $K(u_1, \ldots, u_r)^S$ is contained in a finitely generated purely transcendental extension field of K^S.*

PROOF. Let n denote the order of S, and choose new independent variables v_{ij} over K, where i ranges from 1 to r, and j ranges from 1 to n. Note that

44

the dimension of K over K^S is n, and choose a K^S-basis $(k_1, \ldots, k_n)$ of K. Let U denote the multiplicative group generated by the u_i's. Evidently, there is one and only one group homomorphism π from U to the multiplicative group of $K(v_{11}, \ldots, v_{rn})$ such that

$$\pi(u_i) = \sum_{j=1}^{n} k_j v_{ij}$$

for each i. We extend the action of S on K to an action by field automorphisms on $K(v_{11}, \ldots, v_{rn})$ leaving the v_{ij}'s fixed.

From Galois theory, we know that the K-space spanned by the automorphisms s' of K corresponding to the elements s of S is the space of all K^S-linear endomorphisms of K. Hence, for each index q from $(1, \ldots, n)$, there are elements c_{sq} in K such that the endomorphism $\sum_{s \in S} c_{sq} s'$ sends k_q onto 1 and annihilates every other k_j. Using these in conjunction with the extended action of S on $K(v_{11}, \ldots, v_{rn})$, we obtain

$$v_{iq} = \sum_{s \in S} c_{sq} s(\pi(u_i)).$$

This shows that the rn elements $s(\pi(u_i))$ of $K(v_{11}, \ldots, v_{rn})$ are algebraically independent over K.

Let N denote the multiplicative group generated by the elements $s(\pi(u_i))$. We define a group homomorphism η from U to N by forming the product of the S-conjugates of π, so that

$$\eta(u) = \prod_{s \in S} s(\pi(s^{-1}(u)))$$

for every element u of U. The algebraic independence of the elements $s(\pi(u_i))$ ensures that η is injective, and that the same is true for the K-algebra homomorphism from $K[U]$ to $K(v_{11}, \ldots, v_{rn})$ obtained from η in the evident fashion. By the definition of η, we have $\eta(s(u)) = s(\eta(u))$ for every element s of S and every element u of U. Therefore, our K-algebra homomorphism is also a morphism of S-modules, and so is therefore its unique extension to a field homomorphism

$$\tau : K(u_1, \ldots, u_r) \to K(v_{11}, \ldots, v_{rn}).$$

Now τ restricts to a K^S-linear field homomorphism from $K(u_1, \ldots, u_r)^S$ to $K(v_{11}, \ldots, v_{rn})^S = K^S(v_{11}, \ldots, v_{rn})$. $\qquad \square$

Lemma 1.2. *Let K be a field of characteristic 0, and let $K[[t]]$ be the K-algebra of integral power series in the variable t. Let $(a_1, \ldots, a_q)$ be a subset of K that is linearly independent over the field of rational numbers. Then the elements $t, \exp(a_1 t), \ldots, \exp(a_q t)$ of $K[[t]]$ are algebraically independent over K.*

PROOF. Every polynomial relation among the elements figuring in the lemma may be written in the form

$$\sum c(e_0, \ldots, e_q) t^{e_0} \exp((e_1 a_1 + \cdots + e_q a_q)t) = 0,$$

where the summation goes over a finite set of q-tuples $(e_0, \ldots, e_q)$ of non-negative integers. The assumption on the a_i's ensures that no two of the linear combinations $e_1 a_1 + \cdots + e_q a_q$ are equal. Therefore, the lemma will be established as soon as we have proved that, if $b_1, \ldots, b_n$ are n distinct elements of K, the n power series $\exp(b_i t)$ are linearly independent over the polynomial algebra $K[t]$.

We do this by induction on n, and accordingly suppose that the result holds in the lower cases. Let $p_1, \ldots, p_n$ be elements of $K[t]$ such that

$$\sum_{i=1}^{n} p_i \exp(b_i t) = 0.$$

Multiplying by $\exp(-b_n t)$, and then differentiating k times with respect to t (indicated by $^{(k)}$), we obtain

$$p_n^{(k)} + \sum_{i=1}^{n-1} p_{ik} \exp((b_i - b_n)t) = 0,$$

where the p_{ik}'s are elements of $K[t]$ determined recursively by

$$p_{i0} = p_i; \qquad p_{i(h+1)} = p_{ih}^{(1)} + (b_i - b_n)p_{ih}.$$

If k is large enough, we have $p_n^{(k)} = 0$, and then the inductive hypothesis gives $p_{ik} = 0$ for each i. Hence, the recursion relation yields

$$p_{i(k-1)}^{(1)} + (b_i - b_n)p_{i(k-1)} = 0.$$

Since $b_i - b_n \neq 0$, this gives $p_{i(k-1)} = 0$, and we can repeat the argument until we obtain $p_i = 0$. $\qquad\qquad\square$

2. Let F be a field, G an affine algebraic F-group, τ an element of $\mathscr{L}(G)$. We regard $\mathscr{P}(G)^\circ$ as an F-algebra, with the multiplication obtained by dualizing the comultiplication δ of $\mathscr{P}(G)$. Let J_τ denote the set of all elements f of $\mathscr{P}(G)$ such that $\tau^n(f) = 0$ for every non-negative exponent n, where we agree that τ^0 is the identity element ε of $\mathscr{P}(G)^\circ$. We wish to show that J_τ is a bi-ideal. Since J_τ is evidently an ideal and since $\varepsilon(J_\tau) = (0)$, it suffices to show that

$$\delta(J_\tau) \subset J_\tau \otimes \mathscr{P}(G) + \mathscr{P}(G) \otimes J_\tau.$$

For all non-negative integers p and q, we have

$$\tau^{p+q} = (\tau^p \otimes \tau^q) \circ \delta,$$

whence $\delta(J_\tau)$ is contained in the kernel of the element $\tau^p \otimes \tau^q$ of

$$(\mathscr{P}(G) \otimes \mathscr{P}(G))^\circ.$$

The intersection of the family of these kernels is evidently

$$J_\tau \otimes \mathscr{P}(G) + \mathscr{P}(G) \otimes J_\tau,$$

so that we have the desired conclusion.

Let G_τ denote the annihilator in G of J_τ. Since J_τ is a bi-ideal of $\mathscr{P}(G)$, we know from Proposition I.4.1 that G_τ is an algebraic subgroup of G. In the case where F is of non-zero characteristic p, we have $(f - \varepsilon(f))^p \in J_\tau$ for every element f of $\mathscr{P}(G)$, whence G_τ consists of the neutral element ε alone. On the other hand, if F is of characteristic 0, the group G_τ is significant, by virtue of the following theorem.

Theorem 2.1. *Let F be a field of characteristic 0, and let G be an affine algebraic F-group. For τ in $\mathscr{L}(G)$, the group G_τ is contained in every algebraic subgroup of G whose Lie algebra contains τ. It is irreducible, and τ belongs to $\mathscr{L}(G_\tau)$.*

PROOF. Let H be an algebraic subgroup of G whose Lie algebra contains τ. Let J_H denote the annihilator of H in $\mathscr{P}(G)$. From the assumption that τ belongs to $\mathscr{L}(H)$, we have $\tau_{\lceil}(J_H) \subset J_H$. For every positive integer n, we have $\tau^n = \tau \circ \tau_{\lceil}^{n-1}$. Since τ and ε annihilate J_H, it follows that $J_H \subset J_\tau$, whence $G_\tau \subset H$.

Next, we show that J_τ is a prime ideal. Let a and b be elements of $\mathscr{P}(G) \setminus J_\tau$. Let p be the smallest non-negative exponent such that $\tau^p(a) \neq 0$. Similarly, define q with respect to b. Using that $\tau_{\lceil}$ is a derivation, we obtain

$$\tau^{p+q}(ab) = \varepsilon(\tau_{\lceil}^{p+q}(ab)) = \sum_{u+v=p+q} \frac{(p+q)!}{u!\,v!}\, \tau^u(a)\tau^v(b).$$

The expression on the right reduces to the single non-zero term where $u = p$ and $v = q$. Thus $\tau^{p+q}(ab) \neq 0$, so that ab does not belong to J_τ.

Now it is clear that the theorem will be established as soon as we have shown that the annihilator of G_τ in $\mathscr{P}(G)$ coincides with J_τ. In the case where F is algebraically closed, this follows immediately by applying Theorem II.3.5 to $\mathscr{P}(G)/J_\tau$. In the general case, we shall be able to replace this appeal with an explicit specialization argument based on the fact, to be shown, that $\mathscr{P}(G)/J_\tau$ is contained in a finitely generated purely transcendental extension field of F.

Choose a finite-dimensional left G-stable sub F-space V of $\mathscr{P}(G)$ that generates $\mathscr{P}(G)$ as an F-algebra. Then V is stable under the derivation $\tau_{\lceil}$ of $\mathscr{P}(G)$. There is a finite Galois extension K of F such that the characteristic polynomial of the restriction of $\tau_{\lceil}$ to V splits into a product of linear factors in the polynomial algebra over K. We consider the K-algebra $\mathscr{P}(G) \otimes K$, and the K-linear extensions of ε, τ and $\tau_{\lceil}$, which we continue to denote by the same letters.

By the choice of K, the characteristic roots, $c_1, \ldots, c_n$ say, of the restriction of $\tau_{\lceil}$ to $V \otimes K$ lie in K, so that $V \otimes K$ is the direct sum of $\tau_{\lceil}$-stable K-subspaces $V_1, \ldots, V_n$ such that each V_i is annihilated by some power of $\tau_{\lceil} - c_i$. Now let t be an auxiliary variable, and form the K-algebra $K[[t]]$ of integral formal power series. Then the infinite sum

$$\exp(t\tau_{\lceil}) = \sum_{i \geq 0} \frac{t^i}{i!}\, \tau_{\lceil}^i$$

has a meaning as a map from $\mathscr{P}(G) \otimes K$ to $\mathscr{P}(G) \otimes K[[t]]$. Since $\tau_{[}$ is a K-linear derivation, this exponential map is a homomorphism of K-algebras. If μ_i denotes the restriction of $\tau_{[} - c_i$ to V_i, then the restriction of $\exp(t\tau_{[})$ to V_i is evidently $\exp(tc_i)\exp(t\mu_i)$.

Now we consider the map $\exp(t\tau)$ from $\mathscr{P}(G) \otimes K$ to $K[[t]]$. This coincides with $\varepsilon \circ \exp(t\tau_{[})$, so that it is a homomorphism of K-algebras. Moreover, since some power of μ_i is equal to 0, the above expression for the restriction of $\exp(t\tau_{[})$ to V_i shows that $\exp(t\tau)(V_i)$ is contained in the K-algebra $K[t, \exp(tc_i)]$. Since the V_i's generate $\mathscr{P}(G) \otimes K$ as a K-algebra, it follows that

$$\exp(t\tau)(\mathscr{P}(G) \otimes K) \subset K[t, exp(tc_1), \dots, exp(tc_n)].$$

Since the additive subgroup of K that is generated by the c_i's is torsion-free, it is the free abelian group based on a set $(a_1, \dots, a_q)$ of integral linear combinations of the c_i's. Now it is clear that the field of fractions of

$$K[t, \exp(tc_1), \dots, \exp(tc_n)]$$

is $K(t, \exp(ta_1), \dots, \exp(ta_q))$. By Lemma 1.2, this last field is a purely transcendental extension of K, the displayed elements constituting a transcendence basis.

Let S denote the Galois group of K over F, and let S act coefficient-wise on $K[[t]]$. Since the characteristic polynomial of the restriction of $\tau_{[}$ to $V \otimes K$ has its coefficients in F, the set $(c_1, \dots, c_n)$ of characteristic roots is S-stable. Therefore, S stabilizes the multiplicative group generated by $(t, \exp(ta_1), \dots, \exp(ta_n))$ in $K[[t]]$. The action of S on $K[[t]]$ extends uniquely to an action by field automorphisms on the field of fractions of $K[[t]]$, under which the subfield $K(t, \exp(ta_1), \dots, \exp(ta_q))$ is stable. We have

$$\exp(t\tau)(\mathscr{P}(G)) \subset K(t, \exp(ta_1), \dots, \exp(ta_q))^S$$

and we can apply Lemma 1.1 to conclude that $\exp(t\tau)(\mathscr{P}(G))$ is contained in a finitely generated purely transcendental extension of F. Choose a set $(x_1, \dots, x_m)$ of algebraically independent elements such that this extension field is $F(x_1, \dots, x_m)$. Clearly, the kernel of the restriction of $\exp(t\tau)$ to $\mathscr{P}(G)$ is precisely J_τ, and we may regard $\mathscr{P}(G)/J_\tau$ as a sub F-algebra of $F(x_1, \dots, x_m)$.

Now let J denote the annihilator of G_τ in $\mathscr{P}(G)$. Clearly, $J_\tau \subset J$. Suppose that, contrary to what we wish to prove, we have $J_\tau \neq J$, and choose an element b from $J \setminus J_\tau$. Indicate the canonical homomorphism

$$\mathscr{P}(G) \to \mathscr{P}(G)/J_\tau$$

by $a \mapsto a'$. Let $(p_1, \dots, p_j)$ be a system of representatives in $\mathscr{P}(G) \setminus J_\tau$ for a set of F-algebra generators of $\mathscr{P}(G)/J_\tau$. Write $p_i' = f_i/g_i$ and $b' = f/g$, where the f_i's, g_i's, f and g are non-zero elements of $F[x_1, \dots, x_m]$. There are elements $r_1, \dots, r_m$ in F such that

$$(g_1 \cdots g_m gf)(r_1, \dots, r_m) \neq 0.$$

Clearly, the specialization $x_i \mapsto r_i$ defines an F-algebra homomorphism from $\mathscr{P}(G)/J_\tau$ to F not annihilating b'. This may be viewed as an F-algebra homomorphism from $\mathscr{P}(G)$ to F annihilating J_τ but not b. Thus, we have an element of G_τ not annihilating b, contradicting the assumption that b belongs to J. Therefore, we have the conclusion that $J = J_\tau$. $\square$

We note that G_τ *is commutative*. This is seen as follows. Let s stand for the switch of tensor factors in $\mathscr{P}(G) \otimes \mathscr{P}(G)$. Let p and q be non-negative exponents. We have

$$(\tau^p \otimes \tau^q) \circ s \circ \delta = (\tau^q \otimes \tau^p) \circ \delta = \tau^{p+q} = (\tau^p \otimes \tau^q) \circ \delta,$$

whence

$$(\tau^p \otimes \tau^q) \circ (s \circ \delta - \delta) = 0.$$

This shows that $s \circ \delta - \delta$ sends $\mathscr{P}(G)$ into $J_\tau \otimes \mathscr{P}(G) + \mathscr{P}(G) \otimes J_\tau$. Hence, if x and y are elements of G_τ, we have

$$(x \otimes y) \circ (s \circ \delta - \delta) = 0,$$

which means that $xy = yx$.

Theorem 2.2. *Let G be an algebraic group over a field of characteristic 0. For every sub Lie algebra L of $\mathscr{L}(G)$, let G_L denote the intersection of the family of all algebraic subgroups of G whose Lie algebras contain L. Then G_L is an irreducible algebraic subgroup of G, and $L \subset \mathscr{L}(G_L)$. If H and K are irreducible algebraic subgroups of G, then $H \subset K$ if and only if $\mathscr{L}(H) \subset \mathscr{L}(K)$.*

PROOF. Evidently, G_L is an algebraic subgroup of G. By the first part of Theorem 2.1, we have $G_\tau \subset G_L$ for every element τ of L. This implies that $\mathscr{L}(G_\tau) \subset \mathscr{L}(G_L)$, and now the last part of Theorem 2.1 gives $\tau \in \mathscr{L}(G_L)$. Thus, $L \subset \mathscr{L}(G_L)$. Since $\mathscr{L}((G_L)_1) = \mathscr{L}(G_L)$, this shows also that $G_L \subset (G_L)_1$, so that G_L is irreducible.

Let H be any irreducible algebraic subgroup of G. Since $G_{\mathscr{L}(H)} \subset H$, we have $\mathscr{L}(G_{\mathscr{L}(H)}) \subset \mathscr{L}(H)$. From the above, we have the reversed inclusion, so that $\mathscr{L}(G_{\mathscr{L}(H)}) = \mathscr{L}(H)$. Since H and $G_{\mathscr{L}(H)}$ are irreducible and $G_{\mathscr{L}(H)} \subset H$, we know from the end of Section III.3 that therefore $G_{\mathscr{L}(H)} = H$. Now, if H and K are irreducible algebraic subgroups of G, and $\mathscr{L}(H) \subset \mathscr{L}(K)$, then we have $G_{\mathscr{L}(H)} \subset G_{\mathscr{L}(K)}$, i.e., $H \subset K$. Conversely, if $H \subset K$, then it is clear from the definitions that $\mathscr{L}(H) \subset \mathscr{L}(K)$. $\square$

Theorem 2.3. *Let $\rho \colon G \to H$ be a morphism of algebraic groups over a field of characteristic 0, and let K be the kernel of ρ. Then the kernel of the differential ρ' coincides with $\mathscr{L}(K)$.*

PROOF. In any characteristic, it is clear from the definitions that $\mathscr{L}(K)$ is contained in the kernel of ρ'. Conversely, let τ be an element of the kernel of

$\rho^{\cdot}$. Let f be an element of $\mathscr{P}(H)$, and let x be an element of G. Then we have

$$\tau_{[}(f \circ \rho)(x) = \tau((f \circ \rho) \cdot x) = \tau((f \cdot \rho(x)) \circ \rho)$$
$$= \rho^{\cdot}(\tau)(f \cdot \rho(x)) = 0.$$

Thus, $\tau_{[}(f \circ \rho) = 0$, whence $f \circ \rho - \varepsilon(f \circ \rho)$ belongs to the annihilator J_{τ} of G_{τ}. Since this holds for every element f of $\mathscr{P}(H)$, it follows that $G_{\tau} \subset K$, whence $\mathscr{L}(G_{\tau}) \subset \mathscr{L}(K)$. By Theorem 2.1, this gives $\tau \in \mathscr{L}(K)$. $\qquad\square$

3. Theorem 2.1 yields the following important converse to Proposition III.4.3.

Theorem 3.1. *Let G be an irreducible affine algebraic F-group, where F is a field of characteristic 0. Let ρ be a polynomial representation of G on a finite-dimensional F-space V. Every endomorphism of the form $\rho(x) - i_V$, with x in G, is an F-linear combination of products of elements of $\rho'(\mathscr{L}(G))$.*

PROOF. Let $\tau_1, \ldots, \tau_n$ be elements of $\mathscr{L}(G)$. From Lemma III.4.1, we obtain

$$\rho^* \circ \rho'(\tau_1) \cdots \rho'(\tau_n) = (i_V \otimes (\tau_1)_{[} \cdots (\tau_n)_{[}) \circ \rho^*.$$

Composing this with $i_V \otimes \varepsilon$, we find that

$$\rho'(\tau_1) \cdots \rho'(\tau_n) = (i_V \otimes \tau_1 \cdots \tau_n) \circ \rho^*.$$

Now let v be an element of V° and v an element of V. One verifies directly from the definitions that

$$(v \otimes i_{\mathscr{P}(G)})(\rho^*(v)) = (v/v) \circ \rho.$$

Next, let γ be an element of $\mathscr{P}(G)^{\circ}$. The last relation gives

$$\gamma((v/v) \circ \rho) = (v \otimes \gamma)(\rho^*(v)) = (v/v)((i_V \otimes \gamma) \circ \rho^*).$$

Since the functions v/v span $\mathrm{End}_F(V)^{\circ}$ over F, it follows that, for every α in $\mathrm{End}_F(V)^{\circ}$, we have

$$\gamma(\alpha \circ \rho) = \alpha((i_V \otimes \gamma) \circ \rho^*).$$

Now apply α to our above expression for the product of the $\rho'(\tau_i)$'s, and use the last relation for $\gamma = \tau_1 \cdots \tau_n$. This yields

$$\alpha(\rho'(\tau_1) \cdots \rho'(\tau_n)) = (\tau_1 \cdots \tau_n)(\alpha \circ \rho).$$

Suppose that α annihilates every non-empty product of elements of $\rho'(\mathscr{L}(G))$, and fix an element τ of $\mathscr{L}(G)$. Our last result shows that $\alpha \circ \rho$ is annihilated by every τ^n with $n > 0$. Therefore, the element $\alpha \circ \rho - \varepsilon(\alpha \circ \rho)$ of $\mathscr{P}(G)$ belongs to the annihilator J_{τ} of G_{τ}, which means that α annihilates every endomorphism of the form $\rho(x) - i_V$ with x in G_{τ}. We conclude from this that every such endomorphism is an F-linear combination of products of elements of $\rho'(\mathscr{L}(G))$.

If x_i is an element of G_{τ_i} $(i = 1, \ldots, n)$, we can write the endomorphism $\rho(x_1 \cdots x_n) - i_V$ as an integral linear combination of products of the endomorphisms $\rho(x_i) - i_V$, and it follows from the above that this endomorphism is also an F-linear combination of products of elements of $\rho'(\mathscr{L}(G))$. Thus, we conclude that every element α of $\operatorname{End}_F(V)^\circ$ with the property assumed above annihilates every endomorphism $\rho(x) - i_V$ with x in the group generated by the family of all G_τ's. It follows that α annihilates also the endomorphisms $\rho(x) - i_V$ with x in the closure, H say, of this group. It is clear from Theorem 2.1 that $\mathscr{L}(H) = \mathscr{L}(G)$, whence $H = G$. $\qquad\square$

Corollary 3.2. *Let F be a field of characteristic 0, let G be an irreducible affine algebraic F-group, and let V be a polynomial G-module. Then, with respect to the structure of an $\mathscr{L}(G)$-module given by ρ', the G-fixed part of V coincides with the $\mathscr{L}(G)$-annihilated part, and the family of sub G-modules of V coincides with the family of its sub $\mathscr{L}(G)$-modules.*

4. Let G be an algebraic group. For each element x of G, let c_x denote the conjugation effected by x on G, so that

$$c_x(y) = xyx^{-1}.$$

Since c_x is an automorphism of affine algebraic groups, we have the differential c_x', which is a Lie algebra automorphism of $\mathscr{L}(G)$. Clearly, the map sending each element x of G onto c_x' makes $\mathscr{L}(G)$ into a polynomial G-module. We shall denote this polynomial representation of G by α. It is called the *adjoint representation* of G. Thus,

$$\alpha(x) = c_x'.$$

In order to make this explicit, let us regard the elements of G, as well as those of $\mathscr{L}(G)$, as elements of the F-algebra $\mathscr{P}(G)^\circ$. Using some of the definitions and formal results of Section I.2, we obtain, with x in G, τ in $\mathscr{L}(G)$ and f in $\mathscr{P}(G)$,

$$\alpha(x)(\tau)(f) = \tau(x^{-1} \cdot f \cdot x) = (\tau \circ (x^{-1})_\mathfrak{l} \circ x_\mathfrak{l})(f)$$
$$= (\tau x^{-1} \circ x_\mathfrak{l})(f) = (x\tau x^{-1})(f),$$

whence

$$\alpha(x)(\tau) = x\tau x^{-1}.$$

Next, we calculate the differential of α. Let σ and τ be elements of $\mathscr{L}(G)$ and let f be an element of $\mathscr{P}(G)$. Let f° stand for the element of $\mathscr{L}(G)^\circ$ given by $f^\circ(\rho) = \rho(f)$ for every ρ in $\mathscr{L}(G)$. Then, if f°/τ has its usual meaning as an element of $\operatorname{End}_F(\mathscr{L}(G))^\circ$, the composite $(f^\circ/\tau) \circ \alpha$ is an element of $\mathscr{P}(G)$, and we have

$$\alpha'(\sigma)(\tau)(f) = f^\circ(\alpha'(\sigma)(\tau)) = (f^\circ/\tau)(\alpha'(\sigma))$$
$$= \sigma((f^\circ/\tau) \circ \alpha).$$

Now we obtain an explicit expression for $(f^\circ/\tau) \circ \alpha$ as a function of f. We have, with x in G,

$$((f^\circ/\tau) \circ \alpha)(x) = f^\circ(\alpha(x)(\tau)) = f^\circ(x\tau x^{-1}) = (x\tau x^{-1})(f).$$

In Hopf algebra terms,

$$x\tau x^{-1} = x \circ \tau_[\circ (x^{-1})_[= x \circ \tau_[\circ (i_{\mathscr{P}(G)} \otimes x \circ \eta) \circ \delta$$
$$= x \circ (\tau_[\otimes x \circ \eta) \circ \delta = (x \otimes x) \circ (\tau_[\otimes \eta) \circ \delta$$
$$= x \circ \mu \circ (\tau_[\otimes \eta) \circ \delta.$$

This shows that

$$(f^\circ/\tau) \circ \alpha = (\mu \circ (\tau_[\otimes \eta) \circ \delta)(f).$$

From our first expression for $\alpha'(\sigma)(\tau)(f)$ we see now that

$$\alpha'(\sigma)(\tau) = \sigma \circ \mu \circ (\tau_[\otimes \eta) \circ \delta.$$

Replacing $\sigma \circ \mu$ with $\sigma \otimes \varepsilon + \varepsilon \otimes \sigma$, we find that

$$\alpha'(\sigma)(\tau) = \sigma\tau + \tau(\sigma \circ \eta).$$

Finally, from $\mu \circ (\eta \otimes i_{\mathscr{P}(G)}) \circ \delta = u \circ \varepsilon$, we obtain

$$\sigma \circ \mu \circ (\eta \otimes i_{\mathscr{P}(G)}) \circ \delta = 0,$$

whence, as just above,

$$\sigma \circ \eta + \sigma = 0.$$

Thus, we have

$$\alpha'(\sigma)(\tau) = \sigma\tau - \tau\sigma = [\sigma, \tau].$$

Denoting the derivation effected by τ on $\mathscr{L}(G)$ by D_τ, we summarize our results as follows.

Theorem 4.1. *Let α denote the adjoint representation of the algebraic group G on its Lie algebra $\mathscr{L}(G)$. Then, for every element x of G and every element τ of $\mathscr{L}(G)$, we have*

$$\alpha(x)(\tau) = x\tau x^{-1}.$$

The extended differential of α is given by

$$\alpha'(\tau) = D_\tau.$$

The next three theorems concern basic properties of the adjoint representation, which are decisive only in the case where the base field is of characteristic 0.

Theorem 4.2. *If G is an irreducible algebraic group over a field of characteristic 0, then the kernel of the adjoint representation of G coincides with the center of G.*

PROOF. Let x be an element of the kernel of the adjoint representation. It is evident from Theorem 4.1 that x commutes with every element of $\mathscr{L}(G)$. By Proposition I.2.1, this implies that $x_{\mathfrak{l}}$ commutes with $\tau_{\mathfrak{l}}$ for every element τ of $\mathscr{L}(G)$. Now it follows from Theorem 3.1 that $x_{\mathfrak{l}}$ commutes with $y_{\mathfrak{l}}$ for every element y of G, whence $xy = yx$ for every y in G. $\qquad\square$

A Lie algebra L is called *abelian* if $[x, y] = 0$ for all elements x and y of L.

Theorem 4.3. *Let F be a field, G an irreducible affine algebraic F-group. If G is abelian, so is $\mathscr{L}(G)$. Conversely, if $\mathscr{L}(G)$ is abelian and F is of characteristic 0, then G is abelian.*

PROOF. Suppose that G is abelian. Then the adjoint representation of G is trivial, whence its extended differential is the 0-map. By Theorem 4.1, this implies that $\mathscr{L}(G)$ is abelian.

Now suppose that F is of characteristic 0, and that $\mathscr{L}(G)$ is abelian. This last assumption, in conjunction with Proposition I.2.1, implies that the sub F-algebra of $\mathrm{End}_F(\mathscr{P}(G))$ generated by the elements $\tau_{\mathfrak{l}}$ with τ in $\mathscr{L}(G)$ is commutative. By Theorem 3.1, this implies that the elements $x_{\mathfrak{l}}$, with x in G, commute with each other, whence G is abelian. $\qquad\square$

An *ideal* of a Lie algebra L is a subspace that is stable under every D_x with x in L. Clearly, the ideals are precisely the kernels of Lie algebra homomorphisms.

Theorem 4.4. *Let F be a field, G an irreducible algebraic F-group, K an irreducible algebraic subgroup of G. If K is normal in G, then $\mathscr{L}(K)$ is an ideal of $\mathscr{L}(G)$. Conversely, if $\mathscr{L}(K)$ is an ideal of $\mathscr{L}(G)$, and if F is of characteristic 0, then K is normal in G.*

PROOF. Suppose that K is normal in G. Let I be the annihilator of K in $\mathscr{P}(G)$, let x be an element of G and let τ be an element of $\mathscr{L}(K)$. Then $x^{-1} \cdot I \cdot x \subset I$, i.e., $x_{\mathfrak{l}} x_{\mathfrak{l}}^{-1}$ stabilizes I, whence $\tau \circ x_{\mathfrak{l}} x_{\mathfrak{l}}^{-1}$ annihilates I, i.e., $x\tau x^{-1}$ annihilates I. By Theorem 4.1, this means that $\mathscr{L}(K)$ is G-stable under the adjoint representation α. By Proposition III.4.3, this implies that $\mathscr{L}(K)$ is $\mathscr{L}(G)$-stable under α'. By Theorem 4.1, this means that $\mathscr{L}(K)$ is an ideal of $\mathscr{L}(G)$.

Now suppose that F is of characteristic 0, and that $\mathscr{L}(K)$ is an ideal of $\mathscr{L}(G)$. This means that $\mathscr{L}(K)$ is $\mathscr{L}(G)$-stable under α'. By Corollary 3.2, it follows that $\mathscr{L}(K)$ is G-stable under α. Now let x be an element of G. Then xKx^{-1} is an algebraic subgroup of G whose annihilator in $\mathscr{P}(G)$ is $x \cdot I \cdot x^{-1}$, where I is the annihilator of K. It follows that

$$\mathscr{L}(xKx^{-1}) = \mathscr{L}(K) \circ (x^{-1})_{\mathfrak{l}} \circ x_{\mathfrak{l}} = x\mathscr{L}(K)x^{-1} = \alpha(x)(\mathscr{L}(K)) = \mathscr{L}(K).$$

By Theorem 2.2, this implies that $xKx^{-1} = K$. $\qquad\square$

5. Recall that Theorem II.2.1 provides a finite set E of semi-invariants for an algebraic subgroup H of an algebraic group G that characterizes H. The next result says that E also characterizes $\mathscr{L}(H)$ as a sub Lie algebra of $\mathscr{L}(G)$.

Proposition 5.1. *In the notation of Theorem* II.2.1, $\mathscr{L}(H)$ *consists precisely of those elements* τ *of* $\mathscr{L}(G)$ *which satisfy* $\tau_{\mathfrak{l}}(e) \in Fe$ *for every e in the finite set E of H-semi-invariants constructed in the proof of that theorem.*

PROOF. It is clear from Proposition III.4.3 that every element of $\mathscr{L}(H)$ satisfies the stated condition. Now suppose that τ is an element of $\mathscr{L}(G)$ satisfying this condition. Recall from the proof of Theorem II.2.1 that the elements of E are the functions σ_i/s, where s is an F-space generator of the canonical image S of $\bigwedge^d(V \cap I)$ in $\bigwedge^d(V)$, and the σ_i's are basis elements of the annihilator of S in $\bigwedge^d(V)^\circ$. Since $\tau_{\mathfrak{l}}(\sigma_i/s)$ is a scalar multiple of σ_i/s, while ε annihilates σ_i/s, we have $\tau(\sigma_i/s) = 0$. If $\tau \cdot s$ denotes the transform of s by τ with respect to the extended differential of the representation of G on $\bigwedge^d(V)$, this means that $\sigma_i(\tau \cdot s) = 0$. Thus, $\tau \cdot s$ is annihilated by the annihilator of S in $\bigwedge^d(V)^\circ$, so that $\tau \cdot s$ belongs to S.

Now, if t is an element of $V \cap I$, we have $ts = 0$ in $\bigwedge(V)$. It follows from Proposition III.4.2 that $\mathscr{L}(G)$ acts by derivations on $\bigwedge(V)$. Therefore, we have

$$\tau_{\mathfrak{l}}(t)s + t(\tau \cdot s) = 0$$

because the expression on the left is the transform of ts by τ. By the definition of S, this implies that $\tau_{\mathfrak{l}}(t)$ belongs to $V \cap I$. Since $V \cap I$ generates I as an ideal, it follows that $\tau_{\mathfrak{l}}$ stabilizes the annihilator I of H in $\mathscr{P}(G)$. Therefore, $\tau(I) = (0)$, which means that τ belongs to $\mathscr{L}(H)$. $\qquad\square$

The next result makes a similar addition to Theorem II.2.2, concerning the characterization of the Lie ideal corresponding to a normal algebraic subgroup H of an algebraic group G.

Proposition 5.2. *In the notation of Theorem* II.2.2, $\mathscr{L}(H)$ *consists precisely of those elements* τ *of* $\mathscr{L}(G)$ *which satisfy* $\tau_{\mathfrak{l}}(q) = 0$ *for every element q of the finite set Q of H-invariants constructed in the proof of that theorem.*

PROOF. One sees immediately from Proposition III.4.3 that every element of $\mathscr{L}(H)$ satisfies the stated condition. In order to prove the converse, we must consider the differential of the representation σ of G on U used in the proof of Theorem II.2.2.

Recall that we considered a certain finite-dimensional left G-stable subspace J of $\mathscr{P}(G)$, that U is a sub F-algebra of $\mathrm{End}_F(J)$ stable under the conjugations effected by the elements $\rho(x)$ with x in G, where ρ denotes the representation by left translations on J, and that σ is defined by

$$\sigma(x)(u) = \rho(x)u\rho(x)^{-1}.$$

Let K denote the group of units of $\mathrm{End}_F(J)$, so that ρ is a morphism of affine algebraic F-groups from G to K. Let us identify $\mathscr{L}(K)$ with $\mathrm{End}_F(J)$ in the canonical fashion, so that the adjoint representation of K becomes a representation, γ say, of K on $\mathrm{End}_F(J)$. Then we see from Theorem 4.1 that σ is the representation of G on U that is induced by the representation $\gamma \circ \rho$ of G on $\mathrm{End}_F(J)$. It follows that σ' is the representation of $\mathscr{L}(G)$ on U that is induced by the representation $\gamma' \circ \rho'$ of $\mathscr{L}(G)$ on $\mathrm{End}_F(J)$. Now we see from the second part of Theorem 4.1 that $\sigma'(\tau) = D_{\rho \cdot (\tau)}$ for every element τ of $L(G)$, where the identification of $\mathscr{L}(K)$ with $\mathrm{End}_F(J)$ is to be used in interpreting the expression on the right. This means that

$$\sigma'(\tau)(u) = \rho'(\tau)u - u\rho'(\tau),$$

for every element u of U.

Now suppose that $\tau_{[}(Q) = (0)$. By the definition of Q, this implies that $\tau_{[}$ annihilates the space of representative functions associated with σ, so that τ belongs to the kernel of σ'. By the above, this means that $\rho'(\tau)$ commutes with every element of U. By the proof of Theorem III.2.2, this implies that τ satisfies the condition of Proposition 5.1, whence τ belongs to $\mathscr{L}(H)$. $\square$

Lemma 5.3. *Let F be an algebraically closed field, G an irreducible affine algebraic F-group, H an algebraic subgroup of G. Suppose that the left elementwise fixer of $\mathscr{P}(G)^H$ in G coincides with H. Then $[\mathscr{P}(G)]^H = [\mathscr{P}(G)^H]$.*

PROOF. Let f be an element of $[\mathscr{P}(G)]^H$. We must prove that f belongs to $[\mathscr{P}(G)^H]$. Choose a non-zero element v in $\mathscr{P}(G)$ such that vf lies in $\mathscr{P}(G)$. Now f belongs to the ring of fractions $\mathscr{P}(G)[v^{-1}]$, which is a finitely generated integral domain F-algebra. The F-algebra homomorphisms from $\mathscr{P}(G)[v^{-1}]$ to F are the unique extensions of the F-algebra homomorphisms from $\mathscr{P}(G)$ to F that do not annihilate v, i.e., of the elements x of G such that $v(x) \neq 0$. Let x and y be two such elements of G that coincide on $\mathscr{P}(G)^H$. Then, for every z in G and every g in $\mathscr{P}(G)^H$, we have $g \cdot z$ in $\mathscr{P}(G)^H$, so that

$$(g \cdot z)(x) = (g \cdot z)(y).$$

This may be written

$$g(zx) = ((x^{-1}y) \cdot g)(zx)$$

showing that $(x^{-1}y) \cdot g = g$. By assumption, this implies that $x^{-1}y$ belongs to H, so that $f = (x^{-1}y) \cdot f$. It follows that the unique extensions of x and y to F-algebra homomorphisms from $\mathscr{P}(G)[v^{-1}]$ to F coincide at f. Our conclusion is that homomorphisms from $\mathscr{P}(G)[v^{-1}]$ to F that coincide on $\mathscr{P}(G)^H$ also coincide at f. By Proposition III.2.4, this implies that f is purely inseparably algebraic over $[\mathscr{P}(G)^H]$. If F is of characteristic 0, this means that f belongs to $[\mathscr{P}(G)^H]$.

Now suppose that F is of non-zero characteristic p. Then our result says that there is a non-negative integer n such that f^{p^n} belongs to $[\mathscr{P}(G)^H]$. Thus, there is a non-zero element u in $\mathscr{P}(G)^H$ such that $(uf)^{p^n}$ belongs to $\mathscr{P}(G)^H$. Write w for uf.

We claim that, if $x_1, \ldots, x_m$ are elements of G such that the transforms $x_i \cdot w$ are F-linearly independent, then also the transforms $x_i \cdot w^{p^n}$ are F-linearly independent. In order to see this, let $c_1, \ldots, c_m$ be elements of F such that

$$\sum_{i=1}^{m} c_i x_i \cdot w^{p^n} = 0.$$

We may write this in the form

$$\left(\sum_{i=1}^{m} d_i x_i \cdot w \right)^{p^n} = 0,$$

where the d_i's are the elements of F such that $d_i^{p^n} = c_i$. This gives

$$\sum_{i=1}^{m} d_i x_i \cdot w = 0,$$

whence each $d_i = 0$, so that each $c_i = 0$, and our above claim is established.

Since $\mathscr{P}(G)$ is locally finite as a G-module, it follows from what we have just shown that the F-space spanned by the transforms $x \cdot w$ with x in G is finite-dimensional. Hence, the ideal, J say, of all elements d in $\mathscr{P}(G)$ such that $d(x \cdot w)$ belongs to $\mathscr{P}(G)$ for every element x of G is not the 0-ideal. On the other hand $G \cdot J = J$. Therefore, J has no zero in G. By Theorem II.3.5, it follows that J must coincide with $\mathscr{P}(G)$, which means that w belongs to $\mathscr{P}(G)$. Thus, we have

$$uf \in \mathscr{P}(G) \cap [\mathscr{P}(G)]^H = \mathscr{P}(G)^H.$$

Since u belongs to $\mathscr{P}(G)^H$, this gives $f \in [\mathscr{P}(G)^H]$. $\square$

Theorem 5.4. *Let G be an algebraic group over an algebraically closed field, and let H be a normal algebraic subgroup of G. Then*

$$\dim(G) = \dim(G/H) + \dim(H).$$

PROOF. We have $H_1 \subset G_1$, and H_1 is of finite index in $G_1 \cap H$. Hence, $H_1 = (G_1 \cap H)_1$, so that $\dim(H) = \dim(G_1 \cap H)$. The injection $G_1 \to G$, followed by the canonical map $G \to G/H$, is a morphism of affine algebraic groups from G_1 to G/H. The image is an irreducible normal algebraic subgroup of finite index, so that it coincides with $(G/H)_1$. Thus we have a surjective morphism of affine algebraic groups $G_1 \to (G/H)_1$. By Theorem II.4.4, the induced bijective map

$$G_1/(G_1 \cap H) \to (G/H)_1$$

is a bijective morphism of affine algebraic groups. The transpose of π is an injective algebra homomorphism from $\mathscr{P}((G/H)_1)$ to $\mathscr{P}(G_1/(G_1 \cap H))$. By Proposition III.2.4, the injectiveness of π implies that $[\mathscr{P}(G_1/(G_1 \cap H))]$ is (purely inseparably) algebraic over the field of fractions of the image of $\mathscr{P}((G/H)_1)$. Hence,

$$\dim(G/H) = \dim((G/H)_1) = \dim(G_1/(G_1 \cap H))$$

Now it is clear that we may replace G with G_1 and H with $G_1 \cap H$, which means that we may now suppose that G is irreducible. In this case, Lemma 5.3 applies to show that $[\mathscr{P}(G)]^H = [\mathscr{P}(G)^H]$. By Theorem III.2.3, $[\mathscr{P}(G)]$ is therefore separable over $[\mathscr{P}(G)^H] = [\mathscr{P}(G/H)]$. Let $\rho: G \to G/H$ denote the canonical morphism. Because of the separability just noted, we have from Theorem III.3.3 that $\rho^{\cdot}$ is surjective from $\mathscr{L}(G)$ to $\mathscr{L}(G/H)$. It is clear from Proposition 5.2 that the kernel of $\rho^{\cdot}$ coincides with $\mathscr{L}(H)$. Thus, we have

$$\dim(\mathscr{L}(G)) = \dim(\mathscr{L}(G/H)) + \dim(\mathscr{L}(H)).$$

By Theorem III.3.2, the dimension of an algebraic group is equal to that of its Lie algebra. $\qquad\qquad\square$

Notes

1. Almost exclusively, the original sources for the material of this chapter are [4] and [5].

2. The proof of Theorem 2.1 gives information about the dimension of G_τ, as follows. It is easy to see that each V_i contains a non-zero element v_i such that $\tau_{\mathfrak{l}}(v_i) = c_i v_i$. Replacing V with a translate $V \cdot x$, where x is a suitable element of G, we can ensure that $\varepsilon(v_i) \neq 0$ for each i. Then $\exp(t\tau)(v_i)$ is a non-zero K multiple of $\exp(tc_i)$. Hence,

$$K[\exp(tc_1), \ldots, \exp(tc_n)] \subset \exp(t\tau)(\mathscr{P}(G) \otimes K).$$

Now one can see from the proof of Theorem 2.1 that *the dimension of G_τ is either q or $q + 1$, where q is the rank of the additive group generated by the characteristic values of $\tau_{\mathfrak{l}}$.* In fact, the dimension is q when each μ_i is 0, and $q + 1$ otherwise. The condition that each μ_i be 0 is equivalent to the condition that $\tau_{\mathfrak{l}}$ be a semisimple linear endomorphism.

3. The following example shows how Theorem 4.3 can fail in non-zero characteristic. Let F be an infinite field of non-zero characteristic p. Let G be the group of pairs (r, s) with r and s in F and $s \neq 0$, whose composition is given by

$$(r_1, s_1)(r_2, s_2) = (r_1 + s_1^p r_2, s_1 s_2).$$

Define the functions u and v on G by

$$u(r, s) = r, \qquad v(r, s) = s.$$

Then G has the structure of an irreducible affine algebraic F-group, with $\mathscr{P}(G) = F[u, v, v^{-1}]$. One finds that $\mathscr{L}(G)$ is abelian, although G is not commutative.

4. The bijective morphism

$$\pi\colon G_1/(G_1 \cap H) \to (G/H)_1$$

of the proof of Theorem 5.4 is actually an isomorphism of affine algebraic groups. This can be seen as follows. Using Theorem II.2.3, one may write $\mathscr{P}(G)$ as a direct algebra and G_1-module sum of $\mathscr{P}(G_1)$ and the annihilator of G_1. Then one can show that, for every element u of

$$\mathscr{P}(G_1/(G_1 \cap H)) = \mathscr{P}(G_1)^{(G_1 \cap H)},$$

the sum in $\mathscr{P}(G)$ of the left transforms of u by a set of representatives of the elements of $H/(G_1 \cap H)$ is an element w in $\mathscr{P}(G)^H$ such that $w \circ \pi = u$. Thus, the transpose of π is seen to be surjective, whence it and π are isomorphisms.

Chapter V

Semisimplicity and Unipotency

Section 1 is devoted to the generalities concerning simple and semisimple modules over a ring, and to the theory of a single linear endomorphism. The main result from this second area is the multiplicative *Jordan decomposition* of a linear automorphism, which plays an important role in the structure theory of algebraic groups.

Section 2 introduces *unipotent representations*, which are the extreme opposites of semisimple representations, and the corresponding notion of a *unipotent algebraic group*. Such a group is characterized by the property that all its polynomial representations are unipotent. The main structural result is Theorem 2.3, which says that the multiplicative Jordan components of an element of an algebraic group belong to that group, and that the additive Jordan components of an element of the Lie algebra belong to that Lie algebra.

Section 3 introduces the notion of a *linearly reductive algebraic group*, i.e., a group having the property that all its polynomial representations are semisimple. Over fields of characteristic 0, all the "classical" groups are linearly reductive, and the motivation for this section stems from this fact. Theorem 3.1 is essentially Hilbert's "first main theorem on invariants," in terms of Hopf algebras and comodules.

Section 4 is devoted to semidirect products, preparing the ground for the structure theory of solvable algebraic groups in arbitrary characteristic, and of general algebraic groups in characteristic 0.

Section 5 contains the basic facts concerning the structure of abelian affine algebraic groups.

1. We recall from general algebra that a module is said to be *semisimple* if every submodule is a direct module summand, and that this is so if and only if the module coincides with the sum of its simple submodules.

Proposition 1.1. *Let R be a ring, M a semisimple R-module, $C = \mathrm{End}_R(M)$. Assume that M is finitely generated as a C-module. Then the image of R in $\mathrm{End}(M)$ coincides with $\mathrm{End}_C(M)$.*

PROOF. Choose a finite system $(m_1, \ldots, m_q)$ of C-module generators of M, and let S be the direct sum of q copies of the R-module M, so that this system may be viewed as an element of the R-module S. For $i = 1, \ldots, q$, let

$$\pi_i \colon S \to M \quad \text{and} \quad \sigma_i \colon M \to S$$

denote the ith projection and injection of the direct sum structure of S.

Let us write D for $\mathrm{End}_R(S)$. First, we note that every sub R-module, T say, of S is stable under $\mathrm{End}_D(S)$. In fact, since S is semisimple as an R-module, there is an R-module projection t from S to T. Now t belongs to D, so that we have, for every element e of $\mathrm{End}_D(S)$,

$$e(T) = (et)(S) = (te)(S) \subset t(S) = T.$$

Now let γ be any element of $\mathrm{End}_C(M)$, and let γ' denote the endomorphism of S given by

$$\gamma'(u_1, \ldots, u_q) = (\gamma(u_1), \ldots, \gamma(u_q)).$$

We claim that γ' belongs to $\mathrm{End}_D(S)$. In order to see this, let d be an element of D, and put $d_{ij} = \pi_i d \sigma_j$. We see immediately that d_{ij} belongs to C, whence $\gamma d_{ij} = d_{ij} \gamma$. Noting that

$$d = \sum_{i,j} \sigma_i d_{ij} \pi_j, \qquad \pi_j \gamma' = \gamma \pi_j \quad \text{and} \quad \gamma' \sigma_i = \sigma_i \gamma,$$

we see that $\gamma'd = d\gamma'$, so that γ' belongs to $\mathrm{End}_D(S)$.

Hence, the sub R-module of S that is generated by the element $(m_1, \ldots, m_q)$ is stable under γ'. This means that there is an element r in R such that

$$\gamma'(m_1, \ldots, m_q) = r \cdot (m_1, \ldots, m_q)$$

i.e., $\gamma(m_i) = r \cdot m_i$ for each index i. Since the m_i's generate M as a C-module, it follows that $\gamma(m) = r \cdot m$ for every element m of M. $\qquad\square$

Proposition 1.2. *Let V be a vector space over a field F, and let S be a sub F-algebra of $\mathrm{End}_F(V)$. Let K be an extension field of F. If $V \otimes K$ is semisimple with respect to $S \otimes K$, then V is semisimple with respect to S. Conversely, if V is finite-dimensional and semisimple with respect to S, and F is a perfect field, then $V \otimes K$ is semisimple with respect to $S \otimes K$.*

PROOF. Suppose that $V \otimes K$ is semisimple, and consider a sub S-module U of V. The assumption implies that there is an S-module projection μ from $V \otimes K$ to $U \otimes K$. Choose an F-space complement C of F in K. For v in V, write

$$\mu(v) = \alpha(v) + \beta(v),$$

where $\alpha(v)$ lies in U and $\beta(v)$ in $U \otimes C$. Then one sees immediately that α is an S-module projection from V to U. Our conclusion is that V is semisimple.

In proving the converse, we suppose, without loss of generality, that V is a simple S-module. Moreover, if L is an algebraic closure of K, and if we show that $V \otimes L$ is semisimple with respect to $S \otimes L$, it will follow from what we have already proved that $V \otimes K$ is semisimple with respect to $S \otimes K$. Accordingly, we assume that K is algebraically closed.

Let G be the group of all F-algebra automorphisms of K. Using that F is perfect, and the fact that every automorphism of a subfield of K extends to one of K, one shows that the G-fixed part K^G of K coincides with F. We let G act on $V \otimes K$ by F-algebra automorphisms via the tensor factor K. Clearly, G thus acts by S-module automorphisms that permute the sub $S \otimes K$-modules among themselves.

Since $V \otimes K$ is of finite dimension over K, it contains a non-zero simple sub $S \otimes K$-module, A say. Let B denote the sum in $V \otimes K$ of the family of all transforms $x(A)$, with x in G. Since each $x(A)$ is a simple $S \otimes K$-module, B is semisimple as an $S \otimes K$-module. Therefore, it suffices to prove that $B = V \otimes K$.

Write the non-zero elements of B in the form $\sum_{i=1}^{n} v_i \otimes c_i$, where the v_i's are F-linearly independent elements of V, and the c_i's belong to K. Take such a sum in which n is as small as possible, and multiply it by c_1^{-1} to ensure that then $c_1 = 1$. Apply an element of G to this, and subtract the original from the result. This yields a sum in B with fewer than n summands, so that we must obtain 0. This shows that each c_i must belong to K^G, i.e., to F, so that $B \cap V \neq (0)$. Since V is simple, this implies that $V \subset B$, whence $B = V \otimes K$. $\qquad\qquad\square$

The only purpose of the following lemma is its use in the proof of the next theorem, concerning the Jordan decomposition of a linear endomorphism.

Lemma 1.3. *Let F be a perfect field, x a variable over F. Let f be an element of $F[x] \setminus F$, and let g be a product of mutually inequivalent prime elements of $F[x]$, including the prime factors of f. There is an F-algebra endomorphism π of $F[x]$ satisfying*

(1) $\pi(g) \in F[x]f,$
(2) $\pi(x) - x \in F[x]g.$

PROOF. Since F is perfect, g has no multiple roots in any extension field of F, so that g is relatively prime to its formal derivative g'. Thus, there are elements u and v in $F[x]$ such that

$$ug' + vg = 1.$$

Let σ denote the F-algebra endomorphism of $F[x]$ that sends x onto $x - ug$. Then, for every non-negative exponent e, we have

$$\sigma(x^e) = x^e - ex^{e-1}ug + w_e g^2,$$

where w_e is an element of $F[x]$. It follows that

$$\sigma(g) = g - g'ug + wg^2,$$

where w belongs to $F[x]$. By the choice of u and v, the expression on the right is equal to $vg^2 + wg^2$, so that $\sigma(g)$ lies in $F[x]g^2$. Since some power of g lies in $F[x]f$, it follows that there is a positive integer m such that $\sigma^m(g)$ lies in $F[x]f$.

Put $\pi = \sigma^m$, so that π satisfies (1). Also,

$$\pi(x) - x = \sum_{i=0}^{m-1} (\sigma^{i+1}(x) - \sigma^i(x)) = \sum_{i=0}^{m-1} \sigma^i(\sigma(x) - x)$$

$$= \sum_{i=0}^{m-1} \sigma^i(-ug) \in F[x]g,$$

which shows that π satisfies (2). $\qquad\qquad\square$

Theorem 1.4. *Let F be a perfect field, V a finite-dimensional F-space, e an F-linear endomorphism of V. There are F-linear endomorphisms $e^{(n)}$ and $e^{(s)}$ of V satisfying the following conditions: $e^{(n)}$ is nilpotent, $e^{(s)}$ is semisimple, the sum of these endomorphisms is e, and each is an F-linear combination of positive powers of e. Moreover, if a and b are linear endomorphisms of V such such that a is nilpotent, b is semisimple, $a + b = e$ and $ab = ba$, then $a = e^{(n)}$ and $b = e^{(s)}$.*

PROOF. Let ρ be the F-algebra homomorphism from $F[x]$ to $\mathrm{End}_F(V)$ sending x onto e. The kernel of ρ is a principal ideal $F[x]f$. If f is a multiple of x, we define g as the product of the prime factors of f; otherwise, we define g as this product times x.

Now let π be the F-algebra endomorphism of $F[x]$ that is provided by Lemma 1.3. Let $q = \rho(\pi(x))$. By (2) of Lemma 1.3, q belongs to $eF[e]$. The homomorphism $\rho \circ \pi$ annihilates $F[x]g$, because $\pi(g)$ lies in $F[x]f$. The image $F[q]$ of this homomorphism is therefore a homomorphic image of $F[x]/F[x]g$, and so is a direct F-algebra sum of field extensions of F, because g is squarefree. This shows that q is a semisimple endomorphism of V. We have

$$e - q = \rho(x - \pi(x)) \in \rho(F[x]g),$$

whence it is clear that $e - q$ is nilpotent. Therefore, the elements $e^{(s)} = q$ and $e^{(n)} = e - q$ satisfy the requirements of the theorem.

Finally, let a and b be as described in the theorem. Then a and b commute with e, and therefore also with $e^{(s)}$ and $e^{(n)}$. It follows that $a - e^{(n)}$ is nilpotent and $e^{(s)} - b$ is semisimple. Since these two endomorphisms coincide, each is therefore 0. $\qquad\qquad\square$

The decomposition of Theorem 1.4 is called the *additive Jordan decomposition* of e. The endomorphisms $e^{(s)}$ and $e^{(n)}$ are called the *semisimple* and the *nilpotent component* of e.

In the case where e is an *automorphism* of V, this decomposition gives rise to the *multiplicative Jordan decomposition* of e, as follows. First, we show that, if e is an automorphism, so is $e^{(s)}$. Let W denote the kernel of $e^{(s)}$. Since $e^{(n)}$ commutes with $e^{(s)}$, it stabilizes W. Since $e^{(n)}$ is nilpotent, it follows that, if $W \neq (0)$, it contains a non-zero element w that is annihilated by $e^{(n)}$. But then w lies in the kernel of e, contradicting our assumption. Thus, we must have $W = (0)$, so that $e^{(s)}$ is an automorphism.

Now we define

$$e^{(u)} = i_V + (e^{(s)})^{-1} e^{(n)}.$$

Then $e^{(u)} - i_V$ is nilpotent, which is expressed by saying that $e^{(u)}$ is *unipotent*. We have $e^{(s)}e^{(u)} = e$, and $e^{(s)}$ and $e^{(u)}$ commute with each other. Since e is an automorphism, we have $i_V \in eF[e]$, whence also $e^{(u)} \in eF[e]$. The automorphism $e^{(u)}$ is called the *unipotent component* of e. Finally, as in the situation of Theorem 1.4, one sees readily that the decomposition $e = e^{(s)}e^{(u)}$ is unique, in the sense exactly analogous to the uniqueness of the additive Jordan decomposition.

2. Let M be a module and S a subset of $\mathrm{End}(M)$. We say that S *is nilpotent on* M if there is a positive integer n such that the product of every sequence of n elements of S is 0. Let F be a field, G a group, ρ a representation of G by F-linear automorphisms of a finite-dimensional F-space V. We say that ρ is a *unipotent representation* and that V is a *unipotent G-module* if the set of endomorphisms $\rho(x) - i_V$, with x in G, is nilpotent on V.

Theorem 2.1. *Let F be a field, and let ρ be a representation of a group G by linear automorphisms of a finite-dimensional F-space V. If $\rho(x) - i_V$ is nilpotent for every element x of G, then ρ is a unipotent representation.*

PROOF. If K is an algebraically closed field containing F, we can extend the given structure in the canonical fashion so as to obtain a representation of G by K-linear automorphisms of $V \otimes K$ that still has the property assumed for ρ. Therefore, we suppose without loss of generality that F is algebraically closed. Next, if we proceed by induction on $\dim(V)$, we reduce the theorem to the case where V is simple as a G-module.

In that case, let R denote the sub F-algebra of $\mathrm{End}_F(V)$ that is generated by $\rho(G)$. Since F is algebraically closed and V is simple, we have

$$\mathrm{End}_R(V) = Fi_V.$$

Therefore, we have from Proposition 1.1 that $R = \mathrm{End}_F(V)$. The sub F-space of R that is spanned by the elements $\rho(x) - i_V$ is evidently a two-sided ideal, J say, of R. If $J = (0)$, then there is nothing to prove. Suppose that

$J \neq (0)$. Evidently, $\mathrm{End}_F(V)$ is a simple F-algebra. Therefore, we have $J = \mathrm{End}_F(V)$. On the other hand, since each $\rho(x) - i_V$ is nilpotent, every element of J has trace 0, which is clearly not the case for every element of $\mathrm{End}_F(V)$. Thus, we have a contradiction, and so conclude that $J = (0)$. $\qquad\square$

Proposition 2.2. *Let G be a group of linear automorphisms of a finite-dimensional vector space V. Let S and T be subgroups of G, with T normal in G. If V is unipotent as an S-module and as a T-module, then V is unipotent as an ST-module.*

PROOF. For convenience of notation, let us work with the ordinary group algebra $\mathcal{Z}[G]$ of G over the ring $\mathcal{Z}$ of integers, and let us view V as a $\mathcal{Z}[G]$-module in the evident fashion. If $s \in S$ and $t \in T$ then, in $\mathcal{Z}[G]$,

$$1 - st = s(1 - t) + 1 - s$$

and

$$(1 - t)s = s(1 - s^{-1}ts).$$

Hence, a product $(1 - s_1 t_1) \cdots (1 - s_n t_n)$, where the s_i's belong to S and the t_i's to T, can be written as a sum of products of the form su, where s belongs to S, and u is a product whose factors are either $1 - x$ with x in S or $1 - y$ with y in T, the total number of factors being n. Using that

$$(1 - y)(1 - x) = 1 - y - x(1 - x^{-1}yx),$$

we can rewrite each u as an integral linear combination of products of the form

$$x(1 - x_1) \cdots (1 - x_p)(1 - y_1) \cdots (1 - y_q),$$

where x and the x_i's belong to S, and the y_j's belong to T. Moreover, q is the number of factors $1 - y$ with y in T that occurred in the original expression for u. Therefore, if d is the dimension of V, it follows from the unipotency of V as a T-module that the endomorphism corresponding to u is 0 whenever $q \geq d$. On the other hand, if $q < d$ and $n \geq d^2$, then u must contain at least d successive factors of the form $1 - x$ with x in S, so that the corresponding endomorphism of V is again equal to 0, because V is unipotent as an S-module. $\qquad\square$

A subgroup T of an algebraic group G is called unipotent if the representations of T by translations from the left on the finite-dimensional left T-stable subspaces of $\mathcal{P}(G)$ are unipotent. We express this property by saying that $\mathcal{P}(G)$ is *locally unipotent* as a T-module. It is clear from the definitions that a subgroup T of G is unipotent if and only if the restriction to T of every finite-dimensional polynomial representation of G is unipotent.

Let T be a unipotent subgroup of G, and let V be a finite-dimensional left T-stable subspace of $\mathcal{P}(G)$. Let

$$(0) = V_n \subset \cdots \subset V_0 = V$$

be a composition series for V as a T-module. The unipotency of the representation of T on V is equivalent to the property that the induced representation of T on each V_i/V_{i+1} is trivial. This is expressible by the condition that T annihilate certain elements of $\mathscr{P}(G)$, via evaluation. It follows that *the closure* in G *of a unipotent subgroup is still unipotent*. Taking the above V so that it generates $\mathscr{P}(G)$ as algebra, we see that *every unipotent subgroup of an algebraic group is nilpotent as an abstract group*.

Let G_u denote the subgroup of G that is generated by the family of all normal unipotent subgroups of G. We claim that G_u is a unipotent normal algebraic subgroup of G. In order to see this, let V be any left G_u-stable subspace of $\mathscr{P}(G)$ having finite dimension d, and consider the endomorphism of V that corresponds to a product in $\mathscr{Z}(G)$ of the form

$$(1 - x_1) \cdots (1 - x_d),$$

where each x_i belongs to G_u. There is a subgroup K of G that contains each x_i and is generated by a *finite* family of normal unipotent subgroups of G. By Proposition 2.2, the representation of K on V is unipotent, whence the endomorphism of V corresponding to our product is 0. Our conclusion is that $\mathscr{P}(G)$ is locally unipotent as a G_u-module, i.e., that G_u is a unipotent subgroup of G. Evidently, G_u is normal in G. By the remark just preceding our discussion of G_u, the closure of G_u in G is still unipotent, and it is evidently normal in G. Therefore, G_u must coincide with its closure in G, so that it is an algebraic subgroup of G.

We call G_u the *maximum normal unipotent subgroup* of G. It will play an important role in the general structure theory of algebraic groups.

Let (C, δ, ε) be the structure of a coalgebra over a perfect field F. We wish to examine the left C°-module structure of C with regard to the Jordan decompositions. Recall that the element of $\mathrm{End}_F(C)$ corresponding to an element γ of C° in the left C°-module structure of C is

$$\gamma_{\mathfrak{l}} = (i_C \otimes \gamma) \circ \delta.$$

If V is any finite-dimensional C°-stable sub F-space of C, the restriction of $\gamma_{\mathfrak{l}}$ to V has an additive Jordan decomposition, as described in Theorem 1.4. It is clear from the unicity part of Theorem 1.4, in conjunction with the fact that C is locally finite as a C°-module, that the semisimple and nilpotent components of the restrictions of $\gamma_{\mathfrak{l}}$ to the various V's fit together to yield endomorphisms $\gamma_{\mathfrak{l}}^{(s)}$ and $\gamma_{\mathfrak{l}}^{(n)}$ that commute with each other and whose sum coincides with $\gamma_{\mathfrak{l}}$, such that C is semisimple with respect to $\gamma_{\mathfrak{l}}^{(s)}$ and locally nilpotent with respect to $\gamma_{\mathfrak{l}}^{(n)}$. Since, on each V, each of these coincides with an F-linear combination of powers of $\gamma_{\mathfrak{l}}$, it follows from Proposition I.2.2 that they belong to the image of C° in $\mathrm{End}_F(V)$, i.e., that there are elements $\gamma^{(s)}$ and $\gamma^{(n)}$ in C° whose images under the map $\tau \to \tau_{\mathfrak{l}}$ from C° to $\mathrm{End}_F(C)$ coincide with $\gamma_{\mathfrak{l}}^{(s)}$ and $\gamma_{\mathfrak{l}}^{(n)}$, respectively. Since this map is injective, the elements $\gamma^{(s)}$ and $\gamma^{(n)}$ are determined by this last property. In fact, they are the composites of the corresponding elements of $\mathrm{End}_F(C)$ with ε, as is seen from Proposition I.2.2. We call them the semisimple and nilpotent components of γ.

In the case where γ is an automorphism of C, we obtain the unipotent component $\gamma^{(u)}$ in essentially the same way from the finite-dimensional situation described at the end of Section 1.

Theorem 2.3. *Let F be a perfect field, and let G be an affine algebraic F-group. For every element x of G, the multiplicative Jordan components $x^{(s)}$ and $x^{(u)}$ belong to G. For every element τ of $\mathcal{L}(G)$, the additive Jordan components $\tau^{(s)}$ and $\tau^{(n)}$ belong to $\mathcal{L}(G)$. If ρ is a morphism of affine algebraic F-groups, one has*

$$\rho(x^{(s)}) = \rho(x)^{(s)}, \qquad \rho(x^{(u)}) = \rho(x)^{(u)},$$
$$\rho^{\cdot}(\tau^{(s)}) = \rho^{\cdot}(\tau)^{(s)}, \qquad \rho^{\cdot}(\tau^{(n)}) = \rho^{\cdot}(\tau)^{(n)}.$$

PROOF. Let V be a finite-dimensional left G-stable sub F-space of $\mathcal{P}(G)$. For every linear automorphism e of $V + V^2$ stabilizing both V and V^2, we define the linear automorphism $[e]$ of $\mathrm{Hom}_F(V \otimes V, V^2)$ by

$$[e](h) = e \circ h \circ (e^{-1} \otimes e^{-1}),$$

where we use the same letter for an automorphism and its restriction to V or V^2. It is easy to see that $[e]$ is unipotent whenever e is unipotent. We claim that $[e]$ is semisimple whenever e is semisimple. Clearly, Proposition 1.2 enables us to reduce the proof of this to the case where F is algebraically closed. In that case, V and V^2 are direct sums of 1-dimensional e-stable subspaces, whence $\mathrm{Hom}_F(V \otimes V, V^2)$ is a direct sum of 1-dimensional $[e]$-stable subspaces, showing that $[e]$ is semisimple.

Now we know that $[e^{(s)}]$ is semisimple and that $[e^{(u)}]$ is unipotent. Clearly, they commute with each other, and their product is $[e]$. Therefore, we have from the unicity of the multiplicative Jordan decomposition that

$$[e^{(s)}] = [e]^{(s)}$$

and

$$[e^{(u)}] = [e]^{(u)}.$$

Now let x be an element of G, and let e be the restriction of $x_{\mathfrak{l}}$ to $V + V^2$. Let $m: V \otimes V \to V$ denote the multiplication map. Since $x_{\mathfrak{l}}$ is an F-algebra automorphism of $\mathcal{P}(G)$, we have $[e](m) = m$. It follows that $[e]^{(s)}$ and $[e]^{(u)}$ stabilize Fm, and hence that they also leave m fixed. By the above, this means that $[e^{(s)}]$ and $[e^{(u)}]$ leave m fixed. Since this holds for all finite-dimensional sub G-modules V of $\mathcal{P}(G)$, we conclude that $x_{\mathfrak{l}}^{(s)}$ and $x_{\mathfrak{l}}^{(u)}$ are F-algebra automorphisms of $\mathcal{P}(G)$. This implies that $x^{(s)}$ and $x^{(u)}$ are F-algebra homomorphisms from $\mathcal{P}(G)$ to F, i.e., that they belong to G.

The proof of the corresponding result for Lie algebra elements is the additive analogue of the above. If τ is a linear endomorphism of $V + V^2$ stabilizing V and V^2, we define the linear endomorphism (τ) of

$$\mathrm{Hom}_F(V \otimes V, V^2)$$

by

$$(\tau)(h) = \tau \circ h - h \circ (\tau \otimes i_V + i_V \otimes \tau).$$

As above, we see that $(\tau^{(s)}) = (\tau)^{(s)}$ and $(\tau^{(n)}) = (\tau)^{(n)}$. Then we use the fact that the F-linear derivations of $\mathscr{P}(G)$ are characterized by the property that their restrictions τ to $V + V^2$ satisfy $(\tau)(m) = 0$, for every V as above. This proves that the additive Jordan components of the elements of $\mathscr{L}(G)$ belong to $\mathscr{L}(G)$.

Finally, consider a morphism $\rho: G \to H$ of affine algebraic F-groups. This makes $\mathscr{P}(H)$ into a polynomial G-module in the evident way. Let x be an element of G. We know that every polynomial G-module is locally unipotent with respect to $x^{(u)}$ and semisimple with respect to $x^{(s)}$. In particular, $\mathscr{P}(H)$ is locally unipotent with respect to $\rho(x^{(u)})$ and semisimple with respect to $\rho(x^{(s)})$. Since these automorphisms commute with each other and since their product is $\rho(x)$, they must be the components $\rho(x)^{(s)}$ and $\rho(x)^{(u)}$ of $\rho(x)$.

The corresponding results concerning $\rho^{\cdot}$ are proved in essentially the same way. $\square$

3. We say that a subgroup R of an algebraic group G is *linearly reductive* if $\mathscr{P}(G)$ is semisimple as an R-module. This property is clearly equivalent to the property that every polynomial G-module is semisimple as an R-module. Evidently, if R is linearly reductive, so is the closure of R in G. If S is a normal subgroup of R then every simple R-module is the sum of the family of R-transforms of any simple sub S-module, and hence is semisimple as an S-module. Thus, *a normal subgroup of a linearly reductive subgroup is still linearly reductive.*

It follows that, if G is a linearly reductive algebraic group, then G_u is trivial. The converse is true over fields of characteristic 0, but is not true in general.

Finally, it is easy to see from the definitions that *a direct product of linearly reductive algebraic groups is linearly reductive.*

If F is any field, then the multiplicative group F^* of F, with its standard structure of affine algebraic F-group, as described in Section I.3, is linearly reductive. In fact, we have $\mathscr{P}(F^*) = F[x, x^{-1}]$, where x is the identity map on F^*, and, as an F^*-module, this is the sum of the family of 1-dimensional submodules generated by the powers of x.

Therefore, the direct product of a finite family of copies of F^* is a linearly reductive affine algebraic F-group. When F is an infinite field, this group is irreducible, and it is called an *F-toroid*. If F is algebraically closed and of *non*-zero characteristic, then every irreducible linearly reductive affine algebraic F-group is an F-toroid. On the other hand, over fields of characteristic 0, the supply of linearly reductive groups is ample, including almost all the groups of classical interest.

We wish to describe the role played by linear reductiveness in the context of classical invariant theory. It will be convenient to do this in terms of comodules, especially because the analogous results concerning Lie algebras can thus be covered at the same time.

Let C be a coalgebra over a field F. Suppose there is given the structure of a C-comodule on F; say

$$\gamma: F \to F \otimes C = C.$$

Then γ is determined by F-linearity and the single value $\gamma(1) \in C$. If δ and ε are the comultiplication and counit of C, then the assumption on γ is equivalent to the assumption that $\varepsilon(\gamma(1)) = 1$ and $\delta(\gamma(1)) = \gamma(1) \otimes \gamma(1)$. We identify F with its image $\gamma(F)$ in C. Then our assumption amounts to having the trivial coalgebra F contained in C as a sub coalgebra and direct F-space summand, in such a way that ε is an F-space projection from C to F. Note that, in the case where C is a Hopf algebra, these extra data are contained in the definition of C as such.

If V is a C-comodule, we denote by V^C the subspace of V consisting of the elements v for which $\delta(v) = v \otimes 1$. In the case $C = \mathscr{P}(G)$, where G is an algebraic group, so that V is a polynomial G-module, the subspace V^C is precisely the G-fixed part V^G.

We are concerned with the situation where C has the property that every C-comodule is semisimple. In this case, there is one *and only one* projection of C-comodules from V to V^C. In order to see this, let us recall from Chapter I that the category of C-comodules is naturally isomorphic with the category of locally finite C°-modules of type C. When V is regarded as a C°-module, then V^C is the sub C°-module consisting of all elements v such that, for every α in C°, $\alpha \cdot v = \alpha(1)v$. Since V is semisimple as a C-comodule, it is semisimple as a C°-module, so that there is C°-module complement, W say, for V^C in V. By writing an element v of V as a sum of an element of V^C and an element of W, we see that $\alpha \cdot v - \alpha(1)v$ belongs to W for every element α of C°.

Let V_C denote the subspace of V that is spanned by the elements of the form $\alpha \cdot v - \alpha(1)v$. This is evidently a sub C°-module of W. By semisimplicity, there is a C°-module complement, T say, for V_C in W. It is clear that $T \subset V^C$. Hence,

$$T \subset V^C \cap V_C \subset V^C \cap W = (0).$$

Our conclusion is that V_C is the only C°-module complement for V^C in V, i.e., the only C-comodule complement for V^C in V, so that our above claim is established.

Let H be an F-Hopf algebra, R any F-algebra. Let $\rho: R \otimes R \to R$ denote the multiplication of R. We say that R is an H-*comodule algebra* if it is endowed with a structure $\beta: R \to R \otimes H$ of an H-comodule that is compatible with the algebra structure of R, in the sense that the maps $\beta \circ \rho$ and $(\rho \otimes i_H) \circ (\beta \boxtimes \beta)$ from $R \otimes R$ to $R \otimes H$ coincide. In the case where $H = \mathscr{P}(G)$, this means that G acts by F-algebra automorphisms on R.

Theorem 3.1. *Let F be a field, H an F-Hopf algebra with the property that every H-comodule is semisimple. Suppose that R is a commutative H-comodule* *algebra that is finitely generated as an F-algebra. Then R^H is a finitely generated F-algebra.*

PROOF. It follows directly from the definitions that R^H is a sub F-algebra of R. What must be proved is that it is finitely generated. There is a finite-dimensional sub F-space of R that generates R as an F-algebra. The smallest sub H-comodule of R containing this is still finite-dimensional. Hence, there is a finite-dimensional sub H-comodule V of R that generates R as an F-algebra.

Let S denote the symmetric F-algebra constructed over V. From the H-comodule structure of V, we obtain an H-comodule structure of S in the canonical fashion via the tensor product construction for comodules. It is clear that S thus becomes an H-comodule algebra. From the multiplication of R, we have a surjective map $\sigma: S \to R$, which is evidently a morphism of H-comodule algebras. We have $S = S^H + S_H$, $R = R^H + R_H$, $\sigma(S^H) \subset R^H$ and $\sigma(S_H) \subset R_H$. Since the sums here are direct sums, it follows that $\sigma(S^H) = R^H$. Therefore, it suffices to prove that S^H is finitely generated.

In doing this, we use the grading of S, writing S_m for the homogeneous component of degree m of S. Each S_m is a sub H-comodule of S, so that S^H is the sum of the $(S_m)^H$'s. Let I denote the ideal of S that is generated by the $(S_m)^H$'s with $m > 0$. There are homogeneous elements $u_1, \ldots, u_n$ of strictly positive degrees in S^H such that

$$I = Su_1 + \cdots + Su_n.$$

We shall prove that

$$S^H = F[u_1, \ldots, u_n].$$

It suffices to show that each $(S_m)^H$ is contained in $F[u_1, \ldots, u_n]$. This is evidently true for $m = 0$. Therefore, we suppose that $m > 0$, and that $(S_k)^H$ is contained in $F[u_1, \ldots, u_n]$ for every $k < m$.

Let u be an element of $(S_m)^H$. Then u belongs to I, so that

$$u = s_1 u_1 + \cdots + s_n u_n,$$

with each s_i in S. Moreover, if d_i is the degree of u_i, we may evidently choose s_i from S_{m-d_i}. Now let π be the unique H-comodule projection from S to S^H. Then the restriction of π to S_k is the unique H-comodule projection from S_k to $(S_k)^H$ for every k. Furthermore, if x is a non-zero element of S^H then the maps $sx \to \pi(sx)$ and $sx \to \pi(s)x$ must coincide, because each is an H-comodule projection from Sx to $(Sx)^H$. Hence, we have

$$u = \pi(u) = \pi(s_1)u_1 + \cdots + \pi(s_n)u_n.$$

Now $\pi(s_i)$ belongs to $(S_{m-d_i})^H$, and it follows from our inductive hypothesis that $\pi(s_i)$ belongs to $F[u_1, \ldots, u_n]$. Hence, $u \in F[u_1, \ldots, u_n]$. $\square$

The most important case of Theorem 3.1 is the case where H is the Hopf algebra of polynomial functions of an algebraic group. We state this as follows.

Corollary 3.2. *Let G be a linearly reductive algebraic group, and let R be a finitely generated commutative algebra on which G acts by algebra automorphisms and in such a way that R is a polynomial G-module. Then the G-fixed part of R is a finitely generated algebra.*

4. We discuss the general features of *semidirect products* of algebraic groups. Let F be a field, G an affine algebraic F-group, N a normal algebraic subgroup of G. We know from Theorem II.2.2 that $\mathscr{P}(G)^N$ separates the elements of G/N. By Theorem II.4.3, the sub Hopf algebra $\mathscr{P}(G)^N$ is finitely generated as an F-algebra. Let K denote the group $\mathscr{G}(\mathscr{P}(G)^N)$. This is an affine algebraic F-group with $\mathscr{P}(K) = \mathscr{P}(G)^N$, because the restriction image of G in K separates the elements of $\mathscr{P}(G)^N$. We know, also from Theorem II.4.3, that the restriction morphism from G to K is surjective whenever F is algebraically closed. However, this does not always hold if F is not algebraically closed. We shall say that N is *properly normal* in G if the restriction morphism $G \to \mathscr{G}(\mathscr{P}(G)^N)$ is surjective, so that G/N is an affine algebraic F-group with $\mathscr{P}(G/N) = \mathscr{P}(G)^N$.

Now suppose that N is properly normal in G, and that there is an algebraic subgroup R of G such that the following conditions are satisfied: (1) $G = NR$; (2) $N \cap R = (1)$; (3) the restriction to R of the canonical morphism $G \to G/N$ is an isomorphism of algebraic groups. Under these circumstances we say that G is the *semidirect product* of N and R, and we indicate this by writing $G = N \rtimes R$.

As to condition (3), note that, in any case, the restriction to R of the canonical morphism $G \to G/N$ is a bijective morphism of algebraic groups from R to G/N. The extra assumption is that its inverse is also a morphism of algebraic groups. This is satisfied automatically whenever the base field is algebraically closed and of characteristic 0, by virtue of the following theorem.

Theorem 4.1. *Let F be an algebraically closed field of characteristic 0. Let G be an affine algebraic F-group, and let B be a sub Hopf algebra of $\mathscr{P}(G)$. If the restriction morphism $\rho: G \to \mathscr{G}(B)$ is injective then B coincides with $\mathscr{P}(G)$.*

PROOF. Under the present assumptions, $\mathscr{G}(B)$ is an affine algebraic F-group with $\mathscr{P}(\mathscr{G}(B)) = B$, and ρ is a bijective morphism of affine algebraic F-groups. Clearly, this implies that ρ maps the set of irreducible components of G bijectively onto the set of irreducible components of $\mathscr{G}(B)$. We have the direct F-algebra decompositions of $\mathscr{P}(G)$ and B reflecting the irreducible components, as described in Theorem II.2.3. From what we have just said concerning ρ, it is clear that the transpose of ρ respects these F-algebra decompositions. We see from this that it suffices to prove the theorem in the case where G is irreducible. In that case, we may apply Proposition III.2.4 and conclude that $\mathscr{P}(G)$ is purely inseparably algebraic over $[B]$.

Since F is of characteristic 0, this means that $\mathscr{P}(G)$ is contained in $[B]$. By Lemma II.4.2, this implies that B coincides with $\mathscr{P}(G)$. $\qquad\square$

Now let us examine the above semidirect product $G = N \rtimes R$. For each element f of $\mathscr{P}(G)$, we denote by f_R and f_N the restrictions of f to R and N, so that $f_R \in \mathscr{P}(R)$ and $f_N \in \mathscr{P}(N)$. Let us denote the inverse of the restriction to R of the canonical morphism $\pi\colon G \to G/N$ by $\alpha\colon G/N \to R$. The transpose of α is an isomorphism of Hopf algebras, ρ say, from $\mathscr{P}(R)$ to $\mathscr{P}(G)^N$. On the other hand, define the polynomial projection $\beta\colon G \to N$ by

$$\beta(x) = x\alpha(\pi(x))^{-1}$$

and note that $\beta(xy) = \beta(x)$ whenever y belongs to R. It follows that the transpose of β is an isomorphism of F-algebras, σ say, from $\mathscr{P}(N)$ to $\mathscr{P}(G)^R$, the inverse being the restriction map.

We shall use ρ and σ in showing that, *as an affine algebraic F-set, G is the direct product of N and R.* In fact, we shall show that the multiplication map

$$\mathscr{P}(G)^R \otimes \mathscr{P}(G)^N \to \mathscr{P}(G)$$

is an isomorphism of F-algebras. Then, by preceding this with $\sigma \otimes \rho$, we obtain an isomorphism of F-algebras from $\mathscr{P}(N) \otimes \mathscr{P}(R)$ to $\mathscr{P}(G)$.

From the fact that the set $N \times R$ separates the elements of

$$\mathscr{P}(G)^R \otimes \mathscr{P}(G)^N,$$

it follows that the multiplication map is injective. In order to prove the surjectivity, consider an element f of $\mathscr{P}(G)$, and write

$$\delta(f) = \sum_i u_i \otimes v_i$$

with the u_i's and v_i's in $\mathscr{P}(G)$, so that

$$f(xy) = \sum_i u_i(x)v_i(y).$$

Here, we let x range over N and y over R. Then, from the definitions of ρ and σ, we have

$$v_i(y) = \rho((v_i)_R)(xy) \quad \text{and} \quad u_i(x) = \sigma((u_i)_N)(xy),$$

whence the above shows that

$$f = \sum_i \sigma((u_i)_N)\rho((v_i)_R),$$

so that f belongs to the multiplication image of $\mathscr{P}(G)^R \otimes \mathscr{P}(G)^N$. $\qquad\square$

Theorem 4.2. *Let G be an algebraic group, and suppose there is a linearly reductive subgroup R of G such that $G = G_u R$. Then R is an algebraic subgroup of G, and G is the semidirect product $G_u \rtimes R$.*

Proof. Let R^+ denote the closure of R in G. Then R^+ is still linearly reductive. The intersection $R^+ \cap G_u$ is a normal subgroup of R^+, and therefore is again a linearly reductive subgroup of G. On the other hand, it is clearly a unipotent subgroup of G. It follows that $R^+ \cap G_u = (1)$ and $R = R^+$.

Choose a finite-dimensional sub G-module V of $\mathscr{P}(G)$ that generates $\mathscr{P}(G)$ as an algebra, and let

$$(0) = V_n \subset \cdots \subset V_0 = V$$

be a composition series of V as a G-module. Construct the external direct sum, V' say, of the G-modules V_i/V_{i+1}. Since V is semisimple as an R-module, there is an isomorphism of R-modules $\gamma\colon V' \to V$. Let τ and τ' denote the representations of G on V and V', respectively.

Clearly, there is one and only one map $\alpha\colon G \to R$ such that $xG_u = \alpha(x)G_u$ for every element x of G, and α is evidently a group homomorphism leaving the elements of R fixed. We shall prove that α is a morphism of algebraic groups.

Since G_u acts trivially on each V_i/V_{i+1}, we have $\tau' = \tau' \circ \alpha$. Hence, for every element x of G, we have

$$\gamma\tau'(x)\gamma^{-1} = \gamma\tau'(\alpha(x))\gamma^{-1} = \tau(\alpha(x)).$$

Hence, if v is an element of V, we have

$$v(\alpha(x)) = \varepsilon(\tau(\alpha(x))(v)) = \varepsilon((\gamma\tau'(x)\gamma^{-1})(v)),$$

which shows that $v \circ \alpha$ is a representative function associated with the polynomial representation τ', so that $v \circ \alpha$ belongs to $\mathscr{P}(G)$. Since V generates $\mathscr{P}(G)$ as an algebra, we conclude that $\mathscr{P}(G) \circ \alpha \subset \mathscr{P}(G)$. Since

$$\mathscr{P}(G) \circ \alpha = \mathscr{P}(R) \circ \alpha,$$

this means that α is a morphism of algebraic sets, and hence a morphism of algebraic groups from G to R.

Clearly, $\mathscr{P}(G) \circ \alpha$ is element-wise fixed under the translation action of G_u. Moreover, if f is an element of $\mathscr{P}(G)^{G_u}$, then $f = f \circ \alpha$. We see from this that the transpose of α yields an isomorphism of Hopf algebras

$$\rho\colon \mathscr{P}(R) \to \mathscr{P}(G)^{G_u}.$$

Now define the polynomial map $\beta\colon G \to G_u$ by

$$\beta(x) = x\alpha(x)^{-1}.$$

Then the transpose of β yields an isomorphism of algebras

$$\sigma\colon \mathscr{P}(G_u) \to \mathscr{P}(G)^R$$

Now the argument just preceding Theorem 4.2 applies without change to show that the multiplication map from $\mathscr{P}(G)^R \otimes \mathscr{P}(G)^{G_u}$ to $\mathscr{P}(G)$ is an isomorphism of algebras. From this, we see immediately that the restriction map from G to $\mathscr{G}(\mathscr{P}(G)^{G_u})$ is surjective, i.e., that G_u is properly normal in G.

Our above morphism α has G_u for its kernel, and therefore induces a morphism of algebraic groups from G/G_u to R in the natural fashion. Clearly, this is the inverse of the restriction to R of the canonical morphism from G to G/G_u. Now it is clear that G is the semidirect product $G_u \rtimes R$. $\square$

5. Theorem 5.1. *Let F be a perfect field, and let G be an abelian affine algebraic F-group. There is one and only one linearly reductive algebraic subgroup G_s of G such that G is the direct product $G_u \times G_s$. The elements of G_s are precisely the semisimple elements of G, and every linearly reductive subgroup of G is contained in G_s.*

PROOF. Let R be a linearly reductive subgroup of G. Since F is perfect and $\mathscr{P}(G)$ is locally finite as an R-module, we may apply Proposition 1.2 and conclude that, for every field K containing F, the tensor product $\mathscr{P}(G) \otimes K$ is still semisimple as an R-module. Choose K algebraically closed, and consider a finite-dimensional G-stable sub K-space V of $\mathscr{P}(G) \otimes K$. Since R is abelian, there is a finite family of group homomorphisms ρ from R to K^* and a corresponding decomposition of V into the direct sum of sub K-spaces V_ρ such that each element x of R acts on V_ρ as the scalar multiplication by $\rho(x)$. If R' is another linearly reductive subgroup of G then, since the elements of R' commute with those of R, each V_ρ is R'-stable, and has a decomposition as a direct sum of characteristic subspaces with respect to R'. Thus, by refining the first decomposition, we obtain a decomposition of V into a direct sum of a family of sub K-spaces such that the group generated by R and R' acts by scalar multiplications on each component.

Now let (R_α) be a totally inclusion-ordered family of linearly reductive subgroups of G. Since V is finite-dimensional, the decomposition refinement process obtained by applying the above to successive members of our family (R_α) must terminate, whence we see that V is a sum of 1-dimensional sub K-spaces each of which is stable under the action of the union of our family (R_α). Since this holds for each V, it follows that $\mathscr{P}(G) \otimes K$ is semisimple as a module for the union of the family (R_α). By Proposition 1.2, this implies that this union is a linearly reductive subgroup of G. At the same time, it is clear from the above that the subgroup generated by a pair of linearly reductive subgroups is again a linearly reductive subgroup.

Now we can apply Zorn's lemma and conclude that there exists a maximal linearly reductive subgroup G_s of G. From the last remark, it is clear that every linearly reductive subgroup of G is contained in G_s. In particular, it follows that every semisimple element of G belongs to G_s. On the other hand, it is clear from the situation in $\mathscr{P}(G) \otimes K$ that every element of G_s is a semisimple element.

Now let x be any element of G, and consider its multiplicative Jordan decomposition $x = x^{(u)}x^{(s)}$. We have $x^{(s)} \in G_s$. Since G is abelian, we have $x^{(u)} \in G_u$. Thus, $G = G_u G_s$. By Theorem 4.2, G is therefore the semidirect product of G_u by G_s. Since G is abelian, this is the direct product. $\square$

Theorem 5.2. *If F is an algebraically closed field then every abelian irreducible linearly reductive affine algebraic F-group is an F-toroid.*

PROOF. Let G be such a group, and let V be a finite-dimensional G-stable sub F-space of $\mathscr{P}(G)$ that generates $\mathscr{P}(G)$ as an F-algebra. The assumptions on G imply that V is the direct G-module sum of a finite family of 1-dimensional G-modules. Say

$$V = Fv_1 + \cdots + Fv_q.$$

For each element x of G, we have $x \cdot v_i = \gamma_i(x)v_i$, where γ_i is an element of $\mathscr{P}(G)$ that is a group homomorphism from G to F^*. From now on, we shall refer to such homomorphisms as *polynomial characters*.

We have $v_i = \varepsilon(v_i)\gamma_i$. Since V generates $\mathscr{P}(G)$ as an F-algebra, it follows that, if Σ denotes the multiplicative group generated by the γ_i's, every element of $\mathscr{P}(G)$ is an F-linear combination of elements of Σ. Recall the elementary fact that group homomorphisms from an arbitrary group to the multiplicative group of a field are linearly independent over that field, as functions. It follows from this that every element of $\mathscr{P}(G)$ that is a group homomorphism to F^* belongs to Σ. In other words, Σ is the group of all polynomial characters of G.

Let σ be any polynomial character of G. Evidently, σ is a morphism of affine algebraic F-groups from G to F^*. By Theorem II.4.1, $\sigma(G)$ is therefore an irreducible algebraic subgroup of F^*. By the evident dimension consideration, we have therefore $\sigma(G) = F^*$ or $\sigma(G) = (1)$. In particular, this shows that Σ is torsion-free. Since Σ is finitely generated, it is therefore the free abelian group based on a finite subset $(\sigma_1, \ldots, \sigma_n)$. Since the elements of Σ are F-linearly independent as functions, the σ_i's are algebraically independent over F. We have

$$\mathscr{P}(G) = F[\sigma_1, \ldots, \sigma_n, \sigma_1^{-1}, \ldots, \sigma_n^{-1}].$$

This is the tensor product of the sub Hopf algebras $F[\sigma_i, \sigma_i^{-1}]$, each of which may be identified with $\mathscr{P}(F^*)$. Accordingly, G is the direct product of a family of n copies of F^*. $\square$

Theorem 5.3. *Let F be an algebraically closed field, G an abelian affine algebraic F-group, K an algebraic subgroup of G, and τ a morphism of affine algebraic F-groups from K to an F-toroid T. Then τ extends to a morphism of affine algebraic F-groups from G to T.*

PROOF. Evidently, it suffices to prove the theorem in the case where $T = F^*$. By Theorem 5.1, we have the direct product decompositions $G = G_u \times G_s$ and $K = K_u \times K_s$. Clearly, K_u is an algebraic subgroup of G_u, and K_s is an algebraic subgroup of G_s. Since $\tau(K_u)$ is a *unipotent* subgroup of F^*, it is clear that K_u is contained in the kernel of τ. If we show that the restriction of τ to K_s extends to a morphism of affine algebraic F-groups from G_s to F^*, then we obtain the required extension of τ by composition with the projection morphism from G to G_s with kernel G_u. Therefore, we assume without loss of generality that G and K are linearly reductive.

Next, we reduce the theorem to the case where G is irreducible. Let $\tau_1 : G_1 \cap K \to F^*$ be the restriction of τ. Suppose that there is an extension of τ_1 to a morphism of affine algebraic F-groups $\sigma_1 : G_1 \to F^*$. Consider the group homomorphism

$$\alpha : G_1 \times K \to F^*$$

where

$$\alpha(x, y) = \sigma_1(x)\tau(y).$$

Since σ_1 coincides with τ on $G_1 \cap K$, the kernel of α contains the kernel of the composition morphism from $G_1 \times K$ to $G_1 K \subset G$. Therefore, α yields a group homomorphism β from $G_1 K$ to F^* such that the restriction of β to G_1 coincides with σ_1, and the restriction of β to K coincides with τ. Since F^* is a divisible group, β extends further to a group homomorphism from G to F^*. Since the restriction of β to G_1 is a morphism of affine algebraic F-groups, so is this last group homomorphism from G to F^*.

It remains to deal with the case where G is irreducible. In this case, we know from Theorem 5.2 that G is an F-toroid. Let us regard τ as an element of $\mathscr{P}(K)$, and let us choose an element f of $\mathscr{P}(G)$ whose restriction to K is τ. Since G is an F-toroid, f is an F-linear combination $\sum_{i=1}^{m} c_i h_i$, where each c_i is in F, and each h_i is a polynomial character of G. We choose f so that m is as small as possible. The F-linear combination $\sum_{i=1}^{m} c_i (h_i)_K$ of the restrictions to K of the h_i's is the given group homomorphism τ. Since m is minimal, the $(h_i)_K$'s are mutually distinct. If τ were distinct from each $(h_i)_K$, we would have a contradiction to the fact that distinct group homomorphisms to the multiplicative group of a field are linearly independent as functions. Thus, we must have $m = 1$ and $\tau = (h_1)_K$, so that h_1 is the required extension of τ. $\qquad\qquad\qquad\qquad\qquad\qquad\qquad\qquad\qquad\qquad\qquad\qquad\qquad\square$

Notes

1. Let V be a finite-dimensional real or complex vector space, and let x be a linear endomorphism of V. In the notation of the Jordan decompositions, show that

$$\exp(x)^{(s)} = \exp(x^{(s)}) \quad \text{and} \quad \exp(x)^{(u)} = \exp(x^{(n)}).$$

2. Let F be a field of non-zero characteristic p, and let G be an affine algebraic F-group. Show that, for every unipotent element x of G, there is an exponent $n \geq 0$ such that x is of order p^n. Assuming that F is perfect, deduce from this that, for every element x of G, there is an $n \geq 0$ such that x^{p^n} is semisimple.

3. With regard to our allusions to the classical invariant theory and "classical" linearly reductive groups, see [18].

4. Let F be an algebraically closed field, and let p denote the *characteristic exponent* of F, i.e., the characteristic of F if that is not zero, and 1 otherwise. Let G be an abelian linearly reductive affine algebraic F-group. Let $X(G)$ denote the group of polynomial characters of G. One sees as in the proof of Theorem 5.2 that $X(G)$ is finitely generated. Moreover, it is easily seen that $X(G)$ is p-torsion free. The Hopf algebra $\mathscr{P}(G)$ may be identified with the group algebra $F[X(G)]$, whose comultiplication and antipode are given by $\delta(f) = f \otimes f$ and $\eta(f) = f^{-1}$, for every element f of $X(G)$.

Conversely, let A be any finitely generated abelian group without p-torsion. One shows readily that the group algebra $F[A]$ has no nilpotent elements other than 0, so that $\mathscr{G}(F[A])$ is an affine algebraic F-group with $F[A]$ as its Hopf algebra of polynomial functions. Next, one can show that this is a group G as above, and that $X(G) = A$.

The above yields a functor $X(*)$ from the category of abelian linearly reductive affine algebraic F-groups to the opposite of the category of finitely generated abelian groups with no p-torsion, and a functor $\mathscr{G}(F[*])$ in the opposite direction, by which these two categories become naturally isomorphic.

5. Let F be an algebraically closed field of non-zero characteristic p, and suppose that U is a non-trivial irreducible unipotent affine algebraic F-group. Let M be a finite-dimensional polynomial U-module such that the associated representative functions generate $\mathscr{P}(U)$ as an F-algebra. Let $S^p(M)$ denote the homogeneous component of degree p of the symmetric algebra $S(M)$ built over M, and let T denote the sub U-module of $S^p(M)$ consisting of the pth powers in $S(M)$ of the elements of M.

It has been shown by M. Nagata that T is not a direct U-module summand in $S^p(M)$. This implies that, if G is an irreducible affine algebraic F-group having an infinite unipotent subgroup, then G is not linearly reductive. It will be seen later (cf. Note XIII.3) that the only irreducible affine algebraic F-groups having no infinite unipotent subgroup are the toroids. Thus, one has Nagata's result that the only irreducible linearly reductive affine algebraic F-groups are the toroids.

In order to verify that T has no U-module complement in $S^p(M)$, observe first (using that $[\mathscr{P}(U)]$ is a finitely generated separable field extension of F) that not every element of $\mathscr{P}(U)$ is a p-th power of an element of $\mathscr{P}(U)$. Let Q denote the sub F-algebra of $\mathscr{P}(U)$ consisting of the p-th powers of the elements of $\mathscr{P}(U)$. There is an F-basis $(x_1, \ldots, x_m)$ of M such that, for every index j and every element u of U, one has

$$u \cdot x_j - x_j = \sum_{i > j} f_{ij}(u)x_i$$

with each f_{ij} in $\mathscr{P}(U)$. By assumption, these f_{ij}'s generate $\mathscr{P}(U)$ as an F-algebra. Therefore, it follows from the first observation that there are indices r and s, with $r > s$, such that f_{rs} does not belong to Q, while f_{ij} belongs to Q whenever $j > s$, and f_{is} belongs to Q whenever $i > r$.

Choose the basis $(x_1, \ldots, x_m)$ so that $r - s$ is minimal. Then it follows that f_{rs} is not the sum of an element of Q and an F-linear combination of f_{is}'s with $i \neq r$. In fact, otherwise there are elements α_i in F such that $f_{rs} - \sum_{s < i < r} \alpha_i f_{is}$ belongs to Q. Replacing each x_i where $s < i < r$ with $x_i + \alpha_i x_r$, one obtains a new F-basis of M with the same "triangular" property as $(x_1, \ldots, x_m)$, the same "s", but a smaller "r". This contradicts the minimality of the original $r - s$, so that the claim concerning the "independence" of f_{rs} is established.

Since f_{is} belongs to Q whenever $i > r$, the same is true for the translates $f_{is} \cdot u$, where u is an element of U. Thus, $\sum_j f_{ij}(u) f_{js}$ belongs to Q for every $i > r$ and every element u of U. By virtue of the independence property of f_{rs}, this implies that $f_{ir}(u) = 0$, which means that x_r is U-fixed.

Now suppose that there is a U-module complement, H say, for T in $S^p(M)$. For $i \geq s$ and $i \neq r$, let h_i denote the component of $x_i x_r^{p-1}$ in H. For every element u of U, we have (interpreting f_{ss} as the constant 1)

$$u \cdot (x_s x_r^{p-1}) = \sum_{i \geq s} f_{is}(u) x_i x_r^{p-1}$$

whence

$$u \cdot h_s = \sum_{\substack{i \geq s \\ i \neq r}} f_{is}(u) h_i$$

and

$$u \cdot (x_s x_r^{p-1} - h_s) = f_{rs}(u) x_r^p + \sum_{\substack{i \geq s \\ i \neq r}} f_{is}(u)(x_i x_r^{p-1} - h_i)$$

Since $x_s x_r^{p-1} - h_s$ belongs to T, the expression on the left has the form $\sum_j g_j(u)^p x_j^p$ with each g_j in $\mathscr{P}(U)$. Writing each $x_i x_r^{p-1} - h_i$ as an F-linear combination of x_j^p's, we see that the last equation yields coefficients γ_i in F such that

$$f_{rs} + \sum_{\substack{i \geq s \\ i \neq r}} \gamma_i f_{is} = g_r^p \in Q,$$

which contradicts the independence property of f_{rs}.

Chapter VI

Solvable Groups

This chapter develops the basic structure theory of solvable algebraic groups. For the principal results, it is assumed that the base field is algebraically closed and that the group is irreducible. In the presence of these assumptions, the solvable groups are characterized by the property that their simple polynomial modules are 1-dimensional. This is the Lie–Kolchin Theorem, given here as Theorem 1.1.

Section 2 is devoted to technical preparations for the proof of the main structural result, Theorem 3.2, which gives the decomposition of an irreducible solvable algebraic group G over an algebraically closed field as a semidirect product of G_u by a toroid.

In Section 4, it is shown that irreducible 1-dimensional algebraic groups are commutative, and Section 5 deals with the structure of commutative groups.

1. Theorem 1.1. *Let F be an algebraically closed field, G an irreducible solvable affine algebraic F-group. Then every simple polynomial G-module is 1-dimensional.*

PROOF. By applying Theorem II.4.1 to the polynomial map from $G \times G$ to G that sends each pair (x, y) onto $xyx^{-1}y^{-1}$, we see that the commutator subgroup, K say, of G is an irreducible algebraic subgroup of G. Since G is solvable, we have $K \neq G$ whenever G is non-trivial. Making an induction on the dimension of G, we assume that the theorem holds for K.

Let V be a simple polynomial G-module. From the normality of K in G, it follows that V is a semisimple K-module (cf. the beginning of Section V.3). By our inductive hypothesis, V is therefore a direct sum of a family of 1-dimensional K-stable subspaces. By forming the appropriate partial sums, we

78

obtain a direct K-module decomposition

$$V = V_1 + \cdots + V_q,$$

with corresponding mutually distinct polynomial characters γ_i of K such that each element x of K acts on each V_i as the scalar multiplication by $\gamma_i(x)$. Since K is normal in G, the elements of G permute the V_i's among themselves. Therefore, the stabilizer of V_1 in G is of finite index in G. Since G is irreducible, this implies that the stabilizer of V_1 coincides with G. Since V is simple as a G-module, we have therefore $V_1 = V$. This result means that K acts by scalar multiplications on V.

The determinants of the linear automorphisms of V corresponding to the elements of K must all be equal to one, because these elements are products of commutators. Hence, if n is the dimension of V, we have $\gamma_1(x)^n = 1$ for every element x of K. This shows that the kernel of γ_1 is of finite index in K. Since K is irreducible, this implies that γ_1 is the trivial character, i.e., that every element of V is fixed under the action of K. Therefore, we may view V as a G/K-module. Since G/K is abelian and V is simple as a G/K-module, while F is algebraically closed, it follows that V is 1-dimensional. $\square$

Corollary 1.2. *If G is as in Theorem 1.1 then G/G_u is abelian, and every unipotent element of G belongs to G_u.*

PROOF. By Theorem 1.1, the commutator subgroup of G acts trivially on every simple polynomial G-module. Therefore, it must be contained in G_u, so that G/G_u is abelian.

Now let x be a unipotent element of G, and let H be the subgroup of G that is generated by x and G_u. It follows from Proposition V.2.2 that H is unipotent. Since G/G_u is abelian, H is normal in G. Therefore, we must have $H = G_u$, which means that x belongs to G_u. $\square$

2. Lemma 2.1. *Let F be a field, G a unipotent affine algebraic F-group, e a positive integer not divisible by the characteristic of F. Then the map $x \mapsto x^e$ is bijective from G to G.*

PROOF. First, we deal with the case where F has non-zero characteristic p. Choose a finite-dimensional sub G-module V of $\mathscr{P}(G)$ that generates $\mathscr{P}(G)$ as an F-algebra. Let x be an element of G, and denote the restriction of x_l to V by x'. Since G is unipotent, there is a positive integer n such that the p^n-th power of $x' - i_V$ is 0, so that $(x')^{p^n} = i_V$. Since V generates $\mathscr{P}(G)$ as an F-algebra, this implies that x^{p^n} is the neutral element of G. By the assumption on e, there are integers r and s such that $re + sp^n = 1$. If $y = x^r$, we have $y^e = x$. Moreover, if $z^e = y^e$, and we choose r and s as above for a sufficiently large n, then taking the rth power yields $z = y$.

It remains to deal with the case where F is of characteristic 0. Write i for $i_{\mathscr{P}(G)}$. From the fact that $\mathscr{P}(G)$ is the sum of sub G-modules on each of which $i - x_{\mathrm{f}}$ is nilpotent, it follows that the formal sum

$$\log(x_{\mathrm{f}}) = -\sum_{n>0} n^{-1}(i - x_{\mathrm{f}})^n$$

makes sense as a linear endomorphism of $\mathscr{P}(G)$. Evidently, it is locally nilpotent on $\mathscr{P}(G)$. Similarly, if γ is any locally nilpotent linear endomorphism of $\mathscr{P}(G)$, we interpret the formal exponential series $\exp(\gamma)$ as a linear endomorphism of $\mathscr{P}(G)$. By a straightforward application of the calculus of formal power series, one verifies that, for every integer m,

$$\exp(m \log(x_{\mathrm{f}})) = x_{\mathrm{f}}^m.$$

Using this in conjunction with the fact that x_{f} is an F-algebra homomorphism, we show that $\log(x_{\mathrm{f}})$ is a derivation, as follows.

Let t be an auxiliary variable, and extend the various F-linear endomorphisms of $\mathscr{P}(G)$ to $F[t]$-linear endomorphisms of the polynomial ring $\mathscr{P}(G)[t]$ in the evident way. Clearly, we may extend the above so as to interpret $\exp(t \log(x_{\mathrm{f}}))$ as an $F[t]$-linear endomorphism of $\mathscr{P}(G)[t]$. Now let f and g be elements of $\mathscr{P}(G)$, and consider the expression

$$\exp(t \log(x_{\mathrm{f}}))(fg) - \exp(t \log(x_{\mathrm{f}}))(f)\, \exp(t \log(x_{\mathrm{f}}))(g).$$

This is a polynomial $p(t)$ with coefficients in $\mathscr{P}(G)$. By the above, we have $p(m) = 0$ for every integer m. This implies that every coefficient of $p(t)$ is 0. Equating the coefficient of t to 0 yields

$$\log(x_{\mathrm{f}})(fg) = \log(x_{\mathrm{f}})(f)\, g + f\, \log(x_{\mathrm{f}})(g).$$

Thus, $\log(x_{\mathrm{f}})$ is a derivation of $\mathscr{P}(G)$. A familiar formal argument shows that this implies that $\exp(e^{-1} \log(x_{\mathrm{f}}))$ is an F-algebra automorphism of $\mathscr{P}(G)$. Denoting this by u, we have $u^e = x_{\mathrm{f}}$. Hence, if $y = \varepsilon \circ u$, we have $y \in G$ and $y^e = x$. Finally, if z is any element of G such that $z^e = x$, we have $z_{\mathrm{f}}^e = y_{\mathrm{f}}^e$, from which we get $z = y$ by taking logs, dividing by e, and then applying $\varepsilon \circ \exp$. $\qquad\square$

Let S be a group, M an S-module, f a map from S to M. One says that f is a *cocycle for S in M* if

$$f(xy) = x \cdot f(y) + f(x)$$

for all elements x and y of G. Such a cocycle is called a *coboundary* if there is an element m in M such that

$$f(x) = x \cdot m - m$$

for every element x of S.

Proposition 2.2. *Let F be an algebraically closed field, G an abelian linearly reductive affine algebraic F-group, M a unipotent abelian affine algebraic F-group having a G-module structure such that the defining map from $G \times M$ to M is a polynomial map. Let f be a polynomial (map and) cocycle for G in M. Then f is a coboundary.*

PROOF. By Theorem V.5.2, the irreducible component G_1 of the neutral element in G is an F-toroid. Since F is algebraically closed, G_1 is therefore a divisible group. Since G is abelian, it follows from this that G is the direct product of G_1 and a finite group. With all this information, it is easy to see that there is an increasing sequence $(G(n))$ of finite subgroups of G whose union is dense in G. In the case where F has non-zero characteristic p, the order of every element of each $G(n)$ is not divisible by p, because every element of the abelian linearly reductive group G is semisimple. Therefore, if e_n denotes the order of $G(n)$, then e_n is not divisible by the characteristic of F. Now we have from Lemma 2.1 that, if we write M additively, the multiplication by e_n is bijective from M to M.

Consider the restriction of f to $G(n)$. If we sum the cocycle identity for all y's in $G(n)$, we obtain

$$e_n f(x) = s - x \cdot s,$$

where s is the sum of the $f(y)$'s. Since we can divide by e_n, this shows that the restriction of f to $G(n)$ is a coboundary.

Now let $M(n)$ denote the set for all elements m of M such that

$$f(x) = x \cdot m - m$$

for all elements x of $G(n)$. We have just seen that $M(n)$ is non-empty. Evidently, $M(n + 1) \subset M(n)$ for each n. The assumption on the G-module structure of M implies that each $M(n)$ is closed in M. Since M is Noetherian as a topological space, it follows that the sequence $(M(n))$ is eventually constant. This implies that there is an element m in M such that

$$f(x) = x \cdot m - m$$

for every x in the union of the sequence $(G(n))$. Since this union is dense in G, and since f and the G-module structure are polynomial maps, it follows that $f(x) = x \cdot m - m$ for every element x of G. $\qquad\square$

3. Theorem 3.1. *Let F be an algebraically closed field, G an irreducible nilpotent affine algebraic F-group. The semisimple elements of G constitute a central algebraic subgroup G_s of G such that G is the direct product $G_u \times G_s$. The group G_s is an F-toroid and contains every linearly reductive subgroup of G.*

PROOF. Let C denote the center of G. Since G is nilpotent, C_1 is trivial only if G is trivial, because the descending central series of G terminates at the trivial subgroup and, by Theorem II.4.1, all its members are irreducible algebraic subgroups of G. First, we show by induction on the dimension of G that every semisimple element of G belongs to C. In doing this, we assume that G, and hence C_1, is non-trivial, and that the assertion has been established in the lower cases.

Since C_1 is non-trivial, the dimension of G/C is strictly smaller than that of G. Let x be a semisimple element of G, and let y be an arbitrary element of G. The canonical image of x in G/C is a semisimple element, so that our inductive hypothesis implies that $yxy^{-1} = cx$, with $c \in C$. It is clear that the multiplicative Jordan components of c as an element of G lie in C. It follows that the unipotent component of c is that of yxy^{-1}. Since x is semisimple, so is yxy^{-1}. Therefore, the unipotent component of c is trivial, i.e., c is a semi-simple element of G. On the other hand, c belongs to the commutator sub-group of G, which is contained in G_u, by Corollary 1.2. Therefore, c must be the neutral element, whence x lies in the center of G.

Since the semisimple elements of G belong to C, we conclude from Theorem V.5.1 that they constitute the unique maximum linearly reductive subgroup C_s of C. Thus, the group G_s of the present theorem is C_s. We know from Corollary 1.2 that every unipotent element of G belongs to G_u. There-fore, the Jordan decompositions for the elements of G show that $G = G_u G_s$. By Theorem V.4.2, this implies that G is the semidirect product $G_u \rtimes G_s$. Since G_s is central in G, this means that G is the direct product $G_u \times G_s$. Since G is irreducible, so are therefore G_u and G_s. In particular, by Theorem V.5.2, G_s is therefore an F-toroid.

Finally, let R be any linearly reductive subgroup of G. The image of R under the projection morphism from G to G_u with kernel G_s is a linearly reductive subgroup of G_u, and is therefore trivial. This means that $R \subset G_s$.
$\square$

For subgroups A and B of a group G, we denote by $[A, B]$ the subgroup of G that is generated by the commutators $aba^{-1}b^{-1}$ with a in A and b in B.

Let F be an algebraically closed field, G an affine algebraic F-group. Define a sequence of subgroups of G_u as follows.

$$G_u^{(0)} = G_u; \qquad G_u^{(n+1)} = [G, G_u^{(n)}].$$

Clearly, each $G_u^{(n)}$ is normal in G, and $G_u^{(n+1)} \subset G_u^{(n)}$. The intersection of the family of these $G_u^{(n)}$'s is a normal subgroup of G, and it is denoted by G_u^∞.

Now suppose that G is irreducible and solvable. Define a sequence of subgroups of G as follows. $G^{[0]} = G$; $G^{[n+1]} = [G, G^{[n]}]$. Again, this is a descending sequence of normal subgroups of G. By Theorem II.4.1, each $G^{[n]}$ is an irreducible algebraic subgroup of G, and we know from Corollary 1.2 that $G^{[1]} \subset G_u$. It follows that, for each $n > 1$, we have

$$G_u^{(n)} \subset G^{[n]} \subset G_u^{(n-1)}.$$

Since G is Noetherian as a topological space, the descending sequence of closed subsets $G^{[n]}$ must be eventually constant. Hence, the above shows that

$G_u^\infty = G^{[n]}$ for all sufficiently large n's. In particular, G_u^∞ *is an irreducible normal algebraic subgroup of G whenever G is irreducible and solvable.*

Theorem 3.2. *Let F be an algebraically closed field, G a solvable irreducible affine algebraic F-group, T a maximal F-toroid in G. Then G is the semidirect product $G_u \rtimes T$, and, for every linearly reductive subgroup K of G, there is an element r in G_u^∞ such that $rKr^{-1} \subset T$.*

PROOF. First, we show that there is an F-toroid T in G such that G is the semidirect product $G_u \rtimes T$. By Theorem V.4.2, this will hold provided only that $G = G_u T$. In the case where G is nilpotent, Theorem 3.1 contains the required result. Now suppose that G is not nilpotent. Then G must contain a non-central semisimple element, s say, because otherwise every commutator formed with elements of G is also a commutator formed with unipotent elements, i.e., elements of G_u, which implies that G is nilpotent.

Let G^s denote the centralizer of s in G. Since G is irreducible and $G^s \neq G$, the dimension of G^s is strictly smaller than that of G. Proceeding by induction on the dimension of G, we shall have from the inductive hypothesis that there is an F-toroid T in $(G^s)_1$ such that

$$(G^s)_1 = ((G^s)_1)_u T.$$

Now suppose we have $G = G_u G^s$. Then we have also $G = G_u(G^s)_1$. By Corollary 1.2, $((G^s)_1)_u \subset G_u$. Hence, the above now gives $G = G_u T$. Thus, we shall have the decomposition result as soon as we have shown that $G = G_u G^s$.

Let S denote the smallest algebraic subgroup of G that contains s. Since s is a semisimple element, S is a *linearly reductive* abelian algebraic subgroup of G. Let $G(n)$ denote the nth member of the chain of successive commutator subgroups in G. Each $G(n)$ is an irreducible normal algebraic subgroup of G, and $G(n)$ is trivial for n large enough. We know also that $G(1) \subset G_u$, so that $G(n)$ is unipotent for every $n > 0$. Now $G(n)/G(n+1)$ is a unipotent abelian affine algebraic F-group, and we regard it as an S-module via the conjugation action of S on $G(n)$. Clearly, this module structure satisfies the requirements of Proposition 2.2.

Fix a positive index n, and suppose that x is an element of G with the property that $yxy^{-1}x^{-1}$ belongs to $G(n)$ for every element y of S. Note that every element of G satisfies this condition for $n = 1$. Let $f(y)$ denote the canonical image of $yxy^{-1}x^{-1}$ in $G(n)/G(n+1)$. Then f is a polynomial cocycle for S in $G(n)/G(n+1)$, in the sense of Proposition 2.2. By that proposition, f is therefore a coboundary. When this is written out, one sees that there is an element z in $G(n)$ such that $y(zx)y^{-1}(zx)^{-1}$ belongs to $G(n+1)$ for every element y of S. Since $G(n)$ is eventually trivial, repetition of this process yields an element b of $G(1) \subset G_u$ such that bx commutes with every element of S, and so belongs to G^s. Our conclusion is that

$$G = G_u G^s.$$

As we have seen above, this suffices for establishing the result that there is an F-toroid T such that $G = G_u \rtimes T$.

Now let $G^{[n]}$ be the irreducible normal unipotent algebraic subgroup of G defined just above Theorem 3.2, with $n > 0$. Clearly, the product $G^{[n]}T$ in G is an algebraic subgroup of G and, in fact, is the semidirect product $G^{[n]} \rtimes T$. Let R be any linearly reductive subgroup of G, and suppose that R is contained in $G^{[n]} \rtimes T$ for some $n > 0$. Evidently, $G^{[n+1]} \rtimes T$ is a normal algebraic subgroup of $G^{[n]} \rtimes T$, and the factor group is identifiable with the unipotent affine algebraic F-group $G^{[n]}/G^{[n+1]}$. Since R is linearly reductive, its canonical image in this factor group must be trivial. Therefore, we have $R \subset G^{[n+1]} \rtimes T$. The final conclusion of this argument is that every linearly reductive subgroup of G is contained in $G_u^\infty \rtimes T$.

At this point, we simplify our notation as follows. We write G for $G_u^\infty \rtimes T$, and we write $G(0)$ for G_u^∞. On the other hand, for $n > 0$, we let $G(n)$ stand for the n-th commutator subgroup of the new G. Let K be any linearly reductive subgroup of G. What remains to be proved is that there is an element r in $G(0)$ such that $rKr^{-1} \subset T$. Replacing K with its closure in G, we reduce this to the case where K is an algebraic subgroup of G. From the fact that G/G_u is abelian, it follows that K is abelian. Suppose that, for some $n \geq 0$, we have already found an element r_n of $G(0)$ such that $r_n K r_n^{-1} \subset G(n) \rtimes T$. Note that, for $n = 0$, we may simply take r_0 to be the neutral element of G. Write S for $r_n K r_n^{-1}$, and note that S is a subgroup of G of the same type as K. For every element x of S, let $f(x)$ denote the canonical image of $\gamma(x)$ in $G(n)/G(n+1)$, where γ is the projection from $G(n) \rtimes T$ to $G(n)$ coming from the semidirect product structure. Viewing $G(n)/G(n+1)$ as an S-module via the conjugation action of S on $G(n)$, we see directly that f is a polynomial cocycle as in Proposition 2.2. Therefore, f is a coboundary. Writing this out, we see that there is an element t in $G(n)$ such that $tSt^{-1} \subset G(n+1) \rtimes T$. Now the element $r_{n+1} = tr_n$ takes the place of r_n on the next level. Since $G(n)$ is eventually trivial, repetition of this process leads to the required element r, so that the conjugacy result is established. The fact that the place of T may be taken by any maximal F-toroid in G follows immediately from the conjugacy result. $\Box$

4. The basic example of a solvable algebraic group is the *triangular group*, $T(n)$. This is defined as follows. Let V be a finite-dimensional vector space over a field F, equipped with a basis $(v_1, \ldots, v_n)$. For each index i from $(1, \ldots, n)$, let V_i denote the subspace of V spanned by $(v_1, \ldots, v_i)$, and let us agree that V_0 stands for (0). Now $T(n)$ is defined as the subgroup of $\mathrm{Aut}_F(V)$ consisting of the linear automorphisms that stabilize each V_i. Evidently, $T(n)$ is a solvable affine algebraic F-group, and $T(n)_u$ consists of the elements of $T(n)$ inducing the identity map on each V_i/V_{i-1}. We write $U(n)$ for $T(n)_u$. Clearly, $T(n) = U(n) \rtimes D(n)$, where $D(n)$ consists of the elements stabilizing each line Fv_i. Of course, $D(n)$ is the standard model of an F-toroid.

We analyze $T(n)$ further by introducing a chain of normal algebraic subgroups, as follows. For each pair (i, j) of indices with $1 \le i < j \le n$, let $U(i, j)$ denote the subgroup consisting of the elements of $U(n)$ that fix the elements of V_{j-1} and induce the identity map on V_j/V_i. Evidently, each $U(i, j)$ is a normal algebraic subgroup of $T(n)$, and is contained in $U(n)$. This family is totally ordered by inclusion. We have $U(i, j) \subset U(i', j')$ whenever $j > j'$, and $U(i, j) \subset U(i', j)$ whenever $i \le i'$.

Let f_{ij} denote the function on $U(n)$ whose value at each x is the (i, j) matrix entry of x with respect to our basis of V. It is easy to see that $\mathscr{P}(U(n))$ is generated as an F-algebra by the f_{ij}'s with $1 \le i < j \le n$, and that each $U(r, s)$ is the set of zeros in $U(n)$ of a certain set of f_{ij}'s.

Now let us suppose that F is an infinite field. Then the functions f_{ij} are algebraically independent over F, whence each $\mathscr{P}(U(r, s))$ is an ordinary polynomial algebra. In particular, each $U(r, s)$ is now irreducible. Moreover, it is seen directly from the matrix description that the dimensions of successive members of the chain of $U(r, s)$'s differ by exactly 1. From this, we obtain the following general result.

Theorem 4.1. *Let F be an algebraically closed field, G an irreducible solvable affine algebraic F-group. There is a chain of irreducible normal algebraic subgroups U_i of G, starting with $U_0 = (1)$ and ending with $U_k = G_u$, such that each U_i/U_{i+1} is 1-dimensional.*

PROOF. Let V be a finite-dimensional sub G-module of $\mathscr{P}(G)$ that generates $\mathscr{P}(G)$ as an F-algebra, and let ρ denote the representation of G on V. Then ρ is an isomorphism of affine algebraic F-groups from G to the algebraic subgroup $\rho(G)$ of $\mathrm{Aut}_F(V)$, and we identify G with $\rho(G)$ by means of ρ.

It is clear from Theorem 1.1 that there is a basis $(v_1, \dots, v_n)$ of V such that G is contained in the corresponding triangular group $T(n)$ and G_u is contained in $U(n)$. Put $V(i, j) = (G_u \cap U(i, j))_1$. Then the family of $V(i, j)$'s is totally ordered by inclusion, and each member is an irreducible normal algebraic subgroup of G. Since the dimensions of successive $U(i, j)$'s differ by 1, the corresponding $V(i, j)$'s either coincide or have their dimensions differ by 1. The required chain of U_i's is obtained simply by picking out the distinct terms from the sequence $(V(i, j))$. $\qquad\square$

It is true, but not obvious, that every 1-dimensional irreducible algebraic group is commutative. In proving this, we shall use the following elementary result.

Proposition 4.2. *Let F be an algebraically closed field, M a finite-dimensional F-space, G a group of linear automorphisms of M. Suppose that M is simple as a G-module, and that there is a positive integer e such that $x^e = i_M$ for every element x of G. Then G is finite.*

PROOF. Applying Proposition V.1.1 to the subalgebra of $\mathrm{End}_F(M)$ that is generated by G, we see that G contains an F-basis $(x_1, \ldots, x_q)$ of $\mathrm{End}_F(M)$. Let T denote the trace function on $\mathrm{End}_F(M)$. The map from

$$\mathrm{End}_F(M) \times \mathrm{End}_F(M)$$

to F sending each (x, y) onto $T(xy)$ is a non-degenerate bilinear form, whence the linear map τ from $\mathrm{End}_F(M)$ to F^q defined by

$$\tau(x) = (T(xx_1), \ldots, T(xx_q))$$

is injective. For each element x of G, each $T(xx_i)$ is a sum of e-th roots of 1 in F; namely, characteristic roots of xx_i. The number of summands is $\dim(M)$. This shows that $\tau(G)$ is a finite set. Since τ is injective, G is therefore finite. $\quad\square$

Proposition 4.3. *Let F be an algebraically closed field, G an irreducible affine algebraic F-group such that there is a positive integer e with $x^e = 1_G$ for every element x of G. Then G is unipotent.*

PROOF. Let V be a finite-dimensional sub G-module of $\mathscr{P}(G)$ that generates $\mathscr{P}(G)$ as an F-algebra. Let

$$(0) = V_0 \subset \cdots \subset V_n = V$$

be a composition series of V as a G-module. Let H be the normal algebraic subgroup of G consisting of the elements that fix the elements of each V_i/V_{i-1}. Clearly, H is unipotent. Now let W denote the direct sum of the G-modules V_i/V_{i-1}, so that H is the kernel of the representation of G on W. By Proposition 4.2, the image of G in $\mathrm{Aut}_F(W)$ is finite, which means that G/H is finite. Since G is irreducible, this implies that $G = H$. $\quad\square$

Theorem 4.4. *Every irreducible 1-dimensional algebraic group is commutative.*

PROOF. Let F denote the base field, and let F' be an algebraic closure of F. Let G be an irreducible 1-dimensional affine algebraic F-group. We know from Lemma II.1.2 that $\mathscr{P}(G) \otimes F'$ is an integral domain. The group $\mathscr{G}(\mathscr{P}(G) \otimes F')$ is therefore an *irreducible* affine algebraic F'-group. Evidently, it is of the same dimension, 1, as G. Since G may be identified, in the evident way, with a subgroup of $\mathscr{G}(\mathscr{P}(G) \otimes F')$, this shows that it suffices to prove the theorem in the case where F is algebraically closed, which we now assume to be the case.

First, suppose that G contains an element x that is not of finite order. Then the smallest algebraic subgroup of G containing x has strictly positive dimension, and hence must coincide .with G, giving the result that G is commutative.

It remains to deal with the case where every element of G is of finite order. For every positive integer n, let $G(n)$ be the set of elements x of G such that $x^n = 1_G$. Clearly, $G(n)$ is a closed subset of G. If $G(n)$ is finite then it follows from the irreducibility of G and the stability of $G(n)$ under the conjugation action of G that $G(n)$ is contained in the center of G. Therefore, if every $G(n)$ is finite, then G is commutative. On the other hand, if some $G(n)$ is infinite, then that $G(n)$ must coincide with G, because it has an irreducible component of positive dimension. Hence, we can apply Proposition 4.3 and conclude that G is unipotent. This implies that the commutator subgroup of G does not coincide with G. Since it is irreducible and closed in G, it must therefore be trivial. $\qquad\square$

5. If F is an algebraically closed field, and G is an irreducible 1-dimensional affine algebraic F-group, then we have from Theorem V.5.1 that G is either unipotent or linearly reductive. In the second case, G is the multiplicative group F^*, by Theorem V.5.2. We shall see that, in the case where G is unipotent, it is the additive group of F. However, we obtain this result from a more general theorem on abelian unipotent algebraic groups. This requires some preparation.

Let H be a Hopf algebra. An element h of H is called *primitive* if

$$\delta(h) = h \otimes 1 + 1 \otimes h$$

We denote the subspace of primitive elements by P_H. From $(i_H \otimes \varepsilon) \circ \delta = i_H$, we see that P_H lies in the kernel of ε. In the following lemma, we consider the tensor product of two Hopf algebras A and B, and we shall identify A and B with their canonical images in $A \otimes B$.

Lemma 5.1. *Let A and B be Hopf algebras over a field. Then*

$$P_{A \otimes B} = P_A + P_B.$$

PROOF. Let $u = \sum a \otimes b$ be an element of $P_{A \otimes B}$. Then we have

$$\sum \delta(a) \otimes \delta(b) = s_{23}(\delta(u)) = \sum a \otimes 1_A \otimes b \otimes 1_B + \sum 1_A \otimes a \otimes 1_B \otimes b$$

where s_{23} is the switch of the 2nd and 3rd tensor factors. Now apply the linear map $\varepsilon_A \otimes i_A \otimes i_B \otimes \varepsilon_B$. This yields

$$u = \sum \varepsilon(a)1_A \otimes b + \sum a \otimes \varepsilon(b)1_B.$$

Since u is primitive, it is annihilated by the counit $\varepsilon_A \otimes \varepsilon_B$ of $A \otimes B$, whence we have $\sum \varepsilon(a)\varepsilon(b) = 0$. Using this, we rewrite u by replacing b with $b - \varepsilon(b)1_B$ in the first sum, and a with $a - \varepsilon(a)1_A$ in the second sum. Then we have

$$u = 1_A \otimes s + r \otimes 1_B,$$

with $\varepsilon_B(s) = 0 = \varepsilon_A(r)$. From the primitivity of u, we get (using s_{23} as above)

$$1_A \otimes 1_A \otimes \delta(s) + \delta(r) \otimes 1_B \otimes 1_B = 1_A \otimes 1_A \otimes (s \otimes 1_B + 1_B \otimes s)$$

$$+ (r \otimes 1_A + 1_A \otimes r) \otimes 1_B \otimes 1_B.$$

Applying $\varepsilon_A \otimes \varepsilon_A \otimes i_B \otimes i_B$, we find that $\delta(s) = s \otimes 1_B + 1_B \otimes s$, so that s belongs to P_B. Similarly, r belongs to P_A. $\qquad\square$

Lemma 5.2. *Let F be a field, and let H be the polynomial Hopf algebra $F[x_1, \ldots, x_n]$, where the x_i's are algebraically independent primitive elements. Then P_H consists of the F-linear combinations of the powers $x_i^{p^t}$, where p is the characteristic exponent of F.*

PROOF. Lemma 5.1 reduces the problem to the case $n = 1$. In this case, let us write x for x_1, and let us consider a primitive element u. Using that $\varepsilon(u) = 0$, we write

$$u = \sum_{i=1}^{m} c_i x^i$$

with each c_i in F. Then we have

$$0 = \delta(u) - u \otimes 1 - 1 \otimes u = \sum_{i=2}^{m} \left(c_i \sum_{j=1}^{i-1} \left(\binom{i}{j} x^j \otimes x^{i-j} \right) \right).$$

This shows that, in F, the binomial coefficients $\binom{i}{j}$, with $0 < j < i$, are equal to 0 whenever $c_i \neq 0$. It follows that $c_i = 0$ unless i is a power of p. $\square$

Proposition 5.3. *Let F be a perfect field, and let A be a commutative integral domain Hopf algebra that is generated by a finite set of primitive elements. Then A is generated also by a finite set of algebraically independent primitive elements.*

PROOF. Let $(y_1, \ldots, y_n)$ be a system of primitive elements generating A as an F-algebra. If $n = 1$ and $y_1 \neq 0$, it follows from the assumption that A is an integral domain that y_1 is not algebraic over F, as is seen by noting that $\varepsilon(y_1) = 0$. Now we make an induction on n, assuming that $n > 1$, and that the proposition has been established in the lower cases.

Let H be the polynomial Hopf algebra $F[x_1, \ldots, x_n]$, where the x_i's are algebraically independent primitive elements. Let α be the surjective morphism of Hopf algebras from H to A sending each x_i onto y_i. There is nothing to prove if the kernel of α is (0). Assume this kernel is not (0), and choose a non-zero element h from it having the smallest possible degree. Since $\delta(h) - h \otimes 1 - 1 \otimes h$ is annihilated by $\alpha \otimes \alpha$, it is contained in

$$J \otimes A + A \otimes J,$$

where J denotes the kernel of α. On the other hand, this element has the form $\sum_i a_i \otimes b_i$, where each a_i and each b_i has degree strictly smaller than that of h. Since h has the smallest possible degree among the non-zero elements of J, this element must therefore be 0, i.e., $h \in P_H$.

By Lemma 5.2, h is therefore an F-linear combination of powers $x_i^{p^t}$, where p is the characteristic exponent of F. Since F is perfect, we may therefore write

$$h = h_0 + h_1^p + \cdots + h_k^{p^k},$$

where each h_i is a linear combination of the x_j's. Since h is of minimal degree, we must have $h_0 \neq 0$. It follows that, with suitably relabeled y_i's, we have a relation

$$\sum_{t=0}^{r} c_t y_n^{p^t} = \sum_{i=1}^{n-1} \left(\sum_{u=0}^{s} d_{iu} y_i^{p^u} \right),$$

where $c_0 \neq 0$, $c_r \neq 0$ and at least one $d_{is} \neq 0$. From all relations satisfying these conditions, among all sets of primitive generators $(y_1, \ldots, y_n)$, choose one in which $r + s$ is as small as possible. We show that then $r = 0$.

If $r \neq 0$ and $r \geq s$, we choose an index i with $d_{is} \neq 0$ and replace y_i with $y_i - d_i y_n^{p^{r-s}}$, where $d_i^{p^s} = c_r (d_{is})^{-1}$. Then our relation can be written as a relation among the new y_j's in which r is strictly smaller than before, while s has remained unchanged. This contradicts the minimality of $r + s$ in the original relation, so that we must have $r < s$.

Now we replace y_n with $y_n - \sum_{i=1}^{n-1} d_i y_i^{p^{s-r}}$, where $d_i^{p^r} = (c_r)^{-1} d_{is}$. Then our relation can be written as a relation among the new y_j's in which s is strictly smaller than before, while r has remained unchanged, so that we have again a contradiction to the minimality of $r + s$. Hence, we must have $r = 0$, which means that y_n is an F-linear combination of $y_i^{p^t}$'s with $i < n$. Now the inductive hypothesis applies, because A is generated by $(y_1, \ldots, y_{n-1})$. $\qquad\qquad\square$

If G is an affine algebraic F-group such that $\mathscr{P}(G)$ is generated by a finite set of algebraically independent primitive elements then G is the direct product of a finite family of copies of the additive group of F. We call such a group an *algebraic vector group*.

Theorem 5.4. *Let F be a field, G an irreducible abelian unipotent algebraic F-group. In the case where F has non-zero characteristic p, assume that $x^p = 1_G$ for every element x of G. Suppose also that there is a perfect subfield K of F such that $\mathscr{P}(G)$ has a K-form, in the sense that $\mathscr{P}(G) = A \otimes_K F$, where A is a K-Hopf algebra. Then G is an algebraic vector group, and so is $\mathscr{G}(A)$, with $\mathscr{P}(\mathscr{G}(A)) = A$.*

PROOF. Let us identify G with an algebraic subgroup of $\mathrm{Aut}_F(V)$, where V is a finite-dimensional F-space. Let U denote the subspace of $\mathrm{End}_F(V)$ that is spanned by the elements $x - i_V$ with x in G. Clearly, U is closed under multiplication, the elements of U commute with each other, and every element of U is nilpotent.

More precisely, if $U^{(r)}$ is the ideal of U that is generated by the elements u^r with u in U, there is a non-negative integer q such that $U^{(q+1)} = (0)$ and $q < p$ if F has non-zero characteristic p. We choose q minimal, and we fix an F-basis $(u_1, \ldots, u_n)$ of U such that, for each r with $1 \leq r \leq q$, there is an index r^* such that the u_i's with $i \geq r^*$ constitute an F-basis of $U^{(r)}$.

For u in U, we define the element $\exp(u)$ of $\mathrm{Aut}_F(V)$ by

$$\exp(u) = \sum_{i \le q} (i!)^{-1} u^i.$$

Note that, in the case of non-zero characteristic p, this makes sense, because $q < p$. Now define the map $\alpha \colon U \to \mathrm{Aut}_F(V)$ by

$$\alpha(c_1 u_1 + \cdots + c_n u_n) = \exp(c_1 u_1) \cdots \exp(c_n u_n),$$

where the c_i's belong to F. Clearly, α is a group homomorphism. Since it is also a polynomial map, it is therefore a morphism of affine algebraic F-groups from the algebraic vector F-group U to $\mathrm{Aut}_F(V)$. Let S denote the algebraic subgroup of $\mathrm{Aut}_F(V)$ consisting of the elements $i_V + u$ with u in U. We shall prove that α is an isomorphism of affine algebraic F-groups from U to S.

In doing this, we ignore the origin of U from G, so that we can proceed by induction on q, dealing with nilpotent multiplicatively closed subspaces of $\mathrm{End}_F(V)$. If $q = 0$ then $U = (0)$, and both groups are trivial. If $q = 1$, then $U^{(2)} = (0)$, whence $\alpha(u) = i_V + u$ for every element u of U, and it is clear that our claim is true.

Now suppose that $q > 1$, and that our claim has been established in the lower cases. Let $\gamma_1, \ldots, \gamma_n$ be the coordinate functions on U, so that, for every element u of U,

$$u = \sum_{i=1}^{n} \gamma_i(u) u_i.$$

Let us write t for the index $2^* - 1$. Then we have

$$u - \sum_{i=1}^{t} \gamma_i(u) u_i \in U^{(2)}.$$

It follows that

$$(i_V + u) \prod_{i=1}^{t} \exp(-\gamma_i(u) u_i) \in i_V + U^{(2)}.$$

By inductive hypothesis, the restriction, α_1 say, of α to $U^{(2)}$ is an isomorphism of affine algebraic F-groups from $U^{(2)}$ to $i_V + U^{(2)}$. Hence, we may write

$$i_V + u = \alpha\left[\sum_{i=1}^{t} \gamma_i(u) u_i + \alpha_1^{-1}\left((i_V + u) \prod_{i=1}^{t} \exp(-\gamma_i(u) u_i) \right) \right].$$

If we denote the expression in square brackets on the right by $\beta(i_V + u)$, then β is a polynomial map from S to U such that $\alpha \circ \beta$ is the identity map on S.

We have

$$\alpha(c_1 u_1 + \cdots + c_n u_n) = \left(\prod_{i=1}^{t} \exp(c_i u_i) \right) \alpha_1\left(\sum_{i>t} c_i u_i \right).$$

If this is equal to i_V, then we see that

$$\prod_{i=1}^{t} \exp(c_i u_i) \in i_V + U^{(2)}.$$

It follows from this that $c_i = 0$ for each $i \le t$, and hence that

$$\alpha_1\left(\sum_{i>t} c_i u_i\right) = i_V,$$

so that $c_i = 0$ also for each $i > t$. Thus, α is injective.

With the above, this yields the result that α is bijective, and that β is the inverse of α. Thus, α is indeed an isomorphism of affine algebraic F-groups, whence S is an algebraic vector group.

Since G is an algebraic subgroup of S, it follows that $\mathscr{P}(G)$ is generated as an F-algebra by a finite set of primitive elements. It is easy to see from this that the K-Hopf algebra A is also generated by a finite set of primitive elements. Since K is perfect, we can apply Proposition 5.3 and conclude that $A = K[x_1, \ldots, x_m]$, where the x_i's are algebraically independent primitive elements. Since $\mathscr{P}(G) = A \otimes_K F$, the canonical images in $\mathscr{P}(G)$ of the x_i's constitute a system of algebraically independent primitive F-algebra generators. Clearly, this establishes the theorem. $\square$

Let G be any irreducible unipotent abelian algebraic group. In the case where the base field has non-zero characteristic p, for every non-negative integer e, let $G(e)$ be the closure in G of the subgroup consisting of the elements x^{p^e}. This is clearly an irreducible algebraic subgroup of G, and we know that $G(e)$ is trivial for some e. It follows that, if G is non-trivial, then $G(1) \ne G$. Hence, if G is 1-dimensional, then $G(1)$ is trivial, i.e., $x^p = 1_G$ for every x in G. Now it is clear from Theorem 5.4 that the following result holds.

Corollary 5.5. *Let F be a field, G an irreducible unipotent affine algebraic F-group of dimension 1. Suppose that there is a perfect subfield K of F such that $\mathscr{P}(G)$ has a K-form. Then G is isomorphic with the additive group of F.*

Notes

1. Over non-algebraically closed fields, the structure of solvable algebraic groups is considerably less transparent than the results of this chapter might suggest, even in characteristic 0. For example, let V be the 2-dimensional vector group over the field of real numbers, which we shall here denote by F. Regarding V as the Euclidean plane, let T denote the group of rotations of V. This is a 1-dimensional linearly reductive affine algebraic F-group, with $\mathscr{P}(T) = F[c, s]$, where $c^2 + s^2 = 1$, $\varepsilon(c) = 1$, $\varepsilon(s) = 0$, $\delta(c) = c \otimes c - s \otimes s$ and $\delta(s) = c \otimes s + s \otimes c$. Let G be the semidirect product $V \rtimes T$, the conjugation action of T on V being the rotation action.

Then $G_u = V$, and there is no chain as in Theorem 4.1. The group T is the simplest example of an irreducible linearly reductive group that is not a toroid.

2. In non-zero characteristic p, the higher-dimensional abelian irreducible unipotent algebraic groups need not be vector groups. The simplest example showing this is as follows. For $0 < i < p$, let c_i stand for the integer $p^{-1}\binom{p}{i}$. Let π denote the symmetric polynomial in two variables with integer coefficients given by

$$\pi(x, y) = \sum_{i=1}^{p-1} c_i x^i y^{p-i}.$$

Let F be a field of characteristic p, and define a composition of pairs of elements of F by

$$(a_1, b_1) + (a_2, b_2) = (a_1 + a_2, b_1 + b_2 - \pi(a_1, a_2)).$$

One shows directly that this defines a commutative group, G say (the additive group of Witt vectors of length 2), and that G is a unipotent affine algebraic F-group with $\mathscr{P}(G) = F[u, v]$, where $u(a, b) = a$ and $v(a, b) = b$. The sum in G of p summands, each equal to (a, b), is equal to $(0, a^p)$, showing that G is not an algebraic vector group. If F is infinite, then G is irreducible and of dimension 2.

3. The idea for the proof of Proposition 2.2 is due to Grothendieck (cf. [6], Exposé 6). The proof of Theorem 4.4 was found with the help of David Wigner.

4. Theorem 5.4, as well as the proof given here, is due to M. Rosenlicht [13].

Chapter VII

Elementary Lie Algebra Theory

This chapter establishes fundamental results concerning Lie algebras over fields of characteristic 0. The principal notions involved are semisimplicity, solvability and nilpotency.

The main result of Section 1 is Cartan's solvability criterion. In Section 2, this is used for obtaining the salient features of semisimple Lie algebras, the most important result being that, in characteristic 0, every finite-dimensional module for a finite-dimensional semisimple Lie algebra is semisimple.

Section 3 begins with the theorem that extensions of Lie algebras of finite dimension over a field of characteristic 0 in which the image Lie algebra is semisimple are split. Then, it deals with the radical of a Lie algebra, with reference to the representation theory of the Lie algebra. Section 4 deals with the Levi semidirect sum decomposition of a Lie algebra with respect to its radical.

The results of this chapter will be used in the next chapter, chiefly for proving the semidirect product decomposition theorem for algebraic groups over fields of characteristic 0, which extends Theorem VI.3.2 to the completely general situation.

1. Let F be a field, V a finite-dimensional F-space, e an element of $\mathrm{End}_F(V)$. An easy elementary result (a special case of Fitting's Lemma) says that V is the direct sum of two e-stable subspaces V_0 and V_1 such that the restriction of e to V_0 is nilpotent, while the restriction of e to V_1 is a linear automorphism. Now suppose that F is of characteristic 0 and that the trace of every power of e is equal to 0. Then we can conclude that e is nilpotent, as follows. Our assumption on e implies that, if e_1 is the restriction of e to V_1, then the trace of every power of e_1 is equal to 0. If $V_1 \neq (0)$ then the minimum polynomial of e_1 has a non-zero constant term, and we get a contradiction by applying the trace map to the corresponding relation among the powers of e_1, because F is of characteristic 0. Thus, we have $V_1 = (0)$, which means that e is nilpotent.

Lemma 1.1. *Let F be a field of characteristic 0, and let V be a finite-dimensional F-space. Let $x_1, \ldots, x_n$ and $y_1, \ldots, y_n$ be elements of $\mathrm{End}_F(V)$, and let*

$$e = \sum_{i=1}^{n} (x_i y_i - y_i x_i).$$

If e commutes with each x_i then e is nilpotent.

PROOF. For every positive exponent s, we have

$$e^s = \sum_{i=1}^{n} (x_i(e^{s-1}y_i) - (e^{s-1}y_i)x_i),$$

where e^0 stands for i_V. It is clear from this that the trace of e^s is equal to 0, so that the above remark applies. $\square$

If A and B are subsets of a Lie algebra L, then $[A, B]$ denotes the subspace of L that is spanned by the elements $[a, b]$ with a in A and b in B. The *center* of L is the ideal consisting of the elements x such that $D_x = 0$, i.e., such that $[(x), L] = (0)$. We say that L is *semisimple* if it has no non-zero abelian ideal.

Theorem 1.2. *Let L be a Lie algebra of linear endomorphisms of a finite-dimensional F-space V, where F is a field of characteristic 0. Suppose that V is semisimple as an L-module. If Z is the center of L, then L/Z is a semisimple Lie algebra, $[L, L] \cap Z = (0)$, and every element of Z is semisimple.*

PROOF. Let A denote the (associative) sub F-algebra of $\mathrm{End}_F(V)$ that is generated by L. First, we show that A has no non-zero nilpotent left ideal.

Suppose this is false. Then there is a non-zero left ideal B of A such that $BB = (0)$. Now $B \cdot V$ is a sub A-module of V. Clearly, V is semisimple as an A-module. Therefore, there is a sub A-module W of V such that

$$V = B \cdot V + W$$

and $(B \cdot V) \cap W = (0)$. Now $B \cdot W \subset (B \cdot V) \cap W$, whence $B \cdot W = (0)$. Hence $B \cdot V = B \cdot (B \cdot V) + B \cdot W = (0)$, contradicting $B \neq (0)$.

Let C denote the center of A. By what we have just shown, C has no non-zero nilpotent element. Therefore, every element of C is semisimple, as is seen either quite directly, or from Theorem V.1.4. In particular, every element of Z is semisimple. Moreover, by Lemma 1.1, every element of $[A, A] \cap C$ is nilpotent, so that we must have $[A, A] \cap C = (0)$. In particular, $[L, L] \cap Z = (0)$.

This last result shows that the inverse image in L of an abelian ideal of L/Z is an abelian ideal, J say, of L. Now consider the abelian ideal $[L, J]$.

By Lemma 1.1, all its elements are nilpotent. Let B denote the subspace of A that is spanned by the products of elements of $[L, J]$.

Then the elements of B commute with each other, and $B^m = (0)$ for some positive exponent m. Directly from the definition, we see that $LB \subset BL + B$, whence $AB \subset BA$. Now it follows by induction that $(AB)^s \subset B^sA$ for every positive integer s. Therefore, AB is a *nilpotent* left ideal of A, so that we must have $AB = (0)$. In particular, $[L, J] = (0)$, so that $J \subset Z$. Thus, (0) is the only abelian ideal of L/Z. $\qquad\square$

The ideal $[L, L]$ of a Lie algebra L is called the *commutator ideal*. We say that L is *solvable* if the chain of successive commutator ideals ends with (0). The following result is known as *Lie's Theorem* (cf. Theorem VI.1.1).

Theorem 1.3. *Let L be a solvable Lie algebra over a field of characteristic 0. Then every finite-dimensional semisimple L-module is annihilated by $[L, L]$.*

PROOF. Let V be such an L-module. Let L' denote the image of L in $\operatorname{End}(V)$, and let Z denote the center of L'. By Theorem 1.2, L'/Z has no non-zero abelian ideal. Since L'/Z is solvable, this implies that $L'/Z = (0)$, i.e., that L' is abelian. This means that $[L, L]$ annihilates V. $\qquad\square$

If L is a Lie algebra, V an L-module and S a subset of L then we shall say that S *is nilpotent on* V if there is a positive integer n such that $\rho(S)^n = (0)$, where ρ denotes the given representation of L on V.

Lemma 1.4. *Let L be a Lie algebra, V an L-module, S a subspace of L that is nilpotent on V. Suppose that x is an element of L that is nilpotent on V and such that $[x, S] \subset S$. Then $x + S$ is nilpotent on V.*

PROOF. Without loss of generality, we assume that L is given as a Lie algebra of linear endomorphisms of V. Then the assumption is that there are positive integers p and q such that $S^p = (0)$ and $x^q = 0$. We show that $(x + S)^{pq} = (0)$.

Consider a product $u_1 \cdots u_{pq}$, where each u_i is either an element of S or equal to x. For s in S, we have

$$sx = xs + [s, x] \in xs + S.$$

We see from this that

$$u_1 \cdots u_{pq} \in \sum_{r=0}^{pq} x^r S^t,$$

where t is the number of indices i such that u_i belongs to S. Therefore, this product is 0, unless $t < p$. However, if $t < p$, there must be an index i such that $u_{i+1}, \ldots, u_{i+q}$ are all equal to x, because there are $pq - t > p(q - 1)$ elements x in the $t + 1 \leq p$ intervals not containing elements of S. Consequently, $u_1 \cdots u_{pq} = 0$ in all cases. $\qquad\square$

Now we are in a position to obtain *Engel's Theorem*, which is closely related to Theorem V.2.1.

Theorem 1.5. *Let L be a Lie algebra, and let V be a finite-dimensional L-module. If every element of L is nilpotent on V then L is nilpotent on V.*

PROOF. As above, we assume without loss of generality that L is given as a Lie algebra of linear endomorphisms of V. Let x and y be elements of L, and consider $D_x^m(y)$ for positive integers m. This is a sum of products $\pm x^p y x^q$, where $p + q = m$, so that our assumption gives the result that D_x is nilpotent.

Now we proceed by induction on the dimension of L. There is nothing to prove if $L = (0)$. We suppose that $L \neq (0)$ and that the result has been established in the lower-dimensional cases. Among the sub Lie algebras of L other than L, choose one, H say, having the largest possible dimension. Let us view L as an H-module by the map $x \mapsto D_x$ from H to $\mathrm{End}(L)$. We have just seen that, for this module structure, every element of H is nilpotent on L. By inductive hypothesis, H is therefore nilpotent on L. It follows from this that there is an element y in $L \setminus H$ such that $[H, y] \subset H$. Now we can apply Lemma 1.4 to conclude that $y + H$ is nilpotent on V. Clearly, the subspace of L that is spanned by y and H is a sub Lie algebra of L. By the maximality of H, this subspace therefore coincides with L, and we have just seen that it is nilpotent on V. $\qquad\qquad\square$

Next, we establish *Cartan's solvability criterion*, which is as follows.

Theorem 1.6. *Let F be a field of characteristic 0, V a finite-dimensional F-space, L a Lie algebra of linear endomorphisms of V such that the trace function vanishes on LL. Then L is solvable.*

PROOF. It is easy to see that it suffices to prove this when F is algebraically closed, which we shall now assume to be the case. Making an induction on the dimension of L, we suppose that the result holds for all Lie algebras of lower dimension than L. Then, if $[L, L] \neq L$, it follows that L is solvable. Therefore, we assume that $L = [L, L]$ and derive a contradiction.

Among the sub Lie algebras of L other than L, choose a maximal one, H say. By inductive hypothesis, H is solvable. Consider the H-module L/H, where the module structure comes from the map $x \mapsto D_x$. If U/H is a minimal non-zero sub H-module of this, we know from Theorem 1.3 that the endomorphisms of U/H corresponding to the elements of H commute with each other. Since F is algebraically closed, this implies (via Schur's Lemma) that U/H is 1-dimensional, so that $U = Fy + H$, with $y \notin H$. Since this is a sub H-module of L, there is an element μ in H° such that

$$[x, y] - \mu(x)y \in H$$

for every element x of H. In particular, $Fy + H$ is a sub Lie algebra of L, so that the maximality of H implies that $L = Fy + H$.

Let W be any non-zero simple L-module, and let W_0 be a non-zero simple sub H-module of W. As above for U/H, we see that W_0 is 1-dimensional: $W_0 = Fw_0$. For $i > 0$, define $w_i = y \cdot w_{i-1}$, and let $W_i = Fw_0 + \cdots + Fw_i$. Then, for x in H, we have

$$x \cdot w_{i+1} = x \cdot (y \cdot w_i) = [x, y] \cdot w_i + y \cdot (x \cdot w_i)$$
$$= ([x, y] - \mu(x)y) \cdot w_i + y \cdot (\mu(x)w_i + x \cdot w_i).$$

Since $[x, y] - \mu(x)y$ belongs to H, we see from this inductively that each W_i is a sub H-module of W.

Write $x \cdot w_0 = \sigma(x)w_0$ for every x in H. Then it is easy to verify inductively from the above that, for each index i, the element x of H acts on W_i/W_{i-1} as the scalar multiplication by $\sigma(x) + i\mu(x)$ (where we have written W_{-1} for (0)).

Now let q be the largest index i for which the set $(w_0, \ldots, w_i)$ is linearly independent. Then W_q is a sub L-module of W, so that $W_q = W$. Hence, if x_W denotes the linear endomorphism of W corresponding to the element x of H, and if T denotes the trace function, we have

$$T(x_W) = (q + 1)\sigma(x) + \tfrac{1}{2}q(q + 1)\mu(x).$$

Since $L = [L, L]$, each x is a sum of commutators of elements of L, so that we must have $T(x_W) = 0$, whence

$$\sigma(x) = -\tfrac{1}{2}q\mu(x).$$

Substituting this above, we have that x acts on W_i/W_{i-1} as the scalar multiplication by $(i - \tfrac{1}{2}q)\mu(x)$, which gives

$$T(x_W^2) = \sum_{i=0}^{q} (i - \tfrac{1}{2}q)^2\mu(x)^2.$$

Now let

$$(0) = V_k \subset \cdots \subset V_0 = V$$

be a composition series for the L-module V. Applying our above result to the simple L-modules V_j/V_{j+1}, we obtain

$$T(x_V^2) = \sum_{j=0}^{k-1}\left(\sum_{i=0}^{q_j} (i - \tfrac{1}{2}q_j)^2\right)\mu(x)^2,$$

where $q_j = \dim(V_j/V_{j+1}) - 1$.

If $\mu(x) = 0$ for every x in H, then we have $[L, L] \subset H$, contradicting $[L, L] = L$. Hence there is an element x in H such that $\mu(x) \neq 0$. Since $T(x_V^2) = 0$, it follows that we must have $i = \tfrac{1}{2}q_j$ for all the indices of the above sum. This is possible only if each q_j is 0, i.e., only if each V_j/V_{j+1} is 1-dimensional. But then, since $L = [L, L]$, we have $L \cdot V_j \subset V_{j+1}$ for each j, so that L is nilpotent on V. Evidently, this contradicts $L = [L, L]$. $\qquad\square$

2. Let L be a Lie algebra. The representation of L on L sending each element x of L onto the derivation D_x of L is called the *adjoint representation* of L. If ρ is any finite-dimensional representation of L, then we define the *trace form τ_ρ* on $L \times L$ by

$$\tau_\rho(x, y) = T(\rho(x)\rho(y)),$$

where T again denotes the trace function. Evidently, τ_ρ is bilinear and symmetric. Moreover, one verifies directly that, for all x, y, z in L,

$$\tau_\rho([x, y], z) = \tau_\rho(x, [y, z]).$$

Theorem 2.1. *Let L be a finite-dimensional Lie algebra over a field of characteristic 0. If L is semisimple and ρ is an injective finite-dimensional representation of L, then the trace form τ_ρ is non-degenerate. If the trace form of the adjoint representation of L is non-degenerate then L is semisimple.*

PROOF. Let H denote the set of all elements x of L such that $\tau_\rho(x, y) = 0$ for every y in L. The formal property of τ_ρ noted above shows that H is an ideal of L. Theorem 1.6, applied to $\rho(H)$, shows that H is solvable. If H were not zero, then a member of the sequence of successive commutator ideals of H would be a non-zero abelian ideal of L. Therefore, if L is semisimple, we must have $H = (0)$, which means that τ_ρ is non-degenerate.

Now suppose that the trace form of the adjoint representation of L is non-degenerate. Let I be any abelian ideal of L. If x belongs to I and y to L, then $D_x D_y$ sends L into I and annihilates I, so that $T(D_x D_y) = 0$. Therefore, our assumption implies that $x = 0$, so that $I = (0)$. $\square$

Proposition 2.2. *Let L be a finite-dimensional semisimple Lie algebra over a field of characteristic 0, and let I be an ideal of L. There is one and only one ideal I' of L such that $L = I + I'$ and $I \cap I' = (0)$. Also, $L = [L, L]$.*

PROOF. Since L is semisimple, its center is (0), so that the adjoint representation of L is injective. By Theorem 2.1, the trace form of the adjoint representation of L is therefore non-degenerate. Let I' be the set of all elements x of L such that $T(D_x D_y) = 0$ for every element y of I. The formal property of the trace form noted at the beginning of this section shows that I' is an ideal of L. By Theorem 1.6, $I \cap I'$ is solvable, and hence (0). By definition, I' is the set of zeros in L of a set of linear functions in bijective correspondence with a basis of I. Hence $\dim(I') + \dim(I) \geq \dim(L)$. Since $I \cap I' = (0)$ it follows that $L = I + I'$.

In particular, $L = [L, L] + [L, L]'$, and by applying the endomorphisms D_x we see that $[L, L]'$ must lie in the center of L, which is (0). Thus $L = [L, L]$.

Finally, suppose that J is any ideal of L such that $L = I + J$ and

$$I \cap J = (0).$$

Using that $L = [L, L]$, we find

$$J = [L, J] = [I + I', J] = [I', J] \subset I',$$

and hence $J = I'$. $\square$

Let L be as in Proposition 2.2, and let ρ be an injective representation of L on a finite-dimensional vector space V. There is a linear map

$$\mu : L \otimes L \to \text{End}(L)$$

such that

$$\mu(x \otimes y)(z) = \tau_\rho(z, x)y$$

for all elements x, y, z of L. From the non-degeneracy of τ_ρ, it follows that μ is an isomorphism. In fact, τ_ρ yields an isomorphism from L to L° in the canonical fashion, and μ is the composite of the induced isomorphism $L \otimes L \to L^\circ \otimes L$ with the canonical isomorphism $L^\circ \otimes L \to \text{End}(L)$. We define the *Casimir element u_ρ of ρ* in $L \otimes L$ by $\mu(u_\rho) = i_L$.

The adjoint representation of L yields a representation of L on $L \otimes L$ by the canonical tensor product construction. This representation is characterized by the formula

$$t \cdot (x \otimes y) = [t, x] \otimes y + x \otimes [t, y].$$

On the other hand, the adjoint representation yields a representation of L on $\text{End}(L)$, which is given by

$$t \cdot e = D_t e - e D_t,$$

where $t \in L$ and $e \in \text{End}(L)$. Using the formal property of τ_ρ noted at the beginning of this section, we see that, with respect to these L-module structures, μ is an isomorphism of L-modules. It is clear from this that *u_ρ belongs to the L-annihilated part $(L \otimes L)^L$ of $L \otimes L$.*

Let ρ^2 stand for the linear map $L \otimes L \to \text{End}(V)$ that comes from ρ in the canonical way, so that

$$\rho^2(x \otimes y) = \rho(x)\rho(y).$$

We define the *Casimir operator* of ρ by $\rho^2(u_\rho)$. Clearly, ρ^2 is a morphism of L-modules. It follows that *the Casimir operator is an L-module endomorphism of V* (i.e., that it commutes with every $\rho(x)$).

Let us choose a basis $(y_1, \ldots, y_n)$ of L, and write

$$u_\rho = \sum_{i=1}^{n} x_i \otimes y_i.$$

Then the trace of the Casimir operator is equal to $\sum_{i=1}^{n} \tau_\rho(x_i, y_i)$. From $\mu(u_\rho) = i_L$ we see immediately that $\tau_\rho(x_i, y_i) = 1$ for each i. Thus, the trace of the Casimir operator is equal to $\dim(L)$. In particular, *if $L \neq (0)$ then the Casimir operator of ρ is not nilpotent.*

For any L-module V, let us denote the L-annihilated part of V by V^L, and let $L \cdot V$ denote the L-submodule spanned by the elements $x \cdot v = \rho(x)(v)$ with x in L and v in V.

Lemma 2.3. *Let L be a finite-dimensional semisimple Lie algebra over a field of characteristic* 0, *and let V be a finite-dimension L-module. Then*

$$V = L \cdot V + V^L.$$

PROOF. Let ρ denote the representation of L on V, let I be the kernel of ρ, and let I' be the ideal complementary to I figuring in Proposition 2.2. Clearly, I' is a semisimple Lie algebra, $L \cdot V = I' \cdot V$ and $V^L = V^{I'}$. Therefore, restriction of ρ to I' reduces the problem to the case where ρ is injective. Then, let c_ρ be the Casimir operator of ρ. Let $V = V_1 + V_0$ be the Fitting decomposition of V with respect to c_ρ, so that the restriction of c_ρ to V_1 is a linear automorphism, while the restriction of c_ρ to V_0 is nilpotent. Since c_ρ is an L-module endomorphism, these components V_0 and V_1 are sub L-modules of V. Since $V_1 = c_\rho(V_1)$, we have $V_1 = L \cdot V_1$. If $V \neq (0)$ then c_ρ is not nilpotent, which means that $V_1 \neq (0)$. Therefore, V_0 is of strictly smaller dimension than V. Now it suffices to prove the result for V_0 in the place of V. This is all we need for establishing Lemma 2.3 by induction on the dimension of V. $\square$

Theorem 2.4. *Let L be a finite-dimensional semisimple Lie algebra over a field of characteristic* 0. *Then every finite-dimensional L-module is semisimple.*

PROOF. Let V be a finite-dimensional L-module, and let U be a sub L-module of V. We show that U has an L-module complement in V. Let H be the space of all linear maps $f \colon V/U \to V$. We make H into an L-module by

$$(x \cdot f)(a) = x \cdot f(a) - f(x \cdot a),$$

where $x \in L$ and $a \in V/U$. The elements f of this module with the property that $(x \cdot f)(V/U) \subset U$ for every element x of L clearly constitute a sub L-module, M say, of H. By Lemma 2.3, we have $M = L \cdot M + M^L$.

Now let us choose a linear map $g \colon V/U \to V$ whose composite with the canonical map $V \to V/U$ is the identity map on V/U. Clearly, g belongs to M. Write $g = h + k$, where h belongs to $L \cdot M$ and k belongs to M^L. Then the composite of k with the canonical map $V \to V/U$ is still the identity map on V/U, and k is a morphism of L-modules. Hence $k(V/U)$ is an L-module complement of U in V. $\square$

If L is a Lie algebra and V an L-module, then a *cocycle for L in V* is a linear map f from L to V satisfying the identity

$$f([x, y]) = x \cdot f(y) - y \cdot f(x).$$

Such an f is called a *coboundary* if there is an element v in V such that $f(x) = x \cdot v$ for every element x of L.

Corollary 2.5. *Let L be a finite-dimensional semisimple Lie algebra over the field F of characteristic 0, and let V be a finite-dimensional L-module. Then every cocycle for L in V is a coboundary.*

PROOF. Let f be a cocycle for L in V. We define an L-module structure on the direct sum $V + F$ by setting

$$x \cdot (v, a) = (af(x) + x \cdot v, 0)$$

for every x in L, every v in V and every a in F. In fact, the cocycle identity is exactly what is needed for making this an L-module structure. By Theorem 2.4, this L-module is semisimple, so that the sub L-module V has an L-module complement. If we write the component of $(0, 1)$ in this complement in the form $(-v, 1)$, then we find that $f(x) = x \cdot v$ for every element x of L, because the complement of V in our module must be annihilated by L. $\square$

An application of direct interest is as follows.

Proposition 2.6. *Let L be a finite-dimensional semisimple Lie algebra over a field of characteristic 0. Then every derivation of L is of the form D_x, with x in L.*

PROOF. This is seen immediately from Corollary 2.5 by observing that a derivation is a cocycle for L in L, with respect to the adjoint representation. $\square$

3. Theorem 3.1. *Let E and L be finite-dimensional Lie algebras over a field of characteristic 0, and suppose that L is semisimple. Let $\pi: E \to L$ be a surjective Lie algebra homomorphism. There is a Lie algebra homomorphism $\rho: L \to E$ such that $\pi \circ \rho$ is the identity map on L.*

PROOF. We prove this result by induction on the dimension of the kernel, P say, of π, and thus assume that $P \neq (0)$ and that the theorem has been established in the lower cases. If E is semisimple, we know from Proposition 2.2 that it is the direct Lie algebra sum of P and a complementary ideal P'. Clearly, the restriction of π to P' is an isomorphism from P' to L, and we may take ρ to be the inverse of this isomorphism. Therefore, we assume without loss of generality that E has a non-zero abelian ideal. Let A be a non-zero abelian ideal of E having the smallest possible dimension. Now $(A + P)/P$ is an abelian ideal of the semisimple Lie algebra E/P, and hence is (0). Thus, we have $A \subset P$.

First, suppose that $A \neq P$. Then we consider the surjective Lie algebra homomorphism π' from E/A to L that is induced by π. The kernel of π' is

P/A, and our inductive hypothesis ensures that there is a Lie algebra homomorphism δ from L to E/A such that $\pi' \circ \delta$ is the identity map on L. Write $\delta(L) = M/A$ with $M \subset E$. Since $\dim(A) < \dim(P)$, we can again apply our inductive hypothesis to conclude that there is a Lie algebra homomorphism σ from $\delta(L)$ to M such that $\mu \circ \sigma$ is the identity map on $\delta(L)$, where μ is the canonical map $M \to M/A = \delta(L)$. Clearly, we may take ρ to be $\sigma \circ \delta$.

Now suppose that $P = A$. Since P is now abelian, the adjoint representation of E induces an L-module structure on P such that $\pi(e) \cdot p = [e, p]$ for every e in E and every p in P. The sub L-modules of P are precisely the ideals of E that are contained in P. Since $P = A$, the minimality of A implies that P is a simple L-module. If the representation of L on P is trivial then $[E, P] = (0)$ and the adjoint representation of E induces an L-module structure of E extending that of P. By Theorem 2.4, there is an L-module complement Q for P in E. Clearly, Q is an ideal of E, and the restriction of π to Q is a Lie algebra isomorphism from Q to L whose inverse will serve for ρ in the statement of our theorem.

Now we consider the remaining case where P is a simple non-trivial L-module. Let I be the kernel of the representation of L on P, and let I' be its complementary ideal in L. By our present assumption, $I' \neq (0)$, so that the Casimir element in $I' \otimes I'$ of the representation of I' on P is not 0. Since P is simple as an L-module, the corresponding Casimir operator is an L-module $\textit{auto}$morphism, γ say, and we write our Casimir element as

$$u = \sum_{i=1}^{n} x_i \otimes y_i,$$

where the x_i's and y_i's are elements of L (actually, of I').

Let us choose a linear map f from L to E such that $\pi \circ f$ is the identity map on L. We consider the deviation of f from a Lie algebra homomorphism, i.e., we consider the bilinear map g from $L \times L$ to P that is given by

$$g(x, y) = [f(x), f(y)] - f([x, y]).$$

By the definition of the L-module structure of P, we have $x \cdot p = [f(x), p]$ for every x in L and every p in P. Using this and the Jacobi identity for E, we find that g satisfies the identity

$$x \cdot g(y, z) - y \cdot g(x, z) + z \cdot g(x, y) = g([x, y], z)$$
$$- g([x, z], y) + g([y, z], x).$$

Using this, we obtain the following expression for $y_i \cdot g(x, y)$:

$$x \cdot g(y_i, y) - y \cdot g(y_i, x) + g([y_i, x], y) - g([y_i, y], x) + g([x, y], y_i). \quad (*)$$

Next, we write

$$x_i \cdot (x \cdot g(y_i, y)) = [x_i, x] \cdot g(y_i, y) + x \cdot (x_i \cdot g(y_i, y)). \quad (**)$$

Now we know that the Casimir element u is annihilated by L via the L-module structure of $L \otimes L$ coming from the adjoint representation of L. This means that

$$\sum_{i=1}^{n} ([x, x_i] \otimes y_i + x_i \otimes [x, y_i]) = 0.$$

It follows that, for every bilinear map k from $L \times L$ to P,

$$\sum_{i=1}^{n} (k([x_i, x], y_i) + k(x_i, [y_i, x])) = 0. \qquad (***)$$

Now let us write out $\gamma(g(x, y))$ as $\sum_{i=1}^{n} x_i \cdot (y_i \cdot g(x, y))$ and transform this by applying $(*)$, $(**)$ and the formula obtained by interchanging x and y in $(**)$. The result is

$$\gamma(g(x, y)) = x \cdot h(y) - y \cdot h(x) - h([x, y])$$

$$+ \sum_{i=1}^{n} ([x_i, x] \cdot g(y_i, y) + x_i \cdot g([y_i, x], y))$$

$$- \sum_{i=1}^{n} ([x_i, y] \cdot g(y_i, x) + x_i \cdot g([y_i, y], x)),$$

where h is given by

$$h(z) = \sum_{i=1}^{n} x_i \cdot g(y_i, z).$$

Using $(***)$ with the evident k's, we see that the sums on the second and third lines of the above formula are 0. Now define the map ρ from L to E by

$$\rho(z) = f(z) - \gamma^{-1}(h(z))$$

Then the above formula shows that ρ is a homomorphism of Lie algebras. Also $\pi \circ \rho = \pi \circ f = i_L$. $\qquad \square$

Let A and B be solvable ideals of a Lie algebra L. The ideal $(A + B)/A$ of L/A is a homomorphic image of B and is therefore solvable. Since A is solvable, it follows that $A + B$ is solvable. It is clear from this that every finite-dimensional Lie algebra L has a solvable ideal containing every solvable ideal of L. This unique maximum solvable ideal of L is called the *radical* of L; we denote it by L_r. Clearly, L/L_r is a semisimple Lie algebra.

Theorem 3.2. *Let L be a finite-dimensional Lie algebra over a field of characteristic 0. Then $[L, L] \cap L_r = [L, L_r]$. If V is a finite-dimensional L-module then $[L, L_r]$ is nilpotent on V.*

PROOF. If we apply Theorem 3.1 to the canonical homomorphism $L \rightarrow L/L_r$, we see that there is a semisimple sub Lie algebra S of L that is mapped isomorphically onto L/L_r by the canonical map. Clearly, $L = L_r + S$ and $L_r \cap S = (0)$. Since S is semisimple, we have $[S, S] = S$, whence

$$[L, L] = [L, L_r] + S.$$

It is clear from this that $[L, L] \cap L_r = [L, L_r]$.

We know from Theorem 1.3 that $[L_r, L_r]$ is nilpotent on V. Let T be a subspace of $[L, L_r]$ that contains $[L_r, L_r]$ and is nilpotent on V. If $T \neq [L, L_r]$ there is an element z in L and an element x in L_r such that $[z, x]$ does not belong to T. The sub Lie algebra of L that is spanned by z and L_r is evidently solvable. By Theorem 1.3, its commutator ideal is therefore nilpotent on V. In particular, $[z, x]$ is nilpotent on V. Since $[[z, x], T] \subset T$, it follows from Lemma 1.4 that the space spanned by $[z, x]$ and T is nilpotent on V. Now it is clear that if T is chosen maximal then $T = [L, L_r]$. $\square$

Theorem 3.3. *Let L be a finite-dimensional Lie algebra over a field of characteristic 0, and let V be a finite-dimensional L-module. There is an ideal P of L that is nilpotent on V and contains every ideal of L that is nilpotent on V. Every element of L_r that is nilpotent on V belongs to P.*

PROOF. Let $(0) = V_0 \subset \cdots \subset V_n = V$ be a composition series for the L-module V. Let P_i be the kernel of the induced representation of L on V_i/V_{i-1}, and put $P = \bigcap_{i=1}^n P_i$. Clearly, P is nilpotent on V.

Now let J be any ideal of L that is nilpotent on V, and let W be one of the factor modules V_i/V_{i-1}. From the fact that J is an ideal of L, it follows that $J \cdot W$ is an L-submodule of W. Since W is simple as an L-module, we have therefore either $J \cdot W = W$ or $J \cdot W = (0)$. The first possibility is ruled out because J is nilpotent on V. Therefore, J annihilates W, and we conclude that $J \subset P$.

Now let x be an element of L_r that is nilpotent on V. By Lemma 1.4, the space spanned by x and P is nilpotent on V. We know from Theorem 3.2 that $[L, L_r]$ is nilpotent on V, so that, from what we have just proved,

$$[L, L_r] \subset P.$$

Since $[L, x] \subset [L, L_r] \subset P$, the space spanned by x and P is actually an ideal of L, so that it must be contained in P. $\square$

4. Let L be a Lie algebra. Define $L^0 = L$ and $L^{i+1} = [L, L^i]$. Clearly, the L^i's constitute a descending chain of ideals of L. We say that L is *nilpotent* if this chain ends at (0). If we apply Theorem 3.2 to the adjoint representation, we find that, *if L is a finite-dimensional Lie algebra over a field of characteristic 0, then $[L, L_r]$ is a nilpotent ideal of L.*

Write $L_r^{[0]} = L_r$, and $L_r^{[i+1]}$ for $[L, L_r^{[i]}]$. Assuming that L is finite-dimensional, we have $L_r^{[i+1]} = L_r^{[i]}$ for all sufficiently large indices i, and we denote this limit by $L_r^{[\infty]}$. Clearly, this is a nilpotent ideal of L. If x is an element of $L_r^{[\infty]}$, then D_x is nilpotent, so that $\exp(D_x)$ becomes a polynomial in D_x and has a meaning as a Lie algebra automorphism of L.

Theorem 4.1. *Let L be a finite-dimensional Lie algebra over a field of characteristic 0, and let S be a semisimple sub Lie algebra of L such that $L = L_r + S$ (the existence of S is guaranteed by Theorem 3.1). Let T be any semisimple sub Lie algebra of L. There is an element x in $L_r^{[\infty]}$ such that $\exp(D_x)(T) \subset S$.*

PROOF. Suppose that $T \subset L_r^{[i]} + S$, for some index i. Then, since $T = [T, T]$, it follows that $T \subset L_r^{[i+1]} + S$. Thus, we have $T \subset L_r^{[\infty]} + S$, so that it suffices to prove the theorem when $L_r = L_r^{[\infty]}$. Accordingly, we assume that L_r is nilpotent and then proceed by induction on $\dim(L_r)$. Our inductive hypothesis is now that the theorem has been established in the cases of lower dimensional radical, and we suppose that $L_r \neq (0)$.

Since L_r is nilpotent, its center, Z say, is not (0). It is easy to see that Z is an ideal of L, so that we can apply our inductive hypothesis to L/Z. This gives the existence of an element y of L_r such that $\exp(D_y)$ sends T into $Z + S$. Now suppose that there is an element z in Z such that $\exp(D_z)$ sends $\exp(D_y)(T)$ into S. Then we note that, since D_y and D_z commute with each other, we have

$$\exp(D_z)\exp(D_y) = \exp(D_z + D_y) = \exp(D_{y+z}),$$

so that we may take the required element x to be $z + y$. Thus, it remains only to deal with the case where L_r is abelian.

In this case, we consider the linear maps ρ and σ from T to L_r and S, respectively, such that, for every t in T, $t = \rho(t) + \sigma(t)$. We see directly that σ is a homomorphism of Lie algebras, and that ρ is a cocycle for T in L_r, with respect to the T-module structure of L_r coming from the adjoint representation of L. Since T is semisimple, we have from Corollary 2.5 that there is an element x in L_r such that $\rho(t) = [t, x]$ for every element t of T. Now we have

$$\exp(D_x)(t) = t + D_x(t) = t + [x, t] = t - \rho(t) = \sigma(t),$$

which shows that $\exp(D_x)$ sends T into S. $\qquad\qquad\square$

Notes

1. The most transparent general example of a semisimple Lie algebra is as follows. Let F be a field of characteristic 0, and let V be a finite-dimensional F-space. Let L be the sub Lie algebra of $\mathscr{L}(\mathrm{End}_F(V))$ consisting of the linear endomorphisms of trace 0. It is easy to show directly that L and (0) are the only ideals of L, and that the trace form of the identity representation of L on V is non-degenerate. Using our earlier results on algebraic groups and their Lie algebras, one can show that L is the Lie algebra of the subgroup of $\mathrm{Aut}_F(V)$ consisting of the automorphisms whose determinant is equal to 1.

2. For the basic generalities of Lie algebra theory, see [3]. For a full development of the theory, see [8].

Chapter VIII

Structure Theory in Characteristic 0

The theme of this chapter is the use of Lie algebras in the structural analysis of affine algebraic groups over fields of characteristic 0. In addition to giving far more incisive results than the general structure theory, this Lie theory has the important feature that the basic general results hold over *arbitrary*, not necessarily algebraically closed fields of characteristic 0.

Section 1 reduces the study of unipotent algebraic groups completely to the study of representation-theoretically nilpotent Lie algebras. As a consequence, the theory of factor groups of unipotent algebraic groups is free of any limitations.

Section 2 establishes the result that, in characteristic 0, the tensor product of semisimple group representations is semisimple. This fact is decisive for the role played by linearly reductive groups in the general structure theory. Although the result, Theorem 2.2, has a completely elementary statement, it has never been proved without the use of algebraic groups and their Lie algebras.

Section 3 deals with the "algebraic hulls" $\mathscr{L}(G_L)$ of sub Lie algebras L of the Lie algebra of an algebraic group G. An important result is that if $L = [L, L]$, then L coincides with its algebraic hull. In addition, Section 3 contains the semidirect sum decomposition of $\mathscr{L}(G)$ underlying the semidirect product decomposition of G, which is the main result of Section 4.

1. Let F be a field of characteristic 0, and let G be an affine algebraic F-group. If T is a unipotent algebraic subgroup of G, it is clear from Theorem III.4.3 that $\mathscr{L}(T)$ is locally nilpotent on $\mathscr{P}(G)$. Consequently, $\exp(\tau_{[})$ has a meaning as an element of $\mathrm{End}_F(\mathscr{P}(G))$ for every element τ of $\mathscr{L}(T)$. Since $\tau_{[}$ is a derivation, $\exp(\tau_{[})$ is an F-algebra endomorphism, and $\varepsilon \circ \exp(\tau_{[})$ is an element of G. Clearly, this element of G annihilates the annihilator of T in $\mathscr{P}(G)$, so that it actually belongs to T. We denote it by $\exp(\tau)$. In the statement of Theorem 1.1, we view $\mathscr{L}(T)$ as an affine algebraic F-set, the polynomial functions being the polynomials in the elements of $\mathscr{L}(T)^\circ$.

Theorem 1.1. *Let F be a field of characteristic 0, let G be an affine algebraic F-group and T a unipotent algebraic subgroup of G. Then T is irreducible, and the map sending each element τ of $\mathscr{L}(T)$ onto the element $\exp(\tau)$ of T is an isomorphism of affine algebraic F-sets from $\mathscr{L}(T)$ to T. In this way, the family of all unipotent algebraic subgroups of G is in bijective correspondence with the family of all those sub Lie algebras of $\mathscr{L}(G)$ which are locally nilpotent on $\mathscr{P}(G)$.*

PROOF. Consider the representation of T on $\mathscr{P}(T)^{T_1}$. This factors through the finite group T/T_1. Since F is of characteristic 0, it follows that the representation of T on $\mathscr{P}(T)^{T_1}$ is semisimple. On the other hand, T is locally unipotent on $\mathscr{P}(T)$. Therefore, the representation of T on $\mathscr{P}(T)^{T_1}$ is trivial. By Theorem II.2.2, the element-wise fixer of $\mathscr{P}(T)^{T_1}$ coincides with T_1. Thus, we have $T = T_1$.

Let t be an element of T. Then t_l is locally unipotent on $\mathscr{P}(G)$, so that the expression

$$\log(t_\text{l}) = -\sum_{n>0} n^{-1}(i_{\mathscr{P}(G)} - t_\text{l})^n$$

makes sense as an element of $\mathrm{End}_F(\mathscr{P}(G))$. In proving Lemma VI.2.1, we showed that this is a derivation. Clearly, it stabilizes the annihilator of T in $\mathscr{P}(G)$. Therefore, $\varepsilon \circ \log(t_\text{l})$ is an element of $\mathscr{L}(T)$. We shall denote it by $\log(t)$. From the formal properties of the power series for exp and log, together with what we have just shown, it is clear that the maps

$$\exp\colon \mathscr{L}(T) \to T \quad \text{and} \quad \log T\colon \to \mathscr{L}(T)$$

are mutually inverse polynomial maps.

What remains to be shown is that, if L is any sub Lie algebra of $\mathscr{L}(G)$ that is locally nilpotent on $\mathscr{P}(G)$, then $\exp(L)$ is a unipotent algebraic subgroup of G. In order to see this, we need more information about the exponential map.

Let ρ be a polynomial representation of G on a finite-dimensional F-space V, and consider the extended differential $\rho'\colon \mathscr{L}(G) \to \mathrm{End}_F(V)$. We claim that, for every element τ of $\mathscr{L}(T)$,

$$\rho(\exp(\tau)) = \exp(\rho'(\tau)).$$

In order to prove this, consider the comodule structure

$$\rho^*\colon V \to V \otimes \mathscr{P}(G)$$

corresponding to ρ. By Lemma III.4.1, we have

$$(i_V \otimes \tau_\text{l}) \circ \rho^* = \rho^* \circ \rho'(\tau).$$

It follows directly from this that

$$\exp(i_V \otimes \tau_{\mathfrak{l}}) \circ \rho^* = \rho^* \circ \exp(\rho'(\tau)).$$

The expression on the left is equal to $(i_V \otimes \exp(\tau_{\mathfrak{l}})) \circ \rho^*$, and our claim follows upon applying $i_V \otimes \varepsilon$.

Now let ρ be the adjoint representation of G on $\mathscr{L}(G)$. By Theorem IV.4.1, we have $\rho'(\tau) = D_\tau$ and

$$\rho(x)(\sigma) = x\sigma x^{-1}$$

for every element x of G and every element σ of $\mathscr{L}(G)$. Hence, our above general result gives

$$\exp(\tau)\sigma \exp(\tau)^{-1} = \exp(D_\tau)(\sigma)$$

for every element σ of $\mathscr{L}(G)$ and every element τ of $\mathscr{L}(T)$. In the case where σ also belongs to $\mathscr{L}(T)$, we may apply the exponential map, which yields

$$\exp(\tau)\exp(\sigma)\exp(\tau)^{-1} = \exp(\exp(D_\tau)(\sigma)).$$

Now consider any sub Lie algebra L of $\mathscr{L}(G)$ that is locally nilpotent on $\mathscr{P}(G)$. We wish to prove that $\exp(L)$ is a unipotent algebraic subgroup of G. First, consider the case where L is 1-dimensional; say $L = F\tau$. Evidently, $\exp(L)$ is a unipotent subgroup of G, in this case. Let K denote the closure of $\exp(L)$ in G. Then K is a unipotent algebraic subgroup of G, and

$$L \subset \mathscr{L}(K).$$

Now $\exp(L)$ is the inverse image of L with respect to the polynomial map $\log: K \to \mathscr{L}(K)$, so that $\exp(L)$ is closed in K. Thus, $\exp(L) = K$, so that $\exp(L)$ is a 1-dimensional unipotent algebraic subgroup of G.

Now suppose that $\dim(L) > 1$ and that the result has been established in the lower-dimensional cases. Since L is a nilpotent Lie algebra, we have $L = J + F\tau$, where J is an ideal not containing τ. By our inductive hypothesis, $\exp(J)$ is a unipotent algebraic subgroup of G, and by the above $\exp(F\tau)$ is a 1-dimensional unipotent algebraic subgroup of G.

It is clear from our above result on conjugation of exponentials that $\exp(F\tau)$ normalizes $\exp(J)$. Thus, the product set $\exp(J)\exp(F\tau)$ is a subgroup of G. By Proposition V.2.2, it is a unipotent subgroup of G. Let M denote its closure in G, so that M is a unipotent algebraic subgroup of G. The composition map of G yields a morphism of affine algebraic sets

$$\exp(J) \times \exp(F\tau) \to M$$

whose image, $\exp(J)\exp(F\tau)$, is dense in M. It follows from this that the dimension of M is at most equal to that of $\exp(J) \times \exp(F\tau)$ which is evidently equal to $\dim(L)$. Now we have $L \subset \mathscr{L}(M)$, and $\dim(\mathscr{L}(M)) \leq \dim(L)$, whence $L = \mathscr{L}(M)$. By the part of the theorem we have already proved, $M = \exp(\mathscr{L}(M))$. Hence, $\exp(L) = M$. $\qquad\square$

Theorem 1.2. *Let F and G be as in Theorem 1.1, and let U be a unipotent affine algebraic F-group. Let ρ be a morphism of affine algebraic F-groups from U to G. Then $\rho(U)$ is a unipotent algebraic subgroup of G.*

PROOF. It follows directly from the definitions that $\rho'(\mathscr{L}(U))$ is a sub Lie algebra of $\mathscr{L}(G)$ that is nilpotent on $\mathscr{P}(G)$. By Theorem 1.1, $U = \exp(\mathscr{L}(U))$. From what we have seen in proving Theorem 1.1 concerning the behavior of exp with respect to polynomial representations, it is clear that, for every element τ of $\mathscr{L}(T)$, we have

$$\rho(\exp(\tau)) = \exp(\rho'(\tau)).$$

Hence, we have $\rho(U) = \exp(\rho'(\mathscr{L}(U)))$ which, by Theorem 1.1, is a unipotent algebraic subgroup of G. $\qquad\square$

Theorem 1.3. *Let G be a unipotent affine algebraic F-group, where F is a field of characteristic 0. Let H be a normal algebraic subgroup of G. Then the restriction map from G to $\mathscr{G}(\mathscr{P}(G)^H)$ yields a bijective group homomorphism from G/H to $\mathscr{G}(\mathscr{P}(G)^H)$, so that G/H is an affine algebraic F-group with $\mathscr{P}(G/H) = \mathscr{P}(G)^H$. Moreover, there is a polynomial map $G/H \to G$ whose composite with the canonical map $G \to G/H$ is the identity map on G/H.*

PROOF. Let ρ denote the restriction map from G to $\mathscr{G}(\mathscr{P}(G)^H)$. By Theorem II.2.2, the kernel of ρ coincides with H. By Theorem II.4.3, $\mathscr{P}(G)^H$ is finitely generated as an F-algebra, so that $\mathscr{G}(\mathscr{P}(G)^H)$ is an affine algebraic F-group whose algebra of polynomial functions may be identified with $\mathscr{P}(G)^H$. By Theorem 1.2, $\rho(G)$ is a unipotent algebraic subgroup of $\mathscr{G}(\mathscr{P}(G)^H)$.

We have $\mathscr{L}(\rho(G)) = \rho'(\mathscr{L}(G))$, and we choose a linear map

$$\gamma\colon \mathscr{L}(\rho(G)) \to \mathscr{L}(G)$$

such that $\rho' \circ \gamma$ is the identity map on $\mathscr{L}(\rho(G))$. Now we define the polynomial map α from $\rho(G)$ to G as the composite $\exp \circ \gamma \circ \log$. Using that

$$\rho \circ \exp = \exp \circ \rho',$$

we verify directly that $\rho \circ \alpha$ is the identity map on $\rho(G)$.

Clearly, $\mathscr{P}(\rho(G)) \circ \rho \subset \mathscr{P}(G)^H$. Since $\rho \circ \alpha$ is the identity map, the transpose of ρ is injective from $\mathscr{P}(\rho(G))$ to $\mathscr{P}(G)^H$. We claim that this map is also surjective. In order to see this, let g be an element of $\mathscr{P}(G)^H$ and consider the element $g \circ \alpha$ of $\mathscr{P}(\rho(G))$. We have, for every element σ of $\mathscr{L}(G)$,

$$(g \circ \alpha \circ \rho)(\exp(\sigma)) = (g \circ \alpha)(\exp(\rho'(\sigma))) = g(\exp(\gamma(\rho'(\sigma)))).$$

On the other hand,

$$\rho(\exp(\gamma(\rho'(\sigma)))) = \exp(\rho'(\sigma)) = \rho(\exp(\sigma)).$$

Since the kernel of ρ is H, it follows that $\exp(\gamma(\rho'(\sigma)))$ belongs to the coset $\exp(\sigma)H$. Therefore, the above gives

$$(g \circ \alpha \circ \rho)(\exp(\sigma)) = g(\exp(\sigma)),$$

showing that $g \circ \alpha \circ \rho = g$.

Our conclusion is that the transpose of ρ is an isomorphism of Hopf algebras from $\mathscr{P}(\rho(G))$ to $\mathscr{P}(G)^H$, whence it is clear that ρ is surjective. $\square$

2. Let F be a field, G a group, B a sub Hopf algebra of $\mathscr{R}_F(G)$. By a *B-representation* of G, we mean a representation of G by linear automorphisms of a finite-dimensional F-space whose associated representative functions belong to B. Let K be a normal subgroup of G, and denote the restriction image of B in $\mathscr{R}_F(K)$ by B_K.

Proposition 2.1. *In the above notation, suppose that every B-representation of G whose kernel contains K is semisimple, and that the tensor product of semisimple B_K-representations of K is always semisimple. Then every B-representation of G whose restriction to K is semisimple is semisimple also with respect to G.*

PROOF. Let V be a B-representation space for G that is semisimple as a representation space for K. Let W be a sub G-module of V, and let n denote the dimension of W. Consider the homogeneous component $\bigwedge^n(V)$ of the exterior F-algebra built over V, viewing it as a G-module in the usual way. Clearly, this is a B-representation space of G, and $\bigwedge^n(W)$ may be identified with a 1-dimensional sub G-module of $\bigwedge^n(V)$ in the canonical fashion. For u in $\bigwedge^n(W)$ and x in G, we have $x \cdot u = f(x)u$, where f is a group homomorphism from G to F^*, and $f \in B$. Since $\bigwedge^n(V)$ is a homomorphic image of the n-th tensor power of V, our assumption on K implies that $\bigwedge^n(V)$ is semisimple as a K-module. It follows that there is a direct K-module decomposition

$$\bigwedge{}^n(V) = P + Q,$$

where P consists of all elements p such that $x \cdot p = f(x)p$ for every element x of K, and Q consists of all sums of elements of the form $x \cdot p - f(x)p$ with x in K and p in $\bigwedge^n(V)$. Using that f is a group homomorphism from G to F^* and that K is normal in G, one sees directly that both P and Q are actually sub G-modules of $\bigwedge^n(V)$.

Now we define a new representation of G on P by setting

$$x(p) = f(x^{-1})x \cdot p$$

for every element x of G and every element p of P. Evidently, this is again a B-representation of G, and its kernel contains K. By assumption, it is therefore a semisimple representation. Therefore, the sub G-module $\bigwedge^n(W)$ of P has a G-module complement, R say, in P. However, the G stable subspaces of P for our new representation are clearly the same as those for the original representation. Thus, R is a sub G-module of P also with respect to the original representation. Let $S = R + Q$. Then S is a G-module complement for $\bigwedge^n(W)$ in $\bigwedge^n(V)$.

Let W_1 denote the sub F-space of V consisting of the elements v for which the exterior product $v \wedge^{n-1}(W)$ is contained in S. Since S and $\wedge^{n-1}(W)$ are G-stable, so is W_1. Now observe that S is of codimension 1 in $\wedge^n(V)$, so that it is the space of zeros of some element μ of $(\wedge^n(V))^\circ$. The space $\wedge^{n-1}(W)$ is of dimension n, and if $(u_1, \ldots, u_n)$ is an F-basis for it then an element v of V belongs to W_1 if and only if $\mu(vu_i) = 0$ for each i. This shows that, if m is the dimension of V, then the dimension of W_1 is at least $m - n$.

Let w_1 be any non-zero element of W, and choose elements $w_2, \ldots, w_n$ such that $(w_1, \ldots, w_n)$ is an F-basis of W. Then $w_2 \cdots w_n$ belongs to $\wedge^{n-1}(W)$, and $w_1 \cdots w_n$ is a non-zero element of $\wedge^n(W)$. This shows that w_1 does not belong to W_1, and we have the conclusion that $W \cap W_1 = (0)$. With the above, this implies that V is the direct sum of the sub G-modules W and W_1. Our conclusion is that every sub G-module of V has a G-module complement in V, which means that V is a semisimple G-module. $\qquad\square$

Theorem 2.2. *Let G be an arbitrary group, and let U and V be finite-dimensional semisimple representation spaces for G, over a field of characteristic 0. Then the tensor product G-module $U \otimes V$ is semisimple.*

PROOF. Let S denote the closure of the image of G in the affine algebraic group of all linear automorphisms of the direct sum $U + V$. Clearly, S stabilizes U and V, and the S-stable subspaces of U and V coincide with the G-stable subspaces. Moreover, the images of S and G in the group of all linear automorphisms of $U \otimes V$ have the same closure, so that the S-stable subspaces of $U \otimes V$ coincide with the G-stable subspaces. Therefore, it suffices to prove the theorem for the S-modules U and V. Accordingly, we assume that G is an affine algebraic group, and that U and V are polynomial G-modules.

Since G/G_1 is finite and since our base field is of characteristic 0, every representation of G/G_1 over our base field is semisimple. Therefore, it follows from Proposition 2.1 that no generality is lost in assuming that G is irreducible. In that case, we know from Corollary IV.3.2 that the G-stable subspaces of U, V and $U \otimes V$ coincide with the $\mathcal{L}(G)$-stable subspaces.

Hence, it suffices to prove that if L is a Lie algebra over a field of characteristic 0, and if U and V are finite-dimensional semisimple L-modules, then the tensor product L-module $U \otimes V$ is semisimple. In doing this, we may evidently replace L with its image in $\mathrm{End}(U + V)$. Therefore, we assume that L is given as a Lie algebra of linear endomorphisms of $U + V$ stabilizing U and V, and that $U + V$ is semisimple as an L-module.

By Theorem VII.1.2, $[L, L]$ is semisimple, L is the direct Lie algebra sum of $[L, L]$ and its center, Z say, and every element of Z is a semisimple linear endomorphism of $U + V$. Appealing to Proposition V.1.2, we assume without loss of generality that our base field is algebraically closed. Then U and V are direct sums of 1-dimensional Z-stable subspaces to each of which there corresponds an element μ of Z° such that every element z of Z

acts as the scalar multiplication by $\mu(z)$ on that subspace. Adding together all those 1-dimensional Z-stable subspaces of U, or of V, for which the linear functions μ coincide, we obtain direct L-module decompositions

$$U = U_1 + \cdots + U_p \quad \text{and} \quad V = V_1 + \cdots + V_q$$

such that Z acts on each U_i via a certain element σ_i of Z° and on each V_j by a certain element τ_j of Z°. Now it suffices to show that each $U_i \otimes V_j$ is semisimple as an L-module. Since Z acts on $U_i \otimes V_j$ by scalar multiplications, the sub L-modules of $U_i \otimes V_j$ coincide with the sub $[L, L]$-modules. Since $[L, L]$ is semisimple, we know from Theorem VII.2.4 that $U_i \otimes V_j$ is semisimple as an $[L, L]$-module. Therefore, $U_i \otimes V_j$ is semisimple as an L-module. $\qquad\square$

. Let G be an algebraic group. A sub Lie algebra L of $\mathscr{L}(G)$ is called an *algebraic* sub Lie algebra if there is an algebraic subgroup K of G such that $L = \mathscr{L}(K)$. If the base field is of characteristic 0, there is an algebraic sub Lie algebra of $\mathscr{L}(G)$ containing L and contained in every other such. This is $\mathscr{L}(G_L)$, where G_L is the group of Theorem IV.2.2. We call it the *algebraic hull* of L in $\mathscr{L}(G)$, and we denote it by L^+.

Let W be a polynomial G-module, and let U and V be subspaces of W such that $U \subset V$. The elements x of G with the property that $x \cdot v - v$ belongs to U for every element v of V evidently constitute an algebraic subgroup K of G. If the base field is of characteristic 0, it follows from Proposition III.4.3 and Theorem IV.3.1 that $\mathscr{L}(K)$ consists precisely of those elements of $\mathscr{L}(G)$ which map V into U. Hence, *if L is any sub Lie algebra of $\mathscr{L}(G)$ that maps V into U then L^+ also maps V into U.*

Proposition 3.1. *Let G be an affine algebraic group over a field of characteristic 0, and let L be a sub Lie algebra of $\mathscr{L}(G)$. Then $[L, L] = [L^+, L^+]$.*

PROOF. Consider the adjoint representation of G on $\mathscr{L}(G)$. In the above remark, let $W = \mathscr{L}(G)$, $V = L$ and $U = [L, L]$. The remark shows that $[L^+, L] \subset [L, L]$. Now apply the remark again, with $V = L^+$. This gives $[L^+, L^+] \subset [L, L]$. $\qquad\square$

Theorem 3.2. *Let G be an affine algebraic group over a field of characteristic 0. Every semisimple sub Lie algebra of $\mathscr{L}(G)$ is an algebraic sub Lie algebra.*

PROOF. Let L be a semisimple sub Lie algebra of $\mathscr{L}(G)$. By Theorem VII.2.4, $\mathscr{P}(G)$ is semisimple as an L-module. By the remark just preceding Proposition 3.1, the sub L-modules of $\mathscr{P}(G)$ coincide with the sub L^+-modules. Therefore, $\mathscr{P}(G)$ is semisimple as an L^+-module. In particular, if V is a finite-dimensional sub G-module of $\mathscr{P}(G)$ generating $\mathscr{P}(G)$ as an algebra, then the representation of L^+ on V is injective and semisimple. By Theorem VII.1.2, L^+ is therefore the direct sum of its center, Z say, and $[L^+, L^+]$. By Proposition 3.1, we have $[L^+, L^+] = [L, L]$. Since L is semisimple, we have $[L, L] = L$. Thus L^+ is the direct sum of L and Z, so that it suffices to prove that $Z = (0)$.

Let F' be an algebraic closure of the base field F, and let us make the canonical base field extension, replacing G with $\mathcal{G}(\mathcal{P}(G) \otimes F')$. This replaces $\mathcal{L}(G)$ with $\mathcal{L}(G) \otimes F'$, and we consider the sub Lie algebra $L \otimes F'$. It is clear from Theorem VII.2.1 that $L \otimes F'$ is a semisimple Lie algebra. Its algebraic hull in $\mathcal{L}(G) \otimes F'$ is $L^+ \otimes F'$. The direct decomposition of L^+ has now become the direct decomposition of $L^+ \otimes F'$ as the direct sum of its center, $Z \otimes F'$, and $L \otimes F'$. It is clear from this that we do not lose generality in assuming that F is algebraically closed.

In this case, let z be an element of Z, let r be a characteristic root of the element of $\mathrm{End}_F(V)$ corresponding to z, and let V_r be the corresponding characteristic subspace of V. Evidently, this is stable under the action of L^+. Let $(v_1, \ldots, v_d)$ be an F-basis of V_r, and consider the exterior product $v_1 \cdots v_d$ in the polynomial G-module $\bigwedge^d(V)$. The transform by the endomorphism corresponding to z is

$$z \cdot (v_1 \cdots v_d) = T_{V_r}(z) v_1 \cdots v_d = (dr) v_1 \cdots v_d,$$

where T stands for trace. On the other hand, the sub F-space of $\bigwedge^d(V)$ spanned by $v_1 \cdots v_d$ is L-stable. Since $L = [L, L]$, it is therefore annihilated by L. It follows that it is annihilated also by L^+. Therefore, we must have $r = 0$. Thus, 0 is the only characteristic root of the endomorphism of V that corresponds to z. On the other hand, we know from Theorem VII.1.2 that this endomorphism is semisimple. Therefore, this endomorphism is 0, whence $z = 0$. $\qquad\square$

Theorem 3.3. *Let F be a field of characteristic 0, and let G be an irreducible affine algebraic F-group. Let $[G, G]^+$ denote the closure in G of $[G, G]$. Then $[\mathcal{L}(G), \mathcal{L}(G)] = \mathcal{L}([G, G]^+)$.*

PROOF. First, we deal with the case where F is algebraically closed. By Theorem VII.3.1, applied to the canonical homomorphism from $\mathcal{L}(G)$ to $\mathcal{L}(G)/\mathcal{L}(G)_r$, there is a semisimple sub Lie algebra S of $\mathcal{L}(G)$ such that $\mathcal{L}(G)$ is the semidirect Lie algebra sum of $\mathcal{L}(G)_r$ and S. Hence we have a semidirect sum decomposition

$$[\mathcal{L}(G), \mathcal{L}(G)] = T + S, \qquad \text{where} \quad T = [\mathcal{L}(G), \mathcal{L}(G)_r].$$

It follows from Theorem VII.3.2 that T is locally nilpotent on $\mathcal{P}(G)$. By Theorem 1.1, T is therefore an algebraic sub Lie algebra of $\mathcal{L}(G)$. By Theorem 3.2, S is an algebraic sub Lie algebra of $\mathcal{L}(G)$.

Thus, we have $S = \mathcal{L}(G_S)$ and $T = \mathcal{L}(G_T)$. Since T is an ideal of $\mathcal{L}(G)$, we know from Theorem IV.4.4 that G_T is normal in G. Therefore, $G_T G_S$ is a subgroup of G. Since F is algebraically closed, we may apply Theorem

II.4.1 and conclude that $G_T G_S$ is an irreducible algebraic subgroup of G. Since it is the image of $G_T \times G_S$ by the morphism of affine algebraic sets coming from the composition of G, its dimension is at most

$$\dim(G_T \times G_S) = \dim(T) + \dim(S).$$

Clearly, $T + S \subset \mathscr{L}(G_T G_S)$. Hence, our dimension inequality shows that $T + S = \mathscr{L}(G_T G_S)$, i.e.,

$$[\mathscr{L}(G), \mathscr{L}(G)] = \mathscr{L}(G_T G_S).$$

In particular, $[\mathscr{L}(G), \mathscr{L}(G)]$ is an algebraic sub Lie algebra of $\mathscr{L}(G)$, and so coincides with $\mathscr{L}(G_{[\mathscr{L}(G), \mathscr{L}(G)]})$.

As above for T, we know that $G_{[\mathscr{L}(G), \mathscr{L}(G)]}$ is normal in G. It follows from Theorem III.3.3 and Theorem IV.2.3 that the differential of the canonical morphism from G to $G/G_{[\mathscr{L}(G), \mathscr{L}(G)]}$ induces a Lie algebra isomorphism from $\mathscr{L}(G)/[\mathscr{L}(G), \mathscr{L}(G)]$ to the Lie algebra of $G/G_{[\mathscr{L}(G), \mathscr{L}(G)]}$. Hence, we have from Theorem IV.4.3 that $G/G_{[\mathscr{L}(G), \mathscr{L}(G)]}$ is abelian, which means that $[G, G]$ is contained in $G_{[\mathscr{L}(G), \mathscr{L}(G)]}$. By Theorem II.4.1, $[G, G]$ is an irreducible algebraic subgroup of G. The last inclusion relation implies that

$$\mathscr{L}([G, G]) \subset [\mathscr{L}(G), \mathscr{L}(G)].$$

On the other hand, since $G/[G, G]$ is abelian, so is its Lie algebra. This Lie algebra is the image of $\mathscr{L}(G)$ under the differential of the canonical morphism from G to $G/[G, G]$, whose kernel is $\mathscr{L}([G, G])$. Therefore, we have

$$[\mathscr{L}(G), \mathscr{L}(G)] \subset \mathscr{L}([G, G]).$$

With the above, this gives $[\mathscr{L}(G), \mathscr{L}(G)] = \mathscr{L}([G, G])$.

If F is not algebraically closed, let F' be an algebraic closure of F, and let $G' = \mathscr{G}(\mathscr{P}(G) \otimes F')$. From the above, we know that

$$\mathscr{L}([G', G']) = [\mathscr{L}(G'), \mathscr{L}(G')] = [\mathscr{L}(G), \mathscr{L}(G)] \otimes F'.$$

Clearly, $([G, G]^+)' \subset [G', G']$. On the other hand, $\mathscr{P}(G)^{[G, G]} \otimes F'$ is a Hopf algebra whose comultiplication is commutative, whence the restriction image of $[G', G']$ in $\mathscr{G}(\mathscr{P}(G)^{[G, G]} \otimes F')$ is trivial. It follows that $\mathscr{L}([G', G'])$ is contained in the kernel of the representation of $\mathscr{L}(G')$ on $\mathscr{P}(G)^{[G, G]} \otimes F'$. By Proposition IV.5.2, the kernel of the representation of $\mathscr{L}(G)$ on $\mathscr{P}(G)^{[G, G]}$ is the Lie algebra of the element-wise fixer of $\mathscr{P}(G)^{[G, G]}$ in G, i.e., of $[G, G]^+$. It follows that the kernel of the representation of $\mathscr{L}(G')$ on $\mathscr{P}(G)^{[G, G]} \otimes F'$ is $\mathscr{L}([G, G]^+) \otimes F'$. Thus, we have $\mathscr{L}([G', G']) \subset \mathscr{L}([G, G]^+) \otimes F'$. The reversed inclusion holds because $([G, G]^+)' \subset [G', G']$. Therefore, our conclusion is that

$$\mathscr{L}([G, G]^+) \otimes F' = [\mathscr{L}(G), \mathscr{L}(G)] \otimes F',$$

which clearly yields the conclusion of the theorem. $\square$

Theorem 3.4. *Let F be a field of characteristic 0, and let G be an affine algebraic F-group. Let U and V be irreducible algebraic subgroups of G, and let W be the closure in G of the subgroup generated by U and V. Then $\mathscr{L}(W)$ coincides with the sub Lie algebra of $\mathscr{L}(G)$ that is generated by $\mathscr{L}(U)$ and $\mathscr{L}(V)$.*

PROOF. First, we deal with the case where F is algebraically closed. In this case, we know from Theorem II.4.1 that W is the group generated by U and V. Let S denote the sub Lie algebra of $\mathscr{L}(G)$ that is generated by $\mathscr{L}(U)$ and $\mathscr{L}(V)$. It is easy to see that we must have $G_S = W$, so that $S^+ = \mathscr{L}(W)$. By Proposition 3.1, we have therefore $[S, S] = [\mathscr{L}(W), \mathscr{L}(W)]$. By Theorem II.4.1, $[W, W]$ is an irreducible algebraic subgroup of G, so that Theorem 3.3 gives $[\mathscr{L}(W), \mathscr{L}(W)] = \mathscr{L}([W, W])$. Thus, $[S, S] = \mathscr{L}([W, W])$.

Now let π denote the canonical morphism from W to $W/[W, W]$, and let $\rho\colon U \times V \to W$ be the polynomial map sending each (u, v) onto the product uv in W. Put $\sigma = \pi \circ \rho$. Then σ is a surjective morphism of affine algebraic F-groups from $U \times V$ to $W/[W, W]$. By Theorem III.3.3, the differential $\sigma\dot{}$ is surjective from $\mathscr{L}(U \times V)$ to $\mathscr{L}(W/[W, W])$. Since $\mathscr{L}(U \times V)$ is the direct sum of the canonical images of $\mathscr{L}(U)$ and $\mathscr{L}(V)$, we have

$$\mathscr{L}(W/[W, W]) = \sigma\dot{}(\mathscr{L}(U \times V)) = \pi\dot{}(\mathscr{L}(U)) + \pi\dot{}(\mathscr{L}(V)).$$

By Theorem IV.2.3, the kernel of $\pi\dot{}$ is $\mathscr{L}([W, W]) = [S, S]$. Hence, the above shows that

$$\mathscr{L}(U) + \mathscr{L}(V) + [S, S] = \mathscr{L}(W).$$

Evidently, the sum on the left coincides with S, so that we have the required result in the case where F is algebraically closed.

In the general case, let F' be an algebraic closure of F, and put

$$G' = \mathscr{G}(\mathscr{P}(G) \otimes F'),$$

etc. Then W' is the subgroup of G' that is generated by U' and V'. By the above, the sub Lie algebra of $\mathscr{L}(G')$ that is generated by $\mathscr{L}(U')$ and $\mathscr{L}(V')$ coincides with $\mathscr{L}(W')$. This means that $S \otimes F'$ coincides with $\mathscr{L}(W) \otimes F'$, so that $S = \mathscr{L}(W)$. $\qquad\qquad\square$

If G is an algebraic group, and L is a sub Lie algebra of $\mathscr{L}(G)$, then we say that L is a linearly reductive sub Lie algebra of $\mathscr{L}(G)$ if every polynomial G-module is semisimple as an L-module. In characteristic 0, this condition is equivalent to the condition that G_L be linearly reductive, as is easily seen from the discussion immediately preceding Proposition 3.1. It follows that, if L is linearly reductive, so is L^+.

Theorem 3.5. *Let F be a field of characteristic 0, and let G be an affine algebraic F-group. There is a linearly reductive algebraic sub Lie algebra T of $\mathscr{L}(G)$ such that $\mathscr{L}(G) = \mathscr{L}(G_u) + T$.*

PROOF. We know from Theorem VII.3.1 that there is a semisimple sub Lie algebra S of $\mathscr{L}(G)$ such that $\mathscr{L}(G)_r + S = \mathscr{L}(G)$. By Theorem VII.2.4, S is linearly reductive. Among the linearly reductive sub Lie algebras of $\mathscr{L}(G)$ containing S, choose a maximal one, T say. Since the adjoint representation of $\mathscr{L}(G)$ on $\mathscr{L}(G)$ is the differential of the adjoint representation of G on $\mathscr{L}(G)$, we have that $\mathscr{L}(G)$ is semisimple as a T-module, with respect to the restriction of the adjoint representation. Clearly, $\mathscr{L}(G_u)$ and $\mathscr{L}(G)_r$ are sub T-modules of $\mathscr{L}(G)$, and $\mathscr{L}(G_u) \subset \mathscr{L}(G)_r$. Hence, $\mathscr{L}(G_u)$ has a T-module complement, P say, in $\mathscr{L}(G)_r$. By Theorem VII.3.2, $[\mathscr{L}(G), \mathscr{L}(G)_r]$ is locally nilpotent on $\mathscr{P}(G)$. By Theorem 1.1, this implies that $G_{[\mathscr{L}(G),\,\mathscr{L}(G)_r]}$ is unipotent. Since $[\mathscr{L}(G), \mathscr{L}(G)_r]$ is an ideal of $\mathscr{L}(G)$, this group is also normal in G, and hence is contained in G_u. Therefore, $[\mathscr{L}(G), \mathscr{L}(G)_r]$ is contained in $\mathscr{L}(G_u)$. In particular, it follows that $[T, P] = (0)$. The maximality of T implies that T is an algebraic sub Lie algebra of $\mathscr{L}(G)$.

Evidently, it suffices to show that $\mathscr{L}(G)_r = \mathscr{L}(G_u) + \mathscr{L}(G)_r \cap T$. Suppose this is not the case. Then there is an element σ in P that does not belong to $\mathscr{L}(G_u) + T$. By Theorem V.2.3, the semisimple and nilpotent components $\sigma^{(s)}$ and $\sigma^{(n)}$ belong to $\mathscr{L}(G)_r$, because $\mathscr{L}(G)_r$ is an algebraic sub Lie algebra of $\mathscr{L}(G)$, as one sees easily from the remark just preceding Proposition 3.1. Since $[\mathscr{L}(G), \mathscr{L}(G)_r] \subset \mathscr{L}(G_u)$, it is clear that $\mathscr{L}(G_u) + F\sigma^{(n)}$ is an ideal of $\mathscr{L}(G)$. Since it is locally nilpotent on $\mathscr{P}(G)$, it must therefore coincide with $\mathscr{L}(G_u)$, by the same argument we used just above for $[\mathscr{L}(G), \mathscr{L}(G)_r]$. This means that $\sigma^{(n)}$ belongs to $\mathscr{L}(G_u)$. Since σ does not belong to $\mathscr{L}(G_u) + T$, it follows that $\sigma^{(s)}$ does not belong to T.

Since $[\sigma, T] = (0)$, $\sigma_{[}$ commutes with $\tau_{[}$ for every element τ of T. Therefore, the same is true for $\sigma_{[}^{(s)}$. It follows that $T + F\sigma^{(s)}$ is a sub Lie algebra, U say, of $\mathscr{L}(G)$, and that $\mathscr{P}(G)$ is semisimple as a U-module. Thus, U is a linearly reductive sub Lie algebra of $\mathscr{L}(G)$, in contradiction to the maximality of T. $\qquad\square$

4. Now we are in a position to establish the basic semidirect product decomposition for affine algebraic groups over a field of characteristic 0. This is a substantial extension of Theorem VI.3.2, and it is obtained by strengthening the auxiliary results that entered into the proof of that theorem with the aid of Lie algebras.

Lemma 4.1. *Let G be a linearly reductive algebraic group, and let V be a finite-dimensional polynomial G-module. Then every polynomial cocycle for G in V is a coboundary.*

PROOF. We proceed in exact analogy with the proof of Corollary VII.2.5. Let f be a polynomial cocycle for G in V, and let F denote the base field. We define an action of G on the direct sum $V + F$ by

$$x \cdot (v, a) = (af(x) + x \cdot v, a),$$

where $x \in G$, $v \in V$ and $a \in F$. This is a G-module structure by virtue of the cocycle identity, whence it is clear that $V + F$ thus becomes a polynomial G-module containing V as a sub G-module. Since G is linearly reductive, there is a G-module complement for V in $V + F$. Let v be the element of V such that $(v, 1)$ lies in this complement. Then we must have $f(x) + x \cdot v = v$, because the action of G on the complement of V is trivial. This shows that f is a coboundary. $\qquad\square$

Proposition 4.2. *Let G be an algebraic group over the field F of characteristic* 0, *and suppose that there is a linearly reductive subgroup P of G such that* $G_u P = G$. *Let Q be any linearly reductive subgroup of G. Then there is an element t in G_u such that $tQt^{-1} \subset P$.*

PROOF. By Theorem V.4.2, P is an algebraic subgroup of G, and G is the semidirect product $G_u \rtimes P$. If G_u is trivial, there is nothing to prove. If G_u is non-trivial, then the center of G_u is non-trivial, because G_u is nilpotent. Let C denote this center. We know from Theorem 1.3 that C is properly normal in G_u, in the sense of the beginning of Section V.4. From the semi-direct product decomposition of G, it is clear that C is therefore properly normal in G, and that G/C is identifiable with the semidirect product $(G_u/C) \rtimes P$. Assuming that the proposition has been established in the cases of lower-dimensional G_u, we obtain an element s in G_u such that $sQs^{-1} \subset C \rtimes P$. Thus, proceeding by induction on the dimension of G_u, we reduce the proposition to the case where G_u is abelian.

In that case, we know from Proposition VI.5.4 that G_u is an algebraic vector F-group. Let f denote the restriction to Q of the projection $G \to G_u$ coming from our semidirect product decomposition. Then we see directly that

$$f(xy) = f(x)xf(y)x^{-1}.$$

Replacing Q with its closure in G, we arrange that Q is an algebraic subgroup of G. Writing G_u additively and viewing it as a polynomial Q-module, with Q acting by conjugation, we see that the above identity means that f is a cocycle for Q in G_u. Evidently, f is a polynomial map. By Lemma 4.1, it follows that there is an element t in G_u such that $f(x) = xtx^{-1}t^{-1}$ for every x in Q. This gives $txt^{-1} \in P$, so that $tQt^{-1} \subset P$. $\qquad\square$

Theorem 4.3. *Let F be a field of characteristic* 0, *and let G be an affine algebraic F-group. There is a linearly reductive algebraic subgroup P of G such that G is the semidirect product $G_u \rtimes P$. If Q is any linearly reductive subgroup of G, there is an element t in G_u such that $tQt^{-1} \subset P$.*

PROOF. In view of Theorem V.4.2 and Proposition 4.2, all that remains to be proved is that there is a linearly reductive subgroup P of G such that $G = G_u P$. First, we do this in the case where G is irreducible.

By Theorem 3.5, there is a linearly reductive algebraic sub Lie algebra T of $\mathscr{L}(G)$ such that $\mathscr{L}(G) = \mathscr{L}(G_u) + T$. Consider the corresponding irreducible algebraic subgroup G_T of G. By Corollary IV.3.2, G_T is linearly reductive. The Lie algebra of the closure $(G_u G_T)^+$ of $G_u G_T$ in G contains $\mathscr{L}(G_u)$ and T, and therefore coincides with $\mathscr{L}(G)$. Since G is irreducible, we have therefore $(G_u G_T)^+ = G$. We shall prove that, actually, $G_u G_T = G$.

In order to do this, we extend the base field F to an algebraic closure F' of F. We write G' for $\mathscr{G}(\mathscr{P}(G) \otimes F')$, etc., and we identify G with its canonical image in G'. Now $(G_u)'(G_T)'$ is an algebraic subgroup of G' containing $G_u G_T$, so that $(G_u)'(G_T)' = G'$. Let Σ denote the Galois group of F' relative to F. We let Σ act on $\mathscr{P}(G) \otimes F'$ via the tensor factor F', and we identify the elements of Σ with the corresponding F-algebra automorphisms of $\mathscr{P}(G) \otimes F'$. It is then clear that an element x of G' belongs to G if and only if $\sigma \circ x \circ \sigma^{-1} = x$ for every element σ of Σ.

Let x be an element of G, and write $x = yz$ with y in $(G_u)'$ and z in $(G_T)'$. Then we have

$$yz = (\sigma \circ y \circ \sigma^{-1})(\sigma \circ z \circ \sigma^{-1})$$

whence

$$y^{-1}(\sigma \circ y \circ \sigma^{-1}) = z(\sigma \circ z \circ \sigma^{-1})^{-1}.$$

In the second relation, the element on the left belongs to $(G_u)'$, while that on the right belongs to $(G_T)'$. Thus, each belongs to $(G_u)' \cap (G_T)'$, which is trivial, because it is both linearly reductive and unipotent. Our conclusion is that $y = \sigma \circ y \circ \sigma^{-1}$ and $z = \sigma \circ z \circ \sigma^{-1}$, so that y belongs to

$$(G_u)' \cap G = G_u,$$

and z belongs to $(G_T)' \cap G = G_T$. Thus, $G = G_u G_T$, so that the theorem is established in the case where G is irreducible.

In the general case, we proceed by induction on the dimension of G_u. If G_u is trivial, then we know from what we have just proved that G_1 is linearly reductive. This implies that $\mathscr{P}(G)$ is semisimple as a G_1-module. Since G/G_1 is finite and F is of characteristic 0, every G/G_1-module over F is semisimple. Hence, we can apply Proposition 2.1 and conclude that $\mathscr{P}(G)$ is semisimple as a G-module, i.e., that G is linearly reductive.

Now suppose that G_u is non-trivial, and that the theorem has been established in the cases of lower-dimensional G_u. Let C denote the center of G_u. From the proof of Proposition 4.2, we know that C is properly normal in G_1. Since C is normal in G, it is clear from Theorem II.2.3 that C is therefore properly normal also in G. By inductive hypothesis, there is a linearly reductive algebraic subgroup L of G/C such that G/C is the semidirect product $(G_u/C) \rtimes L$. On the other hand, by what we have already proved, there is a linearly reductive algebraic subgroup P of G_1 such that

$$G_1 = G_u \rtimes P.$$

The canonical image of P in G/C is a linearly reductive subgroup of G/C. By Proposition 4.2, it is contained in a conjugate of L. Therefore, we may choose L so that it contains the canonical image of P. Now we have

$$(G_u/C)L_1 = (G/C)_1 = G_1/C = (G_u/C)((CP)/C),$$

which shows that L_1 coincides with the canonical image $(CP)/C$ of P. Moreover, it is clear from the semidirect product decompositions of G_1 and G/C that the canonical map from P to L_1 is an isomorphism of algebraic groups.

Let M denote the inverse image of L in G, so that $M/C = L$ and

$$M_1/C = L_1.$$

Make a coset decomposition $L = \bigcup_{i=1}^n x_i L_1$, choosing x_1 to be the neutral element. For each i, choose an element y_i in M whose canonical image in L is x_i, taking the neutral element of M for y_1. Define the map $\rho: L \to M$ by making the restriction of ρ to L_1 the inverse of the canonical isomorphism $P \to L_1$ and setting $\rho(x_i u) = y_i \rho(u)$ for every element u of L_1. Then ρ is clearly a polynomial map, and the composite of ρ with the canonical map $M \to L$ is the identity map on L.

We obtain the structure of a polynomial L-module on the algebraic vector F-group C by defining the transform of an element c of C by an element x of L to be

$$x \cdot v = \rho(x)c\rho(x)^{-1},$$

noting that ρ is a group homomorphism mod C, so that the map sending each x onto the conjugation of C effected by $\rho(x)$ is indeed a group homomorphism. Finally, define a map f from $L \times L$ to C by

$$f(x, y) = \rho(x)\rho(y)\rho(xy)^{-1}.$$

Writing this in the form $\rho(x)\rho(y) = f(x, y)\rho(xy)$, we see from the associativity of the group composition of M that f satisfies the identity

$$(x \cdot f(y, z))f(x, yz) = f(x, y)f(xy, z).$$

For each fixed x in L, let f_x denote the map from L to C given by $f_x(y) = f(x, y)$. Then the above identity may be written in the form

$$(x \cdot f_y(z))f_x(yz) = f(x, y)f_{xy}(z).$$

Now write C additively, and identify the elements c of C with the constant maps from L to C with values c. Then the last identity may be written as an identity among maps from L to C, as follows

$$x \cdot f_y + f_x \cdot y = f(x, y) + f_{xy},$$

where the left and right transforms $x \cdot h$ and $h \cdot x$ of a map h from L to C by an element x of L are defined by

$$(x \cdot h)(z) = x \cdot h(z) \quad \text{and} \quad (h \cdot x)(z) = h(xz).$$

All the maps involved above are polynomial maps, and therefore may be viewed as elements of $C \otimes \mathscr{P}(L)$. Since L is linearly reductive, $\mathscr{P}(L)$ is semisimple as a right L-module. Hence, there is a right L-module projection π from $\mathscr{P}(L)$ to $\mathscr{P}(L)^L = F$. Define the map g from L to C by

$$g(x) = (i_C \otimes \pi)(f_x).$$

If we apply $i_C \otimes \pi$ to our above identity for f, we obtain

$$x \cdot g(y) + g(x) = f(x, y) + g(xy).$$

Reverting to the multiplicative notation, we have

$$(x \cdot g(y))g(x) = f(x, y)g(xy).$$

This shows that if h is the map from L to M defined by

$$h(x) = g(x)^{-1}\rho(x)$$

then h is a group homomorphism. Clearly, the composite of h with the canonical map $M \to L$ is the identity map on L. Moreover, h is evidently a polynomial map, and so is a morphism of algebraic groups. We have $M = Ch(L)$, and $h(L)$ is linearly reductive as a subgroup of G. Since

$$(G_u/C)L = G/C,$$

we have $G_u M = G$, i.e., $G_u h(L) = G$. $\qquad\qquad\square$

Theorem 4.4. *Let $\rho: G \to H$ be a morphism of algebraic groups over a field of characteristic 0. Suppose that $\rho(G)$ is dense in H. Then $\rho(G_u) = H_u$.*

PROOF. Clearly, $\rho(G_u)$ is a normal unipotent subgroup of H, so that

$$\rho(G_u) \subset H_u.$$

Let L denote the inverse image of H_u in G. Then L is a normal algebraic subgroup of G containing G_u, whence $L_u = G_u$. By Theorem 4.3, there is a linearly reductive subgroup P of L such that $L = G_u P$. Now $\rho(P)$ is a linearly reductive subgroup of H, and $\rho(P) \subset H_u$. Therefore, $\rho(P)$ is trivial, so that $\rho(L) = \rho(G_u)$. Since $\rho(G)$ is dense in H, we know from Theorem III.3.3 that $\rho'(\mathscr{L}(G)) = \mathscr{L}(H)$. The inverse image of $\mathscr{L}(H_u)$ in $\mathscr{L}(G)$ is $\mathscr{L}(L)$, and $\rho'(\mathscr{L}(L)) = \rho'(\mathscr{L}(G_u))$, because $\rho(L) = \rho(G_u)$. Hence, we have

$$\rho'(\mathscr{L}(G_u)) = \mathscr{L}(H_u).$$

Now we apply Theorem 1.1, obtaining

$$H_u = \exp(\mathscr{L}(H_u)) = \exp(\rho'(\mathscr{L}(G_u))) = \rho(\exp(\mathscr{L}(G_u))) = \rho(G_u). \quad \square$$

Notes

1. In order to see how Theorem 2.2 can fail in non-zero characteristic, consider the following example. Let F be a field of non-zero characteristic p, and let V be an F-space of dimension p. Let J be the kernel of the canonical map from $\otimes^p(V)$ to the 1-dimensional space $\bigwedge^p(V)$. Let G be the group of all F-linear automorphisms of V, and regard $\otimes^p(V)$ and $\bigwedge^p(V)$ as G-modules in the canonical fashion, so that the canonical map is a morphism of G-modules. Now one can show that J has no G-module complement in $\otimes^p(V)$, as follows.

Suppose that U is a G-module complement of J in $\otimes^p(V)$. Then, as a G-module, U is isomorphic with $\bigwedge^p(V)$. Therefore, every element x of G acts on U as the scalar multiplication by the determinant $\det(x)$ of x. Let $(v_1, \ldots, v_p)$ be an F-basis of V, and let S be the subgroup of G consisting of the elements that stabilize our basis. By writing $v_1 \otimes \cdots \otimes v_p$ as the sum of an element of U and an element of J, one sees that

$$\sum_{x \in S} \det(x)x \cdot (v_1 \otimes \cdots \otimes v_p) = 0,$$

which is clearly a contradiction.

2. Theorem 4.3 is due to G. D. Mostow [10]. Note that, in the above example, G_u is trivial, showing that Theorem 4.3 fails in non-zero characteristic.

3. Regarding the conjugacy part of Theorem 4.3, if F is algebraically closed, one can show, as in the proof of Theorem VI.3.2, that t may be chosen from G_u^∞. Actually, this holds even if F is not algebraically closed, but the proof then requires more information on the exponential map than we have at this point (the Campbell–Hausdorff formula).

4. Let G be an affine algebraic group over a field of characteristic 0, and let U be a normal unipotent algebraic subgroup of G. It follows from Theorems 4.3 and 1.3 that U is properly normal in G, in the sense of Section V.4, and that there is a polynomial map from G/U to G whose composite with the canonical map from G to G/U is the identity map on G/U.

Chapter IX

Algebraic Varieties

This chapter is devoted entirely to basic concepts of algebraic geometry. The point of view adopted is that an algebraic variety is a topological space, equipped with a superstructure of functions. Section 1 introduces prevarieties, a preliminary notion slightly more general than that of a variety, which is convenient for developing the basic technical results concerning varieties. Section 2 is devoted to products of prevarieties and the notion of a variety.

In Section 3, we discuss projective varieties, which are fundamental for algebraic group theory, because all the varieties that occur are open subvarieties of projective varieties. The most important example of a projective variety, the Grassmann variety of d-dimensional subspaces of an n-dimensional vector space, is introduced in Section 4. This plays a vital role later on, in the theory of Borel subgroups.

Section 5 deals with the notion of completeness of a variety. In particular, it is shown here that projective varieties are complete. As in all subsequent applications, it is assumed here that the base field is algebraically closed. In the interest of technical clarity, this assumption is made *explicitly* wherever it is meant to be in force.

1. Let S be a topological space, F a field. A *sheaf of F-valued functions* on S is a function $\mathscr{F}_S$ associating with each non-empty open set U of S a sub F-algebra $\mathscr{F}_S(U)$ of the F-algebra F^U of all F-valued functions on U, subject to the following conditions, where we agree that $\mathscr{F}_S(\varnothing) = (0_F)$ and coincides with the restriction image of every $\mathscr{F}_S(U)$.

(1) If U and V are open sets with $U \subset V$ then the restrictions to U of the elements of $\mathscr{F}_S(V)$ belong to $\mathscr{F}_S(U)$.

(2)   Let $\mathscr{V}$ be a family of open sets of S. Suppose that, for each member V of $\mathscr{V}$, there is given an element f_V of $\mathscr{F}_S(V)$ such that the restriction images of f_{V_1} and f_{V_2} in $\mathscr{F}_S(V_1 \cap V_2)$ coincide, for all pairs (V_1, V_2) of members of $\mathscr{V}$. Then the function, with domain the union of the family $\mathscr{V}$, that is defined by the f_V's belongs to $\mathscr{F}_S(\cup (\mathscr{V}))$.

If T is any subspace of S then the restriction maps yield a sheaf of functions on T, from $\mathscr{F}_S$. The resulting $\mathscr{F}_T$ is called the *induced sheaf*.

For our purposes, the basic example of a sheaf of functions is the *sheaf of regular functions* of an irreducible affine algebraic F-set. This is defined as follows.

Let S be an irreducible affine algebraic F-set, and let $[\mathscr{P}(S)]$ denote the field of fractions of the F-algebra $\mathscr{P}(S)$ of polynomial functions on S. We say that an element f of $[\mathscr{P}(S)]$ is *defined at a point* s of S if there are elements u and v of $\mathscr{P}(S)$ such that $f = u/v$ and $v(s) \neq 0$. If (u', v') is another such pair of elements of $\mathscr{P}(S)$ we have $v'u = vu'$ and hence $u(s)/v(s) = u'(s)/v'(s)$. Thus, if f is defined at s, then it determines an element $f(s)$ of F, where

$$f(s) = u(s)/v(s),$$

whenever $f = u/v$ and $v(s) = 0$. This defines f as a *rational function* on S. One says that f is *regular* on a subset U of S if f is defined at every point of U.

Now we define $\mathscr{F}_S$ by making $\mathscr{F}_S(U)$ the F-algebra of the restrictions to U of the rational functions that are regular on U. It is evident that sheaf condition (1) is satisfied. In order to verify (2), consider two non-empty open sets V_1 and V_2 of S, and elements f_{V_1} and f_{V_2} of $\mathscr{F}_S(V_1)$ and $\mathscr{F}_S(V_2)$ whose restrictions to $V_1 \cap V_2$ coincide. Choose u_1, v_1 and u_2, v_2 from $\mathscr{P}(S)$ such that $f_{V_1} = u_1/v_1$ and $f_{V_2} = u_2/v_2$. For $i = 1$ or 2, let P_i be the set of non-zeros of v_i in S. Then, since S is irreducible, $P_1 \cap P_2 \cap V_1 \cap V_2$ is a *non-empty* open set of S, and we have $u_1(s)/u_2(s) = u_2(s)/v_2(s)$ for every point s in this set. Therefore $u_1 v_2 - u_2 v_1$ vanishes on this non-empty open set of S, whence it is 0, so that $u_1/v_1 = u_2/v_2$. This means that f_{V_1} and f_{V_2} are represented by the same element of $[\mathscr{P}(S)]$. It is clear from this that, if (f_V) is a family of functions as described in sheaf condition (2), then there is an element of $[\mathscr{P}(S)]$ that represents each member f_V, showing that condition (2) is satisfied.

The space S, equipped with the sheaf $\mathscr{F}_S$ just defined, is called an *irreducible affine F-variety*.

Definition 1.1. *An irreducible prevariety over the field F is an irreducible Noetherian topological space S, equipped with a sheaf $\mathscr{F}_S$ of F-valued functions satisfying the following condition: S is the union of a finite family of open sets, called affine patches, each of which is an irreducible affine F-variety, with the sheaf induced by $\mathscr{F}_S$ as the sheaf of regular functions.*

By a *prevariety* is meant a Noetherian topological space S, equipped with a sheaf $\mathscr{F}_S$ of functions such that each irreducible component of S becomes an irreducible prevariety when equipped with the induced sheaf of functions.

Clearly, if S is a Noetherian topological space, and if each irreducible component S_i of S is equipped with the structure of an irreducible prevariety such that, on each $S_i \cap S_j$, the sheaves induced by $\mathscr{F}_{S_i}$ and $\mathscr{F}_{S_j}$ coincide, then the $\mathscr{F}_{S_i}$'s fit together so as to yield a sheaf $\mathscr{F}_S$ of functions on S with which S is a prevariety.

An irreducible closed subset of an irreducible affine variety clearly inherits the structure of an irreducible affine variety, the sheaf of regular functions being the induced sheaf. It is easy to see from this that a closed subset of a prevariety becomes a prevariety when equipped with the induced sheaf. It is called a *closed sub prevariety*.

Let S be an irreducible affine variety, and let a be a non-zero element of $\mathscr{P}(S)$. We denote the set of non-zeros of a in S by S_a, and we call this the *principal open set defined by* a. Now form the sub F-algebra $\mathscr{P}(S)[1/a]$ of $[\mathscr{P}(S)]$. Like $\mathscr{P}(S)$, this is a finitely generated integral domain F-algebra, and it is clear that S_a has the structure of an irreducible affine F-variety, with $\mathscr{P}(S_a) = \mathscr{P}(S)[1/a]$, and $\mathscr{F}_{S_a}$ coinciding with the sheaf induced by $\mathscr{F}_S$.

Now consider an arbitrary non-empty open subset U of S. Let J be the annihilator in $\mathscr{P}(S)$ of $S \setminus U$, and let $(a_1, \ldots, a_n)$ be a system of ideal generators of J. Then we have

$$U = S_{a_1} \cup \cdots \cup S_{a_n}.$$

By the above, each S_{a_i} becomes an irreducible affine variety when equipped with the sheaf induced by $\mathscr{F}_S$. Therefore, the sheaf induced on U by $\mathscr{F}_S$ makes U into an irreducible prevariety, the S_{a_i}'s being affine patches.

It follows that every open subset of a prevariety becomes a prevariety when equipped with the induced sheaf. This is called an *open sub prevariety*. Putting this result together with the one above concerning closed subsets, we see that if S is a prevariety, T a closed subset of S and U an open subset of S, then the induced sheaf makes $T \cap U$ into a prevariety. Sets like $T \cap U$ are called *locally closed subsets*. A union of a finite family of locally closed subsets is called a *constructible subset*. This is still a prevariety, when equipped with the induced sheaf.

If U is a non-empty open subset of the irreducible affine variety S, then the rational functions on U, each with its maximum possible domain, constitute a field that is isomorphic, in the evident way, with $[\mathscr{P}(S)]$. This is called the *field of rational functions of* U. For every such U, this field is isomorphic, via restriction, with the field of rational functions of S.

Proposition 1.2. *Let S be an irreducible affine variety over an algebraically closed field F. The rational functions that are regular on all of S are precisely the elements of $\mathscr{P}(S)$, i.e., $\mathscr{F}_S(S) = \mathscr{P}(S)$.*

PROOF. Let f be a rational function that is regular on all of S, and let J_f be the ideal of all elements v of $\mathscr{P}(S)$ such that vf belongs to $\mathscr{P}(S)$. The assumption on f means that J_f has no zero in S. Since F is algebraically closed, this implies that $J_f = \mathscr{P}(S)$, which means that f belongs to $\mathscr{P}(S)$. $\qquad\square$

If S and T are prevarieties then a *morphism of prevarieties* from S to T is a continuous map σ from S to T such that, for every open set V of T, one has $\mathscr{F}_T(V) \circ \sigma \subset \mathscr{F}_S(\sigma^{-1}(V))$. In general terms, this condition is that σ be a *morphism of sheaves*.

Proposition 1.3. *Suppose that σ is a map from a prevariety S to a prevariety T satisfying the following condition. There are open sets $U_1, \ldots, U_n$ of S and $V_1, \ldots, V_n$ of T such that S is the union of the U_i's and T is the union of the V_i's, $\sigma(U_i) \subset V_i$, each V_i is an affine patch of T, and the restrictions to U_i of the elements of $\mathscr{F}_T(V_i) \circ \sigma$ belong to $\mathscr{F}_S(U_i)$. Then σ is a morphism of prevarieties.*

PROOF. Each U_i is the union of a finite family of affine patches. Not requiring that the V_i's be mutually distinct, we enlarge the index set so as to achieve that each new U_i is an affine patch.

Let σ_i denote the restriction of σ to U_i, regarding it as a map from the irreducible affine variety U_i to the irreducible affine variety V_i. By assumption, we have $\mathscr{F}_T(V_i) \circ \sigma_i \subset \mathscr{F}_S(U_i)$. In particular, if f is an element of $\mathscr{P}(V_i)$, and if x is a point of U_i such that $f(\sigma_i(x)) \neq 0$, then we may write $f \circ \sigma_i = a/b$, where a and b are elements of $\mathscr{P}(U_i)$ not vanishing at x. The set of non-zeros of ab in U_i is open, contains x and is mapped by σ_i into the set of non-zeros of f in V_i. It is clear from this that σ_i is continuous. Since the V_i's constitute an open covering of T, it follows that σ is continuous.

Moreover, if V is any open set of T, we may represent the elements of $\mathscr{F}_T(V \cap V_i)$ by fractions formed from elements of $\mathscr{P}(V_i)$. Using that

$$\mathscr{P}(V_i) \circ \sigma_i \subset \mathscr{F}_S(U_i),$$

we conclude that the restrictions to $U_i \cap \sigma^{-1}(V \cap V_i)$ of the elements of $\mathscr{F}_T(V \cap V_i) \circ \sigma$ belong to $\mathscr{F}_S(U_i \cap \sigma^{-1}(V \cap V_i))$. Since

$$\sigma^{-1}(V) \cap U_i \subset \sigma^{-1}(V \cap V_i),$$

this says that the restrictions to $\sigma^{-1}(V) \cap U_i$ of the elements of

$$\mathscr{F}_T(V \cap V_i) \circ \sigma$$

belong to $\mathscr{F}_S(\sigma^{-1}(V) \cap U_i)$. Now it follows from sheaf condition (2) that $\mathscr{F}_T(V) \circ \sigma \subset \mathscr{F}_S(\sigma^{-1}(V))$. $\square$

2. We construct direct products of prevarieties as follows. First, let us consider irreducible prevarieties R and S. If U is an affine patch of R and V is an affine patch of S, then the direct product in the category of affine algebraic sets gives us the structure of an irreducible affine variety on $U \times V$. If P is an open subset of this irreducible affine variety, then P is called an *elementary open subset* of $R \times S$.

If A is an open subset of U and B is an open subset of V, then $(U \setminus A) \times V$ and $U \times (V \setminus B)$ are closed subsets of $U \times V$, and their union is

$$(U \times V) \setminus (A \times B),$$

so that $A \times B$ is open in $U \times V$. It follows that, if Y is open in R and Z is open in S, then $P \cap (Y \times Z)$ is open in $U \times V$, and thus is an elementary open subset of $R \times S$. If we choose Y and Z to be affine patches, we see from this that the intersection of a pair of elementary open subsets of $R \times S$ is again an elementary open subset of $R \times S$. Therefore, we obtain a topology on $R \times S$ by defining the open sets to be the unions of families of elementary open subsets. It is easy to see that this makes $R \times S$ into a *Noetherian* topological space. Moreover, if the sets Y and Z above are non-empty, then $P \cap (Y \times Z)$ is non-empty for every non-empty elementary open subset P. Hence, the intersection of every pair of non-empty open sets of $R \times S$ is non-empty, so that $R \times S$ is irreducible.

Now we define a sheaf $\mathscr{F}_{R \times S}$ of functions on $R \times S$ as follows. Let X be an open subset of $R \times S$. Then we make a function f on X an element of $\mathscr{F}_{R \times S}(X)$ if and only if, for every pair (U, V) as above, the restriction of f to $X \cap (U \times V)$ belongs to $\mathscr{F}_{U \times V}(X \cap (U \times V))$. The verification of the sheaf conditions involves no difficulties. If R is written as the union of a finite family of affine patches U, and if S is written as the union of a finite family of affine patches V, then $R \times S$ is the union of the family of products $U \times V$, each of which is open in $R \times S$, and is an irreducible affine variety. Thus, $R \times S$ is an irreducible prevariety, the $U \times V$'s being affine patches.

We verify that the categorical requirements for a direct product are satisfied. Denote the projections from $R \times S$ to R and S by ρ and σ. Since the projections of a product of irreducible affine varieties are morphisms of affine algebraic sets and hence of prevarieties, an evident application of Proposition 1.3 shows that ρ and σ are morphisms of prevarieties. Now let T be a prevariety, and let α and β be prevariety morphisms from T to R and S. We must show that the map $\alpha \times \beta$ from T to $R \times S$, where

$$(\alpha \times \beta)(t) = (\alpha(t), \beta(t))$$

is a morphism of prevarieties. In the case where R, S and T are all irreducible affine varieties, we know this from the corresponding fact in the category of affine algebraic sets. The general case follows from an evident application of Proposition 1.3.

If R and S are general, not necessarily irreducible prevarieties, let $R_1, \ldots, R_p$ and $S_1, \ldots, S_q$ be the irreducible components of R and S. We define a topology on $R \times S$ by declaring a subset X to be open if and only if

$$X \cap (R_i \times S_j),$$

is open in $R_i \times S_j$ for all i and j. We define a sheaf of functions on $R \times S$ by making a function on an open set X an element of $\mathscr{F}_{R \times S}(X)$ if and only if its restriction to each $X \cap (R_i \times S_j)$ belongs to $\mathscr{F}_{R_i \times S_j}(X \cap (R_i \times S_j))$. clearly, this makes $R \times S$ into a prevariety satisfying the requirements for a direct product in the category of prevarieties.

A *variety* is a prevariety R with the property that the diagonal, $((r, r))_{r \in R}$, is closed in $R \times R$. The significance of this restriction becomes clear with the following proposition.

Proposition 2.1. *A prevariety R is a variety if and only if it satisfies the following condition. For every pair (ρ, σ) of morphisms from a prevariety S to R, the set of points s in S such that $\rho(s) = \sigma(s)$ is closed in S.*

PROOF. Suppose the condition is satisfied. Choose ρ and σ to be the projections from $R \times R$ to the first and second factor. Then the set described in the condition is the diagonal, whence R is a variety.

Conversely, suppose that R is a variety, and let ρ and σ be as in the condition. Consider the morphism $\rho \times \sigma$ from S to $R \times R$. The set described in the condition is the inverse image of the diagonal of $R \times R$ with respect to $\rho \times \sigma$. This is closed in S, because $\rho \times \sigma$ is continuous. $\square$

We remark that there is no terminological conflict: an "irreducible affine variety" in the sense of Section 1 is clearly a variety in the sense of the above definition. A general, not necessarily irreducible *affine variety* is a closed subvariety of an irreducible affine variety. Clearly, its irreducible components are irreducible affine varieties, in both senses.

Proposition 2.2. *If R and S are varieties, so is $R \times S$.*

PROOF. Let ρ and σ denote the projections from $R \times S$ to R and S, and let α and β be morphisms from a prevariety T to $R \times S$. Let $E_{\alpha, \beta}$ denote the set of all points t in T such that $\alpha(t) = \beta(t)$, and use the same notation for other pairs of morphisms. Then we have

$$E_{\alpha, \beta} = E_{\rho \circ \alpha, \rho \circ \beta} \cap E_{\sigma \circ \alpha, \sigma \circ \beta}.$$

Since R and S are varieties, we know from Proposition 2.1 that each of the two sets figuring on the right is closed. Therefore, $E_{\alpha, \beta}$ is closed. By Proposition 2.1, this shows that $R \times S$ is a variety. $\square$

Proposition 2.3. *Suppose that S is a prevariety for which there exists an injective morphism into a variety. Then S is a variety.*

PROOF. Let $\tau: S \to T$ be an injective morphism, where T is a variety. Let α and β be morphisms from a prevariety V to S. Using the same notation as in the last proof, we have from Proposition 2.1 that $E_{\tau \circ \alpha, \tau \circ \beta}$ is closed. Since τ is injective, we have $E_{\tau \circ \alpha, \tau \circ \beta} = E_{\alpha, \beta}$. Now the result follows from Proposition 2.1. $\square$

Proposition 2.4. *Let α be a morphism from a variety S to a variety T. Then the graph of α is closed in $S \times T$.*

PROOF. Define the map δ from $S \times T$ to $T \times T$ by

$$\delta(s, t) = (\alpha(s), t).$$

Then the graph of α is the inverse image, with respect to δ, of the diagonal in $T \times T$. Clearly, δ is a morphism of prevarieties, so that δ is continuous. Since T is a variety, the diagonal is closed, whence also its inverse image is closed. $\qquad\square$

3. Let F be a field, and let A be a finitely generated integral domain F-algebra that is *graded*, in the sense that A is the direct F-space sum of sub F-spaces A_n ($n = 0, 1, \ldots$) such that $A_0 = F$ and $A_r A_s \subset A_{r+s}$ for all r and s. Write A_+ for $\sum_{n>0} A_n$. We assume that $A_+ \neq (0)$, and that the F-algebra homomorphisms from A to F separate the elements of A, so that the elements of A may be regarded as F-valued functions on the set of these homomorphisms.

We consider the F-algebra homomorphisms from A to F not annihilating A_+. Two such homomorphisms, ρ and σ, are said to be *equivalent* if there is a non-zero element c in F such that, for every n and every element a of A_n, one has $\rho(a) = c^n\sigma(a)$. We denote the set of all F-algebra homomorphisms from A to F by $\mathscr{S}(A)$, and the set of equivalence classes of elements of $\mathscr{S}(A)$ not annihilating A_+ by $\mathscr{H}(A)$.

We define a topology on $\mathscr{H}(A)$ by declaring a subset C to be closed if and only if there is a set T of homogeneous elements of A such that the elements of C are the equivalence classes of those homomorphisms which annihilate T. It is equivalent to say that the closed sets are the sets of equivalence classes of the elements of $\mathscr{S}(A)$ annihilating homogeneous ideals of A, but not A_+. It is clear from this that our definition makes $\mathscr{H}(A)$ into a Noetherian topological space.

We show that $\mathscr{H}(A)$ is irreducible, as follows. Suppose that X and Y are closed sets in $\mathscr{H}(A)$ such that $X \cup Y = \mathscr{H}(A)$. For every subset S of $\mathscr{H}(A)$, let I_S denote the ideal of A that is generated by those homogeneous elements of A which are annihilated by the representatives in $\mathscr{S}(A)$ of the elements of S. Then we have $I_X \cap I_Y \subset I_{\mathscr{H}(A)}$. If f is an element of $I_{\mathscr{H}(A)}$ then fA_+ is annihilated by every element of $\mathscr{S}(A)$. Since $\mathscr{S}(A)$ separates the points of A, this implies that $fA_+ = (0)$. Since $A_+ \neq (0)$ and A is an integral domain, this gives $f = 0$. Thus, $I_{\mathscr{H}(A)} = (0)$, so that $I_X \cap I_Y = (0)$. Since A is an integral domain, it follows that one of I_X or I_Y is (0); say $I_X = (0)$. By the definition of a closed set of $\mathscr{H}(A)$, this means that $X = \mathscr{H}(A)$.

Let $[A]_0$ denote the subfield of $[A]$ consisting of the fractions a/b, where a and b are homogeneous elements of the same degree. Each such fraction determines an F-valued function on a certain non-empty open subset of $\mathscr{H}(A)$, as in the definition of a rational function on an affine algebraic set, because $\rho(a)/\rho(b) = \sigma(a)/\sigma(b)$ whenever ρ and σ are equivalent elements of $\mathscr{S}(A)$ not annihilating b. Now we define a sheaf of F-valued functions on $\mathscr{H}(A)$, as follows.

If U is a non-empty open subset of $\mathscr{H}(A)$, then $\mathscr{F}_{\mathscr{H}(A)}(U)$ consists of the restrictions to U of those elements of $[A]_0$ which are defined at every point of U. As before, we agree that $\mathscr{F}_{\mathscr{H}(A)}(\varnothing) = (0_F)$. Evidently, sheaf condition

(1) is satisfied. The verification of sheaf condition (2) is almost identical with the verification of this condition in the case of an irreducible affine variety, as carried out in Section 1.

Now we *assume that A is generated as an F-algebra by A_1*. Since A is finitely generated as an F-algebra, A_1 is finite-dimensional, and since $A_+ \neq (0)$ we have $A_1 \neq (0)$. Let d be any non-zero element of A_1, and let $A_{(d)}$ denote the sub F-algebra of $A[d^{-1}]$ consisting of the sums of elements of the form a/d^n, where a is an element of A_n ($n = 0, 1, \ldots$). If $(a_1, \ldots, a_m)$ is an F-basis of A_1, then $A_{(d)}$ is generated as an F-algebra by the elements $a_1/d, \ldots, a_m/d$. Thus, $A_{(d)}$ is a finitely generated integral domain F-algebra. Let $\mathcal{H}(A)_d$ denote the complement in $\mathcal{H}(A)$ of the closed set determined by d. If r is a point of $\mathcal{H}(A)_d$, and if ρ is a representative of r in $\mathcal{S}(A)$, then $\rho(d) \neq 0$, so that ρ defines an element of $\mathcal{S}(A_{(d)})$ by canonical extension. Clearly, this element depends only on r, not on the particular choice of the representative ρ. Thus, we have a map

$$\delta \colon \mathcal{H}(A)_d \to \mathcal{S}(A_{(d)}).$$

We regard $\mathcal{S}(A_{(d)})$ as an irreducible affine variety whose defining algebra of polynomial functions is $A_{(d)}$. Then it is clear from the definitions that δ is a morphism of sheaves when $\mathcal{H}(A)_d$ is equipped with the sheaf induced from $\mathcal{F}_{\mathcal{H}(A)}$.

We wish to prove that δ is actually an *isomorphism* of sheaves. In order to construct the inverse of δ, choose an F-basis $(a_1, \ldots, a_m)$ of A_1 such that $a_1 = d$. Then the elements of $A_{(d)}$ may be written in the form $p(a_2/d, \ldots, a_m/d)$, where p is a polynomial in $m - 1$ variables with coefficients in F. Let e be the degree of p, and let p^* denote the polynomial in m variables $x_1, \ldots, x_m$ given by

$$p^*(x_1, \ldots, x_m) = x_1^e p(x_2/x_1, \ldots, x_m/x_1).$$

Then, if $p(a_2/d, \ldots, a_m/d) = 0$, the element $p^*(a_1, \ldots, a_m)$ of A is 0. Using this, we see that, given an element σ of $\mathcal{S}(A_{(d)})$, there is an element σ' of $\mathcal{S}(A)$ such that $\sigma'(a_1) = 1$ and $\sigma'(a_i) = \sigma(a_i/d)$ for every $i > 1$. If s is the equivalence class of σ' in $\mathcal{H}(A)$, then s belongs to $\mathcal{H}(A)_d$ and $\delta(s) = \sigma$. In this way, we obtain a map

$$\gamma \colon \mathcal{S}(A_{(d)}) \to \mathcal{H}(A)_d$$

such that $\delta \circ \gamma$ is the identity map on $\mathcal{S}(A_{(d)})$. Also, it is clear from the above construction that, for every r in $\mathcal{H}(A)_d$, the element $\delta(r)'$ of $\mathcal{S}(A)$ is a representative of r, so that $\gamma \circ \delta$ is the identity map on $\mathcal{H}(A)_d$. Finally, one sees directly from the definitions that γ is a morphism of sheaves.

Thus, $\mathcal{H}(A)_d$, with the sheaf of functions induced by $\mathcal{F}_{\mathcal{H}(A)}$, is isomorphic with the irreducible affine variety $\mathcal{S}(A_{(d)})$. Since $\mathcal{H}(A)$ is the union of the family of open subsets $\mathcal{H}(A)_{a_i}$, this proves that $\mathcal{H}(A)$ is an irreducible prevariety. In order to show that, actually, $\mathcal{H}(A)$ is a variety, we use the following lemma.

Lemma 3.1. *Suppose that V is a prevariety such that, for every pair (x, y) of points of V, there is an affine patch of V containing x and y. Then V is a variety.*

PROOF. Let ρ and σ be morphisms from a prevariety W to V, and let $E_{\rho,\sigma}$ be the set of all points w in W such that $\rho(w) = \sigma(w)$. By Proposition 2.1, it suffices to show that $E_{\rho,\sigma}$ is closed in W. Let w be a point of the closure of $E_{\rho,\sigma}$. By assumption, there is an affine patch S of V containing $\rho(w)$ and $\sigma(w)$. Now $\rho^{-1}(S) \cap \sigma^{-1}(S)$ is an open subset of W containing w. Let ρ' and σ' be the restrictions of ρ and σ to this subset. These are morphisms from the prevariety $\rho^{-1}(S) \cap \sigma^{-1}(S)$ to the variety S. By Proposition 2.1, the set $E_{\rho',\sigma'}$ is therefore closed in $\rho^{-1}(S) \cap \sigma^{-1}(S)$, i.e., $E_{\rho,\sigma} \cap \rho^{-1}(S) \cap \sigma^{-1}(S)$ is closed in $\rho^{-1}(S) \cap \sigma^{-1}(S)$. Every open subset of $\rho^{-1}(S) \cap \sigma^{-1}(S)$ containing w has a non-empty intersection with $E_{\rho,\sigma}$, because it is open also in W. Therefore, w belongs to $E_{\rho,\sigma}$. $\qquad\qquad\square$

Now let r and s be points of $\mathcal{H}(A)$. Choose representatives ρ and σ for r and s in $\mathcal{S}(A)$. There are elements a and b in A_1 such that $\rho(a) \neq 0$ and $\sigma(b) \neq 0$, because ρ and σ do not annihilate A_+. If $\sigma(a) = 0$ and $\rho(b) = 0$, we have $\rho(a + b) \neq 0$ and $\sigma(a + b) \neq 0$. Thus, in every case, there is an element c in A_1 such that $\rho(c) \neq 0$ and $\sigma(c) \neq 0$. This shows that both r and s belong to the affine patch $\mathcal{H}(A)_c$. By Lemma 3.1, it follows that $\mathcal{H}(A)$ is a variety. We call $\mathcal{H}(A)$ the *projective variety defined by A*. We know that it is irreducible. A general, not necessarily irreducible projective variety is a closed subvariety of an $\mathcal{H}(A)$. Its irreducible components are again $\mathcal{H}(A)$'s, as is easy to see from the definitions.

Next, we show that *the direct product $\mathcal{H}(A) \times \mathcal{H}(B)$ of two irreducible projective varieties is again an irreducible projective variety.* Let $A \cdot B$ denote the subalgebra of $A \otimes B$ that is generated by the subspaces $A_n \otimes B_n$. We regard this as a graded algebra, with $(A \cdot B)_n = A_n \otimes B_n$. Clearly, this algebra satisfies all the conditions we imposed above in defining projective varieties. Accordingly, we have the projective variety $\mathcal{H}(A \cdot B)$. The canonical map from $\mathcal{S}(A) \times \mathcal{S}(B)$ to $\mathcal{S}(A \otimes B)$ evidently induces a map δ from $\mathcal{H}(A) \times \mathcal{H}(B)$ to $\mathcal{H}(A \cdot B)$, which we shall prove to be an isomorphism of varieties.

Let $(a_1, \ldots, a_m)$ be an F-basis of A_1, and let $(b_1, \ldots, b_n)$ be an F-basis of B_1. Then $(a_1 \otimes b_1, \ldots, a_m \otimes b_n)$ is an F-basis of $(A \cdot B)_1$. Each product $\mathcal{H}(A)_{a_i} \times \mathcal{H}(B)_{b_j}$ is open in $\mathcal{H}(A) \times \mathcal{H}(B)$, and the restriction of δ to this is a morphism of varieties to the affine patch $\mathcal{H}(A \cdot B)_{a_i \otimes b_j}$ of $\mathcal{H}(A \cdot B)$. By Proposition 1.3, this implies that δ is a morphism of varieties.

Next, we show that δ is injective. Let (r, s) and (r', s') be points of $\mathcal{H}(A) \times \mathcal{H}(B)$ such that $\delta(r, s) = \delta(r', s')$. Choose representatives ρ and ρ' of r and r' in $\mathcal{S}(A)$, and representatives σ and σ' of s and s' in $\mathcal{S}(B)$. Then we have, for all indices i and j,

$$\rho(a_i)\sigma(b_j) = c\rho'(a_i)\sigma'(b_j),$$

where c is a non-zero element of F. There is a pair (p, q) of indices such that $\rho(a_p) \neq 0$ and $\sigma(b_q) \neq 0$. Then the above relations give

$$\rho(a_i) = (c\sigma'(b_q)/\sigma(b_q))\rho'(a_i)$$

and

$$\sigma(b_j) = (c\rho'(a_p)/\rho(a_p))\sigma'(b_j)$$

for all indices i and j. This shows that $r = r'$ and $s = s'$, and we have the conclusion that δ is injective.

Now let u be a point of $\mathscr{H}(A \cdot B)$, and let μ be a representative of u in $\mathscr{S}(A \cdot B)$. There is a pair (p, q) of indices such that $\mu(a_p \otimes b_q) \neq 0$. We claim that there is an element ρ in $\mathscr{S}(A)$ such that $\rho(a_i) = \mu(a_i \otimes b_q)$ for each i. In order to see this, consider a polynomial f in m variables with coefficients in F such that $f(a_1, \ldots, a_m) = 0$. Write $f = f_0 + \cdots + f_k$, where each f_i is homogeneous of degree i. Then $f_i(a_1, \ldots, a_m) = 0$ for each i. Hence

$$f(a_1 \otimes b_q, \ldots, a_m \otimes b_q) = \sum_{i=0}^{k} f_i(a_1, \ldots, a_m) \otimes b_q^i = 0.$$

It is clear from this that the required element ρ of $\mathscr{S}(A)$ exists. Similarly, there is an element σ in $\mathscr{S}(B)$ such that $\sigma(b_j) = \mu(a_p \otimes b_j)$ for each j. Now ρ and σ do not annihilate all of A_+ or B_+, respectively. Therefore, they represent elements r and s of $\mathscr{H}(A)$ and $\mathscr{H}(B)$. If we define the element $\rho \cdot \sigma$ of $\mathscr{S}(A \cdot B)$ so that $(\rho \cdot \sigma)(a \otimes b) = \rho(a)\sigma(b)$ then $\rho \cdot \sigma$ represents $\delta(r, s)$. We have

$$(\rho \cdot \sigma)(a_i \otimes b_j) = \rho(a_i)\sigma(b_j) = \mu(a_i \otimes b_q)\mu(a_p \otimes b_j)$$

$$= \mu(a_i a_p \otimes b_q b_j)$$

$$= \mu(a_p \otimes b_q)\mu(a_i \otimes b_j)$$

for all indices i and j. Since $\mu(a_p \otimes b_q) \neq 0$, this shows that $\rho \cdot \sigma$ is equivalent to μ, so that $\delta(r, s) = u$. Thus, δ is surjective.

Our conclusion is that δ is a bijective morphism of varieties. Moreover, the above shows that, for each index pair (p, q), the restriction of δ^{-1} to $\mathscr{H}(A \cdot B)_{a_p \otimes b_q}$ is a morphism of varieties to $\mathscr{H}(A)_{a_p} \times \mathscr{H}(B)_{b_q}$. It follows, by Proposition 1.3, that δ^{-1} is a morphism of varieties. Thus, δ is an isomorphism of varieties from $\mathscr{H}(A) \times \mathscr{H}(B)$ to $\mathscr{H}(A \cdot B)$.

4. For our purposes, the most important projective variety is the *Grassmann variety* $\mathscr{G}_d(V)$, whose points are the d-dimensional subspaces of an n-dimensional vector space V over a field F. We consider the exterior F-algebra $\bigwedge(V^\circ)$, and we let V act on this by homogeneous derivations of degree -1.

Explicitly, if $\rho_1 \cdots \rho_m$ is the product in $\bigwedge^m(V^\circ)$ of m elements ρ_i of V°, and if v is an element of V, then the transform is given by

$$v \cdot (\rho_1 \cdots \rho_m) = \sum_{i=1}^{m} (-1)^{i-1} \rho_i(v) \rho_1 \cdots \hat{\rho}_i \cdots \rho_m,$$

where the $\hat{}$ indicates omission. From this action, we have the canonical homomorphism of F-algebras

$$\pi \colon \bigwedge(V) \to \operatorname{End}_F(\bigwedge(V^\circ)).$$

For each index d with $0 < d \leq n$, this map π induces a linear isomorphism from $\bigwedge^d(V)$ to $(\bigwedge^d(V^\circ))^\circ$.

Let A denote the symmetric algebra built over the F-space $\bigwedge^d(V^\circ)$, so that A is graded, and generated as an F-algebra by $A_1 = \bigwedge^d(V^\circ)$. We *assume that F is infinite*, so that we have the irreducible projective variety $\mathscr{H}(A)$. Via the above linear isomorphism, the 1-dimensional subspaces of $\bigwedge^d(V)$ may be identified with the points of $\mathscr{H}(A)$.

The d-dimensional subspaces of V are in bijective correspondence with those 1-dimensional subspaces of $\bigwedge^d(V)$ which are spanned by *decomposable* elements, i.e., products of d-tuples of elements of V. We shall show that they constitute a closed subvariety of $\mathscr{H}(A)$. This comes from the following criterion.

Let p be a non-zero element of the 1-dimensional F-space $\bigwedge^n(V^\circ)$. Then a non-zero element z of $\bigwedge^d(V)$ is decomposable if and only if $\pi(z)(p)\pi(zx)(p) = 0$ for every x in $\bigwedge^{n-d-1}(V)$.

In order to establish this, let us first suppose that z is decomposable; say $z = v_1 \cdots v_d$. The criterion is trivially satisfied if $z = 0$. Therefore, we assume that $z \neq 0$, so that the set $(v_1, \ldots, v_d)$ is linearly independent. We complete it to an F-basis $(v_1, \ldots, v_n)$ of V, and we let $(\rho_1, \ldots, \rho_n)$ be the dual basis of V°. We may take $p = \rho_1 \cdots \rho_n$. If $x = v_{i_1} \cdots v_{i_{n-d-1}}$ then $zx = 0$ unless each $i_k > d$. If each $i_k > d$ and $zx \neq 0$ there is exactly one v_j different from each v_{i_k} and with $j > d$, and we have $\pi(zx)(p) = \pm \rho_j$. On the other hand, $\pi(z)(p) = \pm \rho_{d+1} \cdots \rho_n$. Therefore, $\pi(z)(p)\pi(zx)(p) = 0$ in every case, and hence for every x in $\bigwedge^{n-d-1}(V)$.

Now suppose that z satisfies the above vanishing conditions, and write z as an F-linear combination of products of basis elements $v_1, \ldots, v_n$ of V, as follows

$$z = \sum_{i_1 < \cdots < i_d} c(i_1, \ldots, i_d) v_{i_1} \cdots v_{i_d}.$$

We may suppose that $c(1, \ldots, d) \neq 0$, and hence that $c(1, \ldots, d) = 1$. For $j = d+1, \ldots, n$, put $x_j = v_{d+1} \cdots \hat{v}_j \cdots v_n$. Then, for each $i > d$, we have $zx_j v_i = \pm \delta_{ij} v_1 \cdots v_n$, whence we see that the elements zx_j are linearly independent. Therefore, the elements $\pi(zx_j)$ of $\operatorname{End}_F(\bigwedge(V^\circ))$ are linearly independent. They belong to $\pi(\bigwedge^{n-1}(V))$, and it is easy to see that the evaluation at p is injective from $\pi(\bigwedge^{n-1}(V))$ to V°. Therefore, the elements

$\pi(zx_j)(p)$ of V° are linearly independent. For each $j > d$, put $\rho_j = \pi(zx_j)(p)$, and choose $\rho_1, \ldots, \rho_d$ from V° such that $(\rho_1, \ldots, \rho_n)$ is an F-basis of V°. Let $(w_1, \ldots, w_n)$ be the dual basis of V. Then our assumption on z gives $\pi(z)(p)\rho_j = 0$ for every $j > d$. This implies that $\pi(z)(p)$ is an F-multiple of $\rho_{d+1} \cdots \rho_n$, whence z is an F-multiple of $w_1 \cdots w_d$.

It is evident from the criterion we have just established that z is decomposable if and only if it is a zero of a certain set of elements of A_2. Therefore, the set of d-dimensional subspaces of V is identified, via π, with a closed subset of $\mathscr{H}(A)$.

Finally, we show that this is an irreducible subset, so that, *if the base field is infinite, $\mathscr{G}_d(V)$ is an irreducible projective variety.*

Let G denote the irreducible affine algebraic F-group $\mathrm{Aut}_F(V)$. In the evident fashion, G acts transitively on our variety of d-dimensional subspaces of V. Fix a d-dimensional subspace S of V. It is easy to see that the map associating with each element σ of G the subspace $\sigma(S)$ of V is a morphism of varieties from G to $\mathscr{G}_d(V)$. Since it is surjective, and since G is irreducible, it follows that $\mathscr{G}_d(V)$ is irreducible, by continuity.

5. Definition 5.1. *A variety X is said to be complete if, for all varieties W, the canonical projection from $X \times W$ to W is a closed map.*

The following proposition contains suggestions regarding the significance of the completeness property.

Proposition 5.2. *Let X and Y be varieties.*

(1)   *If X is complete and Y is a closed subvariety of X then Y is complete.*
(2)   *If X and Y are complete, so is $X \times Y$.*
(3)   *If X is complete and γ is a morphism of varieties from X to Y then $\gamma(X)$ is closed in Y and complete.*
(4)   *A complete irreducible affine variety consists of a single point.*

PROOF. Evidently, (1) follows immediately from the definition.

As for (2), it suffices to observe that the projection from $X \times Y \times W$ to W is the composite of the projection to $Y \times W$ with the projection from $Y \times W$ to W, each of which is a closed map.

In order to establish (3), consider the graph, G_γ say, of γ in $X \times Y$. By Proposition 2.4, G_γ is closed in $X \times Y$. Since X is complete, the projection image of G_γ in Y is therefore closed, i.e., $\gamma(X)$ is closed in Y. If π denotes the canonical projection from $X \times W$ to W, and π' denotes the canonical projection from $\gamma(X) \times W$ to W, then $\pi = \pi' \circ (\gamma \times i_W)$. If C is a closed subset of $\gamma(X) \times W$, then $(\gamma \times i_W)^{-1}(C)$ is closed in $X \times W$. Since X is complete, it follows that $\pi((\gamma \times i_W)^{-1}(C))$ is closed in W, i.e., that $\pi'(C)$ is closed in W. This shows that $\gamma(X)$ is complete.

Finally, suppose that V is a complete irreducible affine variety, and let f be an element of $\mathscr{P}(V)$. If the base field is finite then V consists of a single point, because it is finite and irreducible. Therefore, we assume without loss of generality that our base field is an infinite field F. Then we regard F as a 1-dimensional irreducible affine algebraic variety in the usual way. Our element f of $\mathscr{P}(V)$ is a morphism of varieties from V to F. By part (3) above, $f(V)$ is closed in F and complete. Since $f(V)$ is an irreducible subset of F, it follows that either $f(V) = F$, or $f(V)$ consists of a single point. In the first case, it would follow that F is a complete variety. However, this is not the case, because the closed subset of $F \times F$ consisting of the points (a, b) with $ab = 1$ projects onto the non-closed subset $F \setminus (0)$ of F. Therefore, $f(V)$ consists of a single point. Our conclusion is that the constants are the only elements of $\mathscr{P}(V)$, so that V consists of a single point. $\qquad\square$

Theorem 5.3. *Every projective variety over an algebraically closed field is complete.*

PROOF. Evidently, it suffices to prove that every irreducible projective variety $P = \mathscr{H}(A)$ is complete. We must show that, for every variety W, the canonical projection from $P \times W$ to W is a closed map. Now W is the union of a finite family of affine patches U. A subset of W is closed in W if and only if its intersection with each U is closed in U, and a subset of $P \times W$ is closed in $P \times W$ if and only if its intersection with each $P \times U$ is closed in $P \times U$. Therefore, it suffices to deal with the case where W is an irreducible affine variety, $\mathscr{S}(B)$ say.

Let $(a_1, \ldots, a_n)$ be a basis of A_1, so that P is the union of the affine patches $P_{a_i} = \mathscr{S}(A_{(a_i)})$, in the notation of Section 3. Accordingly, $P \times W$ is the union of the family of open subvarieties $P_{a_i} \times W$, each of which is an irreducible affine variety $\mathscr{S}(A_{(a_i)} \otimes B)$.

Consider a closed subset C of $P \times W$, and put $C_i = C \cap (P_{a_i} \times W)$. Let $I(C_i)$ denote the annihilator ideal of C_i in $A_{(a_i)} \otimes B$. Regarding $A \otimes B$ as a graded algebra with the grading coming from that of A, let I be the homogeneous ideal of $A \otimes B$ whose homogeneous components I_e consist of those elements

$$\sum_{e_1 + \cdots + e_n = e} (a_1)^{e_1} \cdots (a_n)^{e_n} \otimes b(e_1, \ldots, e_n)$$

for which the corresponding elements

$$\sum_{e_1 + \cdots + e_n = e} (a_1/a_i)^{e_1} \cdots (a_n/a_i)^{e_n} \otimes b(e_1, \ldots, e_n)$$

belong to $I(C_i)$ for each i. Working in $A[a_1^{-1}, \ldots, a_n^{-1}] \otimes B$, consider an element f of $I(C_i)$. There is an exponent $e \geq 0$ such that $((a_i)^e \otimes 1)f$ belongs to $A \otimes B$. Then, for each j,

$$((a_i/a_j)^e \otimes 1)f \in A_{(a_j)} \otimes B$$

and this element vanishes on

$$C_i \cap (P_{a_j} \times W) = C_j \cap (P_{a_i} \times W).$$

On the other hand, the element

$$((a_i/a_j)^{e+1} \otimes 1)f = ((a_i/a_j) \otimes 1)((a_i/a_j)^e \otimes 1)f$$

vanishes on the part of C_j that does not lie in $P_{a_i} \times W$. Since j is arbitrary, this shows that the element $((a_i)^{e+1} \otimes 1)f$ belongs to I_{e+1}.

Now let w be a point of W not belonging to the projection image of C. Then C_i and $P_{a_i} \times (w)$ are disjoint closed subsets of the irreducible affine variety $P_{a_i} \times W$. Let J be the annihilator of w in B. Then $A_{(a_i)} \otimes J$ is the annihilator of $P_{a_i} \otimes (w)$ in $A_{(a_i)} \otimes B$. Since F is algebraically closed, it follows from the disjointness of the sets C_i and $P_{a_i} \times (w)$ that we must have

$$I(C_i) + A_{(a_i)} \otimes J = A_{(a_i)} \otimes B.$$

This means that there are elements f_i in $I(C_i)$, g_{ij} in $A_{(a_i)}$ and m_{ij} in J such that

$$f_i + \sum_j g_{ij} \otimes m_{ij} = 1.$$

By the above, there is an exponent e such that $((a_i)^e \otimes 1)f_i$ belongs to I_e for each i, and $(a_i)^e g_{ij}$ belongs to A for all i and j. With this, the above expression for 1 yields the result that $(a_i)^e \otimes 1$ belongs to $I_e + (A \otimes J)_e$. By enlarging e, if necessary, we obtain the result that all monomials of total degree e in $a_1, \ldots, a_n$ belong to $I_e + (A \otimes J)_e$. This means that

$$(A \otimes B)_e = I_e + (A \otimes J)_e.$$

Now consider the finitely generated B-module $R = (A \otimes B)_e/I_e$. Our last result means that $J \cdot R = R$. If $(r_1, \ldots, r_k)$ is a system of B-module generators of R, we have therefore relations

$$r_p = \sum_{q=1}^{k} u_{pq} r_q$$

with each u_{pq} in J. The determinant, f say, of this system is of the form $1 + u$, with $u \in J$, and $f \cdot R = (0)$. Thus, we have $A_e \otimes Bf \subset I_e$ and $f \notin J$. In particular, $(a_i)^e \otimes f$ belongs to I_e, so that f vanishes on the projection image of C in W, while $f(w) \neq 0$. Thus, W_f is an open subset of W containing w and not meeting the projection image of C. Our conclusion is that the complement in W of the projection image of C is open. $\qquad\square$

Notes

1. The simplest example of an open subvariety of an affine variety that is not an affine variety is as follows. Let F be an algebraically closed field, x and y independent variables over F, and S the irreducible affine F-variety

$\mathscr{S}(F[x, y])$. Identifying the points s of S with the corresponding pairs $(s(x), s(y))$ of elements of F, let U denote the open subvariety $S \setminus (0, 0)$ of S. One can show that every rational function of S that is defined at every point of U belongs to $F[x, y]$. If U were affine, $\mathscr{P}(U)$ would therefore be the restriction image of $F[x, y]$, by virtue of Proposition 1.2. But then $(0, 0)$ would belong to U. [In order to prove the assertion concerning the rational functions, let f be an element of $\mathscr{F}_S(U)$, and consider the ideal J_f of all elements u of $F[x, y]$ such that uf belongs to $F[x, y]$. Then the only zero of J_f in S is $(0, 0)$, which implies that J_f contains some power of x, as well as some power of y. Using unique factorization in $F[x, y]$, one sees from this that f must belong to $F[x, y]$].

2. The recipe we have used for constructing the direct product of projective varieties is due to M. E. Sweedler.

3. The proof of Theorem 5.3 is due to Grothendieck. Generally, the above treatment of the completeness property comes from D. Mumford [11].

Chapter X

Morphisms of Varieties
and Dimension

This chapter is concerned mainly with the dimension-theoretical analysis of morphisms between varieties. This involves substantially more commutative algebra than has been used up to now. Thus, Section 1 establishes Noether's Normalization Theorem, which is used for reducing some of the required ideal theoretical considerations to the situation of an ordinary polynomial algebra. The remaining results of Section 1 concern the connections between the dimensions of irreducible closed subvarieties of irreducible affine varieties and the generation of their annihilating ideals.

Section 2 contains the "Going Down Theorem" of Cohen–Seidenberg for prime ideals with respect to integral ring extensions, which is used in proving Proposition 2.4. This proposition is an important step in the examination of inverse images with respect to morphisms.

Section 3 deals with the commutative algebra underlying the notion of a normal variety. Every variety has a non-empty open normal subvariety, and the morphisms to a normal variety are far more tractable than morphisms are in general. This fact appears in Section 4, especially in Theorem 4.3, which enables one to exploit the vital criterion of Theorem 4.5 for a morphism to be an open map.

1. Proposition 1.1. *Let F be a field, and let $A = F[a_1, \ldots, a_n]$ be a finitely generated commutative F-algebra. Let I be an ideal of A, other than A. There are elements $b_1, \ldots, b_n$ in A, and an index $r \le n$, such that b_i belongs to I for each $i < r$, $I \cap F[b_r, \ldots, b_n] = (0)$ and A is integral over $F[b_1, \ldots, b_n]$.*

PROOF. We make an induction on n. If $n = 0$ we have $A = F$ and $I = (0)$, so that the proposition holds trivially. Now suppose that $n > 0$ and that the proposition has been established in the lower cases. If $I = (0)$ we may evidently take $r = 1$ and each $b_i = a_i$. Therefore, we assume that $I \ne (0)$.

Choose any non-zero element b_1 from I, and let $x_1, \ldots, x_n$ be independent variables over F. Write

$$b_1 = f(a_1, \ldots, a_n),$$

where $f \in F[x_1, \ldots, x_n]$, and let d denote the total degree of f. Since $I \neq A$, we must have $d > 0$. For each $i > 1$, put

$$c_i = a_i - a_1^{(d+1)^{i-1}}.$$

Then we have

$$b_1 = f(a_1, c_2 + a_1^{d+1}, \ldots, c_n + a_1^{(d+1)^{n-1}}).$$

The term of the highest degree in x_1 in the expansion of

$$(x_1)^{e_1}(x_2 + x_1^{d+1})^{e_2} \cdots (x_n + x_1^{(d+1)^{n-1}})^{e_n},$$

is x_1^e, where

$$e = e_1 + e_2(d+1) + \cdots + e_n(d+1)^{n-1}.$$

As the e_i's range over the natural numbers from 0 to d, the exponent e of x_1 here is never repeated. It follows that, if we arrange the full expansion of our above expression for b_1 according to the powers of a_1, we obtain

$$b_1 = u(a_1)^e + \sum_{i=0}^{e-1} f_i(c_2, \ldots, c_n)(a_i)^i,$$

where u is a non-zero element of F, the exponent e is greater than 0 and the f_i's belong to $F[x_2, \ldots, x_n]$. Hence, a_1 is integral over $F[b_1, c_2, \ldots, c_n]$.

Now write B for $F[c_2, \ldots, c_n]$ and J for $I \cap B$. By our inductive hypothesis, there are elements $b_2, \ldots, b_n$ in B and an index $r \leq n$ such that b_i belongs to J for each i with $2 \leq i < r$, $J \cap F[b_r, \ldots, b_n] = (0)$ and B is integral over $F[b_2, \ldots, b_n]$. Clearly, b_i belongs to I for each $i < r$, and

$$I \cap F[b_r, \ldots, b_n] = (0).$$

Finally, each a_i is integral over $F[b_1, c_2, \ldots, c_n]$, while $F[b_1, c_2, \ldots, c_n]$ is integral over $F[b_1, \ldots, b_n]$. Therefore, each a_i is integral over $F[b_1, \ldots, b_n]$. $\qquad\square$

The following almost immediate consequence is known as *Noether's Normalization Theorem*.

Theorem 1.2. *Let F be a field, and let R be a finitely generated commutative F-algebra. There is a subset $(z_1, \ldots, z_s)$ of R that is algebraically free over F and such that R is integral over $F[z_1, \ldots, z_s]$.*

PROOF. Write $R = F[x_1, \ldots, x_n]/I$, where the x_i's are independent variables over F. Let $(b_1, \ldots, b_n)$ and r be as obtained from Proposition 1.1, with $F[x_1, \ldots, x_n]$ in the place of A. Let z_i denote the canonical image in R of

b_{i+r-1} $(i = 1, \ldots, n + 1 - r)$. Since $F[x_1, \ldots, x_n]$ is integral over $F[b_1, \ldots, b_n]$, it follows that R is integral over the canonical image of $F[b_1, \ldots, b_n]$ in R. This image is $F[z_1, \ldots, z_{n+1-r}]$, because b_i belongs to I for each $i < r$. Clearly, the set $(b_1, \ldots, b_n)$ must be algebraically free over F. Since $I \cap F[b_r, \ldots, b_n] = (0)$, it follows that the set $(z_1, \ldots, z_{n+1-r})$ is algebraically free over F. $\square$

Let X be an irreducible algebraic variety, U and V affine patches of X. By restriction to an affine patch W contained in $U \cap V$, each, the field of rational functions of U, and the field of rational functions of V, is isomorphic with the field of rational functions of W. Thus, we may identify all the fields of rational functions of affine patches of X, and so arrive at the notion of *the field of rational functions of X*. The degree of transcendence of this field over the base field is the *dimension* of X, denoted $\dim(X)$. This is equal to the dimension of every affine patch of X.

Proposition 1.3. *Let X be an irreducible variety, and let Y be a closed irreducible subvariety of X, other than X. Then $\dim(Y) < \dim(X)$.*

Proof. There is an affine patch U of X such that $Y \cap U \neq \varnothing$. Now $Y \cap U$ is closed in U, and $Y \cap U \neq U$, because otherwise Y contains the closure of U, which is X. Also, $Y \cap U$ is open in Y, and hence is irreducible. We have $\dim(X) = \dim(U)$ and $\dim(Y) = \dim(Y \cap U)$. Since U is an irreducible affine variety, we can apply Lemma II.3.8 in the evident way to conclude that $\dim(Y \cap U) < \dim(U)$. $\square$

Corollary 1.4. *Let X be an irreducible affine variety, and let Y be a closed irreducible subset of X such that $\dim(Y) = \dim(X) - 1$. Then, for every non-zero element f of $\mathscr{P}(X)$ such that $f(Y) = (0)$, Y is an irreducible component of the set of zeros of f in X.*

Proof. Evidently, the irreducible set Y is contained in some irreducible component, Z say, of the set of zeros of f. By Proposition 1.3, we have $\dim(Z) < \dim(X)$, and $\dim(Y) \leq \dim(Z)$. Since $\dim(X) - \dim(Y) = 1$, this gives $\dim(Y) = \dim(Z)$. By Proposition 1.3, this implies that $Y = Z$. $\square$

Proposition 1.5. *Let X be an irreducible affine variety, and let f be a non-zero element of $\mathscr{P}(X)$. Suppose that Y is an irreducible component of the set of zeros of f in X. Then $\dim(Y) = \dim(X) - 1$.*

Proof. Let $Y_1, \ldots, Y_t$ be the irreducible components of the set of zeros of f, with $Y_1 = Y$. Let J_i denote the annihilator of Y_i in $\mathscr{P}(X)$. For any ideal I, let I^* denote the radical of I. Then we have $(\mathscr{P}(X)f)^* = J_1 \cap \cdots \cap J_t$. Choose an element g from the non-empty set $(J_2 \cap \cdots \cap J_t) \setminus J_1$ (if $t = 1$, we interpret this expression as $\mathscr{P}(X) \setminus J_1$), and consider the principal open subset X_g of X. This is an irreducible affine variety, and $Y \cap X_g$ is precisely the set of zeros of f in X_g. Moreover $Y \cap X_g = Y_{g'}$, where g' is the restriction of g to Y, so that $Y \cap X_g$ is an irreducible affine variety of the same dimension as Y. Since $\dim(X_g) = \dim(X)$, it suffices to prove that

$$\dim(Y \cap X_g) = \dim(X_g) - 1.$$

Thus, we may replace X with X_g and Y with $Y \cap X_g$. Then we have the simplified situation where Y is the set of zeros of f in X, and the annihilator of Y in $\mathscr{P}(X)$ is $(\mathscr{P}(X)f)^*$, for which we shall simply write J. In this situation, we proceed as follows.

If $d = \dim(X)$, we have from Theorem 1.2 that there is a subset $(z_1, \ldots, z_d)$ of $\mathscr{P}(X)$ that is algebraically free over the base field F and such that $\mathscr{P}(X)$ is integral over $F[z_1, \ldots, z_d]$. Let us write A for $\mathscr{P}(X)$ and B for $F[z_1, \ldots, z_d]$, and let us consider the finite algebraic extension $[A]$ of $[B]$. Since f is integral over B, and since B is integrally closed in $[B]$, the monic minimum polynomial for f relative to $[B]$ has all its coefficients in B. Let μ denote the norm map for $[A]$ relative to $[B]$. Then $\mu(f)$ is, up to sign, a power of the constant term of the minimum polynomial for f, so that $\mu(f)$ lies in $(Bf) \cap B \subset J \cap B$. Hence we have $(B\mu(f))^* \subset J \cap B$. Conversely, let g be an element of $J \cap B$. There is a positive integer m such that $g^m = hf$, with some h in A. Applying μ, we obtain $g^{em} = \mu(h)\mu(f)$, where e is the degree of $[A]$ relative to $[B]$. As with $\mu(f)$, we find $\mu(h) \in B$, so that our last equation shows that g lies in $(B\mu(f))^*$. Thus, we have $J \cap B = (B\mu(f))^*$.

Since B is a unique factorization domain, and since $(B\mu(f))^*$ is the *prime* ideal $J \cap B$, the element $\mu(f)$ must be a non-zero F-multiple of a power of a prime element p of B, so that $(B\mu(f))^* = Bp$. Now p is an irreducible polynomial in $z_1, \ldots, z_d$ with coefficients in F. By relabeling, if necessary, we arrange to have z_d actually occur in p.

Now $\dim(Y)$ is the transcendence degree of $[A/J]$ over F, which is equal to the transcendence degree of $[B/(J \cap B)]$ over F. If y_i denotes the canonical image of z_i in $B/(J \cap B)$, we have

$$F[y_1, \ldots, y_{d-1}] \subset B/(J \cap B).$$

If the elements $y_1, \ldots, y_{d-1}$ were not algebraically independent over F there would be a non-zero polynomial q in $d - 1$ variables with coefficients in F such that $q(y_1, \ldots, y_{d-1}) = 0$. This means that $q(z_1, \ldots, z_{d-1})$ belongs to Bp. Since z_d actually occurs in p, this contradicts the algebraic independence of $z_1, \ldots, z_d$. Therefore, the transcendence degree of $B/(J \cap B)$ over F is at least equal to $d - 1$, whence $\dim(Y) \geq d - 1$. By Proposition 1.3, we have $\dim(Y) < d$. Hence, $\dim(Y) = d - 1$. $\square$

Corollary 1.6. *Suppose X is an irreducible affine variety and Y is a closed irreducible subset of X, with $\dim(Y) = \dim(X) - r$, where $r > 0$. There are closed irreducible subsets Y_i of X such that $Y = Y_r \subset \cdots \subset Y_1$ and*

$$\dim(Y_i) = \dim(X) - i,$$

for each i.

PROOF. If $r = 1$ there is nothing to prove. Suppose $r > 1$ and the corollary established in the lower cases. Since $Y \neq X$ there is a non-zero element f in $\mathscr{P}(X)$ that vanishes on Y. Now Y is contained in some irreducible component Y_1 of the set of zeros of f in X. By Proposition 1.5, $\dim(Y_1) = \dim(X) - 1$. Now apply the inductive hypothesis with Y_1 in the place of X. $\qquad\square$

Corollary 1.7. *Let X be an irreducible affine variety, and let $f_1, \ldots, f_r$ be elements of $\mathscr{P}(X)$. Suppose Y is an irreducible component of the set $\mathscr{V}(f_1, \ldots, f_r)$ of common zeros in X of the f_i's. Then $\dim(Y) \geq \dim(X) - r$.*

PROOF. This follows by induction on r, using Proposition 1.5 for the inductive step. In fact, Y is an irreducible closed subset of some irreducible component, Z say, of $\mathscr{V}(f_1, \ldots, f_{r-1})$. Since Y is a maximal irreducible subset of $\mathscr{V}(f_1, \ldots, f_r)$, it follows that Y is an irreducible component of $Z \cap \mathscr{V}(f_r)$. By inductive hypothesis, $\dim(Z) \geq \dim(X) - r + 1$. If the restriction of f_r to Z is 0, then $Y = Z$. If the restriction of f_r to Z is not 0, we have from Proposition 1.5 that $\dim(Y) = \dim(Z) - 1 \geq \dim(X) - r$. $\qquad\square$

Lemma 1.8. *Let T be a commutative ring, $P_1, \ldots, P_k$ prime ideals of T, and K a subset of T that is closed under addition and multiplication. Then, if the union of the family of P_i's contains K, one of the P_i's contains K.*

PROOF. Making an induction on k, we may suppose that $k > 1$ and that K is not contained in the union of any proper subset of the set of P_i's, and show that then K is not contained in $\bigcup_{i=1}^{k} P_i$. Choose a_j from $K \setminus \bigcup_{i \neq j} P_i$ for each j, and put

$$b_j = a_1 \cdots \hat{a}_j \cdots a_k, \qquad b = b_1 + \cdots + b_k.$$

If K is contained in the union of the set of P_i's, we must have $a_j \in P_j$, so that $b_i \in P_j$ for each i other than j. On the other hand, $b_j \notin P_j$, because none of its factors a_i belongs to P_j. It follows that b does not belong to P_j. Since b belongs to K, this contradicts the assumption that the union of the family of P_i's contains K. $\qquad\square$

Theorem 1.9. *Let X be an irreducible affine variety. Suppose that $Y_1, \ldots, Y_r$ are irreducible closed subsets of X such that $Y_r \subset \cdots \subset Y_1$ and*

$$\dim(Y_i) = \dim(X) - i,$$

for each i. Then there are elements $f_1, \ldots, f_r$ in $\mathscr{P}(X)$ satisfying the following conditions, for each i with $1 \leq i \leq r$:

(1)   *Y_i is an irreducible component of $\mathscr{V}(f_1, \ldots, f_i)$;*
(2)   *every irreducible component of $\mathscr{V}(f_1, \ldots, f_i)$ has dimension $\dim(X) - i$.*

PROOF. By Corollary 1.4, there is an element f_1 in $\mathscr{P}(X)$ such that Y_1 is an irreducible component of $\mathscr{V}(f_1)$. By Proposition 1.5, every irreducible component of $\mathscr{V}(f_1)$ has dimension $\dim(X) - 1$.

Now suppose we have already found $f_1, \ldots, f_i$ satisfying (1) and (2). Write Z_1 for Y_i, and let $Z_2, \ldots, Z_m$ be all the other irreducible components of $\mathcal{V}(f_1, \ldots, f_i)$. The dimension of each of these components is $\dim(X) - i$, which is greater than the dimension of Y_{i+1}. Therefore, none of $Z_1, \ldots, Z_m$ is contained in Y_{i+1}. Let P_j denote the annihilator of Z_j in $\mathcal{P}(X)$, and let K denote the annihilator of Y_{i+1} in $\mathcal{P}(X)$. By our last remark, K is not contained in any P_j. By Lemma 1.8, this implies that K is not contained in the union of the set of P_j's. Thus, there is an element f_{i+1} in $\mathcal{P}(X)$ that vanishes on Y_{i+1} but not on any Z_j. Let Z be an irreducible component of $\mathcal{V}(f_1, \ldots, f_{i+1})$. Then $Z \subset Z_j$ for some j and $Z \subset \mathcal{V}(f_{i+1})$. By Proposition 1.5, every irreducible component of $Z_j \cap \mathcal{V}(f_{i+1})$ is of dimension

$$\dim(Z_j) - 1 = \dim(X) - (i + 1).$$

Therefore, we have $\dim(Z) \leq \dim(X) - (i + 1)$. On the other hand, by Corollary 1.7, we have $\dim(Z) \geq \dim(X) - (i + 1)$. Thus,

$$\dim(Z) = \dim(X) - (i + 1).$$

Since $Y_{i+1} \subset \mathcal{V}(f_{i+1}) \cap Y_i$, we have $Y_{i+1} \subset \mathcal{V}(f_1, \ldots, f_{i+1})$. Since $\dim(Y_{i+1}) = \dim(X) - (i + 1)$, which is the dimension of every irreducible component of $\mathcal{V}(f_1, \ldots, f_{i+1})$, it follows that Y_{i+1} is an irreducible component of $\mathcal{V}(f_1, \ldots, f_{i+1})$. Thus, the set $(f_1, \ldots, f_{i+1})$ satisfies conditions (1) and (2) for $i + 1$ in the place of i. $\qquad\square$

2. Let σ be a morphism from an irreducible variety X to a variety Y. We say that σ is *dominant* if $\sigma(X)$ is dense in Y. If X is not irreducible, we say that σ is dominant if the restriction of σ to each irreducible component of X is a dominant morphism from that component to some irreducible component of Y and $\sigma(X)$ is dense in Y.

Theorem 2.1. *Suppose $\sigma\colon X \to Y$ is a dominant morphism between irreducible varieties. Let W be a closed irreducible subset of Y, and let Z be an irreducible component of $\sigma^{-1}(W)$ such that $\sigma(Z)$ is dense in W. Then*

$$\dim(Z) \geq \dim(W) + \dim(X) - \dim(Y).$$

PROOF. There is an affine patch U of Y such that $U \cap W \neq \varnothing$. Then $U \cap W$ is a closed irreducible subset of the irreducible affine variety U, and $\dim(U \cap W) = \dim(W)$. Now σ induces a morphism of varieties from $\sigma^{-1}(U)$ to U, and $Z \cap \sigma^{-1}(U)$ is an irreducible component of $\sigma^{-1}(U \cap W)$ whose image is dense in $U \cap W$. We have $\dim(Z \cap \sigma^{-1}(U)) = \dim(Z)$, $\dim(U \cap W) = \dim(W)$, $\dim(U) = \dim(Y)$ and $\dim(\sigma^{-1}(U)) = \dim(X)$. Therefore, it suffices to prove the theorem in the case where Y is affine ($Y = U$), which we shall now assume.

Let $r = \dim(Y) - \dim(W)$. By Corollary 1.6 and Theorem 1.9, there are elements $f_1, \ldots, f_r$ in $\mathcal{P}(Y)$ such that W is an irreducible component of

$\mathscr{V}(f_1,\ldots,f_r)$. Let $g_i = f_i \circ \sigma$, so that g_i is an everywhere regular function on X. Then we have $Z \subset \mathscr{V}(g_1,\ldots,g_r)$. Since Z is irreducible, it is contained in some irreducible component Z_0 of $\mathscr{V}(g_1,\ldots,g_r)$. If T' denotes the closure of a set T, we have

$$W = \sigma(Z)' \subset \sigma(Z_0)' \subset \mathscr{V}(f_1,\ldots,f_r).$$

Since W is an irreducible component of $\mathscr{V}(f_1,\ldots,f_r)$ and since $\sigma(Z_0)'$ is irreducible, it follows that $W = \sigma(Z)' = \sigma(Z_0)'$, whence $Z_0 \subset \sigma^{-1}(W)$. Since Z is an irreducible component of $\sigma^{-1}(W)$, we must therefore have $Z = Z_0$, so that Z is an irreducible component of $\mathscr{V}(g_1,\ldots,g_r)$. Intersecting with a suitable affine patch of X and applying Corollary 1.7, we see that therefore $\dim(Z) \geq \dim(X) - r = \dim(W) + \dim(X) - \dim(Y)$. $\square$

A morphism $\sigma\colon X \to Y$ between affine varieties is called *finite* if $\mathscr{P}(X)$ is integral over $\mathscr{P}(Y) \circ \sigma$.

Proposition 2.2. *Let $\sigma\colon X \to Y$ be a finite dominant morphism between irreducible affine varieties over an algebraically closed field F. Then σ is surjective.*

PROOF. Let y be a point of Y, and view y as an F-algebra homomorphism from $\mathscr{P}(Y)$ to F. Since σ is dominant, its transpose is injective from $\mathscr{P}(Y)$ to $\mathscr{P}(X)$. Consequently, y defines an F-algebra homomorphism y' from $\mathscr{P}(Y) \circ \sigma$ to F. By Proposition II.3.2, y' extends to an F-algebra homomorphism y'' from some valuation subring S of $[\mathscr{P}(X)]$ containing $\mathscr{P}(Y) \circ \sigma$ to F. Since $\mathscr{P}(X)$ is integral over $\mathscr{P}(Y) \circ \sigma$, we have $\mathscr{P}(X) \subset S$. If x is the restriction of y'' to $\mathscr{P}(X)$, then $\sigma(x) = y$. $\square$

Theorem 2.3. *Let R be an integral domain that is integrally closed in its field of fractions. Let S be an integral domain containing R and integral over R. Let Q be a prime ideal of S, and let P_0 be a prime ideal of R that is contained in Q. There is a prime ideal Q_0 in S such that $Q_0 \subset Q$ and $Q_0 \cap R = P_0$.*

PROOF. Let D denote the multiplicatively closed subset of S consisting of the products rs with r in $R \setminus P_0$ and s in $S \setminus Q$. First, we show that it suffices to prove that $(SP_0) \cap D = \varnothing$, where SP_0 denotes the ideal of S generated by P_0.

If this is the case, we can apply Zorn's Lemma in the evident way to show that there is an ideal Q_0 in S that is maximal in the set of all ideals of S containing SP_0 and not meeting D. We show that Q_0 is a prime ideal. Since Q_0 does not meet D, we have $Q_0 \neq S$. Let u and v be elements of $S \setminus Q_0$. Then, by the maximality of Q_0, both $Q_0 + Su$ and $Q_0 + Sv$ meet D. Therefore, $Q_0 + Suv$ also meets D, so that uv cannot lie in Q_0. Thus, Q_0 is a prime ideal of S. We have

$$P_0 \subset (SP_0) \cap R \subset Q_0 \cap R.$$

On the other hand,

$$(Q_0 \cap R)\backslash P_0 \subset Q_0 \cap D = \varnothing,$$

so that $Q_0 \cap R \subset P_0$. Thus, $Q_0 \cap R = P_0$. Finally,

$$Q_0 \subset S \backslash D \subset S \backslash (S \backslash Q) = Q,$$

so that Q_0 satisfies the requirements of Theorem 2.3.

It remains to be proved only that $(SP_0) \cap D = \varnothing$. Suppose this is false, and choose r from $R \backslash P_0$ and s from $S \backslash Q$ such that rs belongs to SP_0. Then

$$rs = \sum_i s_i p_i,$$

with each s_i in S and each p_i in P_0. Now s_i is a root of a monic polynomial with coefficients in R, whence $s_i p_i$ is a root of a monic polynomial whose coefficients other than the leading one lie in P_0. We show that the sum of two elements a and b having this property also has this property. Suppose m and n are the degrees of polynomials as described just above for a and b, respectively. Let $u_1, \ldots, u_{mn}$ be the products $a^i b^j$, with $0 \le i < m$ and $0 \le j < n$.

Then every monomial $a^e b^f$ with $e + f \ge m + n - 1$ is a linear combination with coefficients in P_0 of the u_k's. Hence, for $t = m + n - 1$ and $i = 1, \ldots, mn$,

$$(a + b)^t u_i = \sum_{j=1}^{mn} p_{ij} u_j,$$

with each p_{ij} in P_0. It follows that, if T is the determinant with entries $(a + b)^t \delta_{ij} - p_{ij}$, we have $Tu_i = 0$ for each i. Since one of the u_i's is 1, this gives $T = 0$, showing that $a + b$ has the property in question.

Now we conclude that rs is a root of a monic polynomial, h say whose coefficients other than the leading one lie in P_0. Let f be the monic minimum polynomial of rs with respect to $[R]$. Then $h = fg$, where g is a monic polynomial with coefficients in $[R]$. The roots (in an algebraic closure of $[R]$) of f and g are roots of h, and thus are integral over R. Therefore the coefficients of f and g are integral over R. Since R is integrally closed in $[R]$, it follows that the coefficients of f and g belong to R. We claim that these coefficients, except for the leading ones, actually belong to P_0.

In order to see this, let m and n denote the degrees of f and g, respectively, and write $f = x^m + f_1$, $g = x^n + g_1$, where f_1 is a polynomial of degree less than m, and g_1 is a polynomial of degree less than n. Indicating the canonical homomorphism from the polynomial ring $R[x]$ to the polynomial ring $(R/P_0)[x]$ by $'$, we have, from $h = fg$,

$$(x^m + (f_1)')(x^n + (g_1)') = x^{m+n},$$

whence

$$x^m(g_1)' + x^n(f_1)' + (f_1)'(g_1)' = 0.$$

This implies that $(f_1)'(g_1)' = 0$, because otherwise the product of the non-zero terms of lowest degree in $(f_1)'$ and $(g_1)'$ cannot cancel out. Moreover, the above relation shows that the vanishing of one of $(f_1)'$, $(g_1)'$ implies that of the other. Therefore, we have $(f_1)' = 0 = (g_1)'$, which establishes our claim.

Thus, we have

$$f = x^m + p_1 x^{m-1} + \cdots + p_m,$$

with each p_i in P_0. Now

$$x^m + \left(\frac{p_1}{r}\right) x^{m-1} + \cdots + \left(\frac{p_m}{r^m}\right),$$

is the monic minimum polynomial for s relative to $[R]$. Since s is integral over R, it follows, by a now familiar argument, that p_i/r^i belongs to R for each i. But $r^i(p_i/r^i)$ belongs to P_0, while r^i does not belong to P_0. Hence p_i/r^i belongs to P_0. But this gives $s^m \in SP_0 \subset Q$, contradicting the fact that s does not belong to Q. $\qquad\square$

Proposition 2.4. *Let $\sigma: X \to Y$ be a finite dominant morphism between affine varieties over an algebraically closed field. Let W be a closed irreducible subset of Y, and let Z be an irreducible component of $\sigma^{-1}(W)$. Then $\sigma(Z)$ is closed in Y, and $\dim(\sigma(Z)) = \dim(Z)$. For at least one such Z, one has $\sigma(Z) = W$. If X and Y are irreducible and $\mathscr{P}(Y)$ is integrally closed in its field of fractions then $\sigma(Z) = W$ for every irreducible component Z of $\sigma^{-1}(W)$.*

PROOF. Clearly, Z and the closure, $\sigma(Z)'$ say, of $\sigma(Z)$ are irreducible, and the restriction of σ to Z is a dominant morphism $\sigma_1: Z \to \sigma(Z)'$. Let J be the annihilator of Z in $\mathscr{P}(X)$, so that $\mathscr{P}(Z) = \mathscr{P}(X)/J$. Let I be the annihilator of $\sigma(Z)$ in $\mathscr{P}(Y)$. Then $\mathscr{P}(\sigma(Z)') = \mathscr{P}(Y)/I$, and $\mathscr{P}(\sigma(Z)') \circ \sigma_1$ may be identified with $(\mathscr{P}(Y) \circ \sigma)/(I \circ \sigma)$. Now $I \circ \sigma = J \cap (\mathscr{P}(Y) \circ \sigma)$, so that

$$\mathscr{P}(\sigma(Z)') \circ \sigma_1 = (\mathscr{P}(Y) \circ \sigma)/(J \cap (\mathscr{P}(Y) \circ \sigma)).$$

Since $\mathscr{P}(X)$ is integral over $\mathscr{P}(Y) \circ \sigma$, this shows that $\mathscr{P}(Z)$ is integral over $\mathscr{P}(\sigma(Z)') \circ \sigma_1$, i.e., that σ_1 is a finite morphism. Now we have from Proposition 2.2 that σ_1 is surjective, which means that $\sigma(Z)$ is closed in Y. From the fact that $\mathscr{P}(Z)$ is integral over $\mathscr{P}(\sigma(Z)) \circ \sigma_1$, it is clear that $\dim(\sigma(Z)) = \dim(Z)$.

Now let P be the annihilator of W in $\mathscr{P}(Y)$, and Q the annihilator of $\sigma^{-1}(W)$ in $\mathscr{P}(X)$. Then $P \circ \sigma = Q \cap (\mathscr{P}(Y) \circ \sigma)$. Write $Q = J_1 \cap \cdots \cap J_k$, where the J_i's are the annihilators in $\mathscr{P}(X)$ of the irreducible components, Z_i say, of $\sigma^{-1}(W)$. Then $P \circ \sigma$ is the intersection of the family of ideals $J_i \cap (\mathscr{P}(Y) \circ \sigma)$. Since $P \circ \sigma$ is a prime ideal, there is an index i such that $P \circ \sigma = J_i \cap (\mathscr{P}(Y) \circ \sigma)$. We claim that $W = \sigma(Z_i)$. In order to see this, let g be an element of $\mathscr{P}(Y)$ and suppose that $g(\sigma(Z_i)) = (0)$. Then $g \circ \sigma$ belongs to $J_i \cap (\mathscr{P}(Y) \circ \sigma)$, i.e., $g \circ \sigma \in P \circ \sigma$. Since σ is dominant, it follows that g belongs to P, i.e., that $g(W) = (0)$. Since $\sigma(Z_i)$ is closed in Y, this shows that $W = \sigma(Z_i)$.

Now suppose that X and Y are irreducible and that $\mathscr{P}(Y)$ is integrally closed in its field of fractions. For each i, we have

$$P \circ \sigma = Q \cap (\mathscr{P}(Y) \circ \sigma) \subset J_i \cap (\mathscr{P}(Y) \circ \sigma).$$

By Theorem 2.3, there is a prime ideal T in $\mathscr{P}(X)$ such that $T \subset J_i$ and $T \cap (\mathscr{P}(Y) \circ \sigma) = P \circ \sigma$. Now $\sigma(\mathscr{V}(T)) \subset W$, so that $\mathscr{V}(T)$ is an irreducible subset of $\sigma^{-1}(W)$. Since $Z_i \subset \mathscr{V}(T)$, we must therefore have $\mathscr{V}(T) = Z_i$, whence $T = J_i$. Thus we have $J_i \cap (\mathscr{P}(Y) \circ \sigma) = P \circ \sigma$ for each i. As we have seen above, this implies that $\sigma(Z_i) = W$ for each i. $\qquad\square$

3. Lemma 3.1. *Let A be a Noetherian integral domain that is integrally closed in its field of fractions $[A]$. Let L be a finite separable algebraic field extension of $[A]$. Then the integral closure of A in L is Noetherian as an A-module.*

PROOF. Let τ denote the trace map $L \to [A]$. Since L is separable over $[A]$, the $[A]$-bilinear trace form $(u, v) \mapsto \tau(uv)$ on $L \times L$ is non-degenerate. Let A^L denote the integral closure of A in L. Clearly, A^L contains an $[A]$-basis $(u_1, \ldots, u_n)$ of L. Because of the non-degeneracy of the trace form, we can find elements $t_1, \ldots, t_n$ in L such that $\tau(u_i t_j) = \delta_{ij}$ for all i and j. Evidently, $(t_1, \ldots, t_n)$ is also an $[A]$-basis of L.

Now let u be any element of A^L, and write

$$u = \sum_{i=1}^{n} a_i t_i,$$

with each a_i in $[A]$. Multiplying by u_j and then applying τ, we find $a_i = \tau(uu_j)$. Since uu_j belongs to A^L, this shows that a_i is integral over A. Since A is integrally closed in $[A]$, we have therefore $a_i \in A$. Thus, A^L is contained in $At_1 + \cdots + At_n$. Since A is Noetherian, it follows that A^L is Noetherian as an A-module. $\qquad\square$

Theorem 3.2. *Let F be a field, A a finitely generated integral domain F-algebra. Let L be a finite algebraic field extension of $[A]$. Then the integral closure of A in L is Noetherian as an A-module.*

PROOF. By Theorem 1.2, there is a finite subset $(z_1, \ldots, z_s)$ of A that is algebraically free over F and such that A is integral over $F[z_1, \ldots, z_s]$. Write B for $F[z_1, \ldots, z_s]$, and note that the integral closure B^L of B in L coincides with the integral closure A^L of A. If B^L is Noetherian as a B-module then, a fortiori, it is finitely generated as an A-module, and therefore Noetherian as an A-module, because A is Noetherian. Therefore, it suffices to prove Theorem 3.2 in the case where A is an ordinary polynomial algebra $F[x_1, \ldots, x_n]$. We assume this from here on.

Let L' be an algebraic closure of L. Then L' contains a purely inseparable algebraic field extension P of $[A]$ such that P is a perfect field. Let $(y_1, \ldots, y_m)$ be a set of field generators for L over $[A]$. The coefficients of the monic minimum polynomials of the y_j's relative to P generate a finite purely inseparable algebraic field extension S of $[A]$, and the subfield $S[L]$ of L' is a finite separable algebraic field extension of S. Write $S = [A][u_1, \ldots, u_k]$. There is a power, q say, of the characteristic of F such that u_j^q belongs to $[A]$ for each j; if the characteristic of F is 0, we define q as 1.

Now each u_j^q is a fraction formed with two polynomials in the x_i's with coefficients in F. Let $a_1, \ldots, a_s$ be all these coefficients. Working in L', put

$$C = A[x_1^{1/q}, \ldots, x_n^{1/q}, a_1^{1/q}, \ldots, a_s^{1/q}].$$

Then C is integral over A. Let K denote the field extension of F that is generated by the elements $a_i^{1/q}$. Then $(x_1^{1/q}, \ldots, x_n^{1/q})$ is algebraically free over K, and C is the polynomial algebra over K that is generated by this set. Hence, C is integrally closed in its field of fractions, which is

$$[C] = [A][x_1^{1/q}, \ldots, x_n^{1/q}, a_1^{1/q}, \ldots, a_s^{1/q}].$$

Clearly, $S \subset [C]$, and the integral closure A^S of A in S is contained in C. Since C is finitely generated as an A-module and A is Noetherian, A^S is therefore Noetherian as an A-module. A fortiori, A^S is a Noetherian ring. Moreover, A^S is integrally closed in its field of fractions, because this field of fractions coincides with S.

Now we can apply Lemma 3.1 to conclude that the integral closure of A^S in the finite separable field extension $S[L]$ of S is Noetherian as an A^S-module. Since A^S is Noetherian as an A-module, it follows that the integral closure of A^S in $S[L]$ is Noetherian also as an A-module. Evidently, this integral closure contains A^L. Since A is Noetherian, it follows that A^L is Noetherian as an A-module. $\qquad\Box$

If R and S are commutative rings with $R \subset S$ then the *conductor* $\mathscr{C}(R, S)$ *of R in S* is defined as the largest ideal of S that is contained in R. It consists of the elements r of R for which $rS \subset R$.

Let us say that an integral domain A is *normal* if it is integrally closed in its field of fractions. If A is normal, and S is a multiplicatively closed subset of A, then $A[S^{-1}]$ is normal, as is easy to verify. Conversely, suppose that the integral closure A^* of A in $[A]$ is finitely generated as an A-module and that $A[(A \setminus P)^{-1}]$ is normal for every maximal ideal P of A. Then A is normal. Indeed, otherwise we have $\mathscr{C}(A, A^*) \neq A$, so that there is a maximal ideal P in A containing $\mathscr{C}(A, A^*)$. Since $A[(A \setminus P)^{-1}]$ is normal and A^* is finitely generated as an A-module, there is an element u in $A \setminus P$ such that $uA^* \subset A$. But this means that u belongs to $\mathscr{C}(A, A^*)$ and hence to P, so that we have a contradiction.

Let X be a variety, p a point of X. The *local ring at p* is the direct limit of the system of rings $\mathscr{F}_X(U)$, where U ranges over the open subsets of X containing p, with respect to the restriction maps $\mathscr{F}_X(U) \to \mathscr{F}_X(V)$ for the pairs (U, V) with $V \subset U$. This is also called the *stalk at p* of the sheaf of regular functions, and is denoted by $\mathscr{F}_X(p)$. Its elements are the equivalence classes of functions regular on a neighborhood of p, two such functions being called equivalent if their restrictions to some neighborhood of p coincide. If U is an affine patch of X containing p, and if P is the annihilator of p in $\mathscr{P}(U)$, then $\mathscr{F}_X(p)$ may evidently be identified with the ring of fractions $A[(A \setminus P)^{-1}]$, where $A = \mathscr{P}(U)$. The point p of X is called a *normal point* if $\mathscr{F}_X(p)$ is normal. The variety X is said to be a *normal variety* if every point of X is normal.

Let X be an irreducible affine variety. By Theorem 3.2, the integral closure of $\mathscr{P}(X)$ in its field of fractions is Noetherian as a $\mathscr{P}(X)$-module. If the base field is algebraically closed then every maximal ideal of $\mathscr{P}(X)$ is the annihilator of a point of X. Hence we see from our above discussion of normal integral domains that *an irreducible affine variety X over an algebraically closed field is normal if and only if $\mathscr{P}(X)$ is normal.* Of course, the sufficiency of this condition holds even if the base field is not algebraically closed.

Let X be an irreducible affine variety, and write A for $\mathscr{P}(X)$. Let A^* denote the integral closure of A in $[A]$. Let $(u_1, \ldots, u_n)$ be a system of A-module generators for A^*. There is a non-zero element a in A such that au_i belongs to A for each i, whence $aA^* \subset A$, i.e., $a \in \mathscr{C}(A, A^*)$. Consider the principal open subset $X_a = \mathscr{S}(A[a^{-1}])$ of X. We have

$$(A[a^{-1}])^* = A^*[a^{-1}].$$

Since $aA^* \subset A$, we have $A^* \subset A[a^{-1}]$, whence $(A[a^{-1}])^* = A[a^{-1}]$. Thus, X_a is a normal variety.

We summarize as follows, for reference.

Proposition 3.3. *Let V be a variety, and let $X_1, \ldots, X_n$ be affine patches of V covering V. If $\mathscr{P}(X_i)$ is integrally closed in its field of fractions for each i then V is a normal variety. The converse holds whenever the base field is algebraically closed. Every variety has a non-empty normal open subvariety.*

4. Proposition 4.1. *Let A and B be commutative rings, with $A \subset B$. Suppose that f is an element of the polynomial ring $B[x]$ that is integral over $A[x]$. Then the coefficients of f are integral over A.*

PROOF. By assumption, there are elements $p_1, \ldots, p_m$ in $A[x]$ such that

$$f^m + p_1 f^{m-1} + \cdots + p_m = 0.$$

Let r be an integer greater than the maximum of the degrees of f and the p_i's. Put $g = f - x^r$. In the above, substitute $g + x^r$ for f and write the resulting relation for g over $A[x]$ in the form

$$(-g)[g^{m-1} + q_1 g^{m-2} + \cdots + q_{m-1}] = q_m,$$

where the q_i's are elements of $A[x]$. By the choice of r, the polynomial $-g$ is monic. On the other hand,

$$q_m = x^{rm} + p_1 x^{r(m-1)} + \cdots + p_{m-1} x^r + p_m,$$

showing that q_m is monic. Therefore, the above relation shows that

$$g^{m-1} + q_1 g^{m-2} + \cdots + q_{m-1}$$

is monic.

Generally, if u and v are monic polynomials in $B[x]$ such that the coefficients of uv are integral over A, then the coefficients of u and v are integral over A. This is proved by constructing a commutative ring C containing B such that u and v are products of polynomials $x - c$ with c in C. Then, since the coefficients of uv are integral over A, so are all these c's, whence the same holds for the coefficients of u and v.

Applying this to the above, since the coefficients of q_m lie in A, we see that the coefficients of g are integral over A, which means that the coefficients of f are integral over A. $\qquad\square$

Lemma 4.2. *Let A be an integral domain, B a subring of A such A is finitely generated as a B-algebra. Then there is a finite subset $(x_1, \ldots, x_r)$ of A that is algebraically free over $[B]$, and a non-zero element b of B, such that $A[b^{-1}]$ is integral over $B[b^{-1}][x_1, \ldots, x_r]$.*

PROOF. Let A' be the sub $[B]$-algebra of $[A]$ that is generated by A. Then A' is a finitely generated $[B]$-algebra to which we can apply Theorem 1.2. This yields a finite subset $(x_1, \ldots, x_r)$ of A' that is algebraically free over $[B]$ and such that A' is integral over $[B][x_1, \ldots, x_r]$. Each x_i is a fraction a_i/b_i where a_i lies in A and b_i in B. Multiplying each x_i by b_i does not disturb the relevant properties. Thus, we see that the x_i's may be chosen from A. Since A is finitely generated as a B-algebra, and since every element of A is a zero of a monic polynomial with coefficients in $[B][x_1, \ldots, x_r]$, we can clearly find a non-zero element b in B such that $A[b^{-1}]$ is integral over $B[b^{-1}][x_1, \ldots, x_r]$. $\qquad\square$

Theorem 4.3. *Let F be an algebraically closed field, and let $\sigma\colon X \to Y$ be a dominant morphism between irreducible varieties over F. Then $\sigma(X)$ contains a non-empty open subset U of Y. Moreover, U may be so chosen that, if W is a closed irreducible subset of Y meeting U, and Z is an irreducible component of $\sigma^{-1}(W)$ meeting $\sigma^{-1}(U)$, then $\dim(Z) = \dim(W) + \dim(X) - \dim(Y)$.*

PROOF. Appealing to Proposition 3.3, we reduce the theorem to the case where Y is a normal variety. As in the beginning of the proof of Theorem 2.1, we see that no generality is lost in assuming that Y is affine. Now write X as a union of affine patches $X_1, \ldots, X_n$. Then, if U_i is an open subset of Y satisfying the requirements of the theorem for the restriction of σ to X_i, the intersection of the family of U_i's satisfies the requirements for σ. Therefore, it suffices to deal with the case where X is affine. Accordingly, we assume that both X and Y are affine, and that Y is normal.

Write A for $\mathcal{P}(X)$, and B for $\mathcal{P}(Y) \circ \sigma$. Let $(x_1, \ldots, x_r)$ and b be as obtained in Lemma 4.2, so that $A[b^{-1}]$ is integral over the polynomial algebra $B[b^{-1}][x_1, \ldots, x_r]$. We have $b = c \circ \sigma$, with c in $\mathcal{P}(Y)$, and

$$B[b^{-1}] = \mathcal{P}(Y_c) \circ \sigma.$$

We shall show that Y_c satisfies the requirements for U in Theorem 4.3.

We have $\sigma^{-1}(Y_c) = X_b$, and $\mathcal{P}(X_b) = A[b^{-1}]$. Let us regard the polynomial algebra $F[x_1, \ldots, x_r]$ as the algebra $\mathcal{P}(F^r)$ of polynomial functions on the direct product F^r of r copies of F, with its standard structure of an affine F-variety. Consider the affine F-variety $Y_c \times F^r$. We have

$$\mathcal{P}(Y_c \times F^r) = \mathcal{P}(Y_c) \otimes F[x_1, \ldots, x_r].$$

Using the composition with σ on the factor $\mathcal{P}(Y_c)$, we obtain an F-algebra injection $\mathcal{P}(Y_c \times F^r) \to \mathcal{P}(X_b)$ whose image is $B[b^{-1}][x_1, \ldots, x_r]$. This defines a dominant morphism $\tau: X_b \to Y_c \times F^r$. Since $\mathcal{P}(X_b)$ is integral over $B[b^{-1}][x_1, \ldots, x_r]$, this morphism τ is a finite morphism. By Proposition 2.2, τ is therefore surjective. Clearly, the restriction of σ to X_b is the composite of τ with the canonical projection $Y_c \times F^r \to Y_c$. Hence

$$Y_c = \sigma(X_b) \subset \sigma(X).$$

Now let W and Z be as in the statement of Theorem 4.3. We may replace W with $W \cap Y_c$, and Z with $Z \cap X_b$, and σ with its restriction to X_b. Thus, we may assume that $W \subset Y_c$ and that Z is an irreducible component of $\sigma^{-1}(W)$ in $X_b = \sigma^{-1}(Y_c)$. Then Z is an irreducible component of $\tau^{-1}(W \times F^r)$. Now $W \times F^r$ is a closed irreducible subset of $Y_c \times F^r$. The integral closure of $\mathcal{P}(Y_c \times F^r)$ in its field of fractions is contained in

$$[\mathcal{P}(Y_c)][x_1, \ldots, x_r].$$

By Proposition 4.1, the elements of this integral closure, when written as polynomials in $x_1, \ldots, x_r$, have their coefficients integral over $\mathcal{P}(Y_c)$. Since Y is a normal variety, so is Y_c. Therefore, we have from Proposition 3.3 that $\mathcal{P}(Y_c)$ is integrally closed in $[\mathcal{P}(Y_c)]$. Thus, $\mathcal{P}(Y_c \times F^r)$ is integrally closed in its field of fractions. Now we can apply Proposition 2.4 to conclude that $\tau(Z) = W \times F^r$ and $\dim(\tau(Z)) = \dim(Z)$. Therefore, we have $\dim(Z) = \dim(W) + r$. Finally, it is clear that r is equal to the transcendence degree of $[\mathcal{P}(X)]$ over $[\mathcal{P}(Y) \circ \sigma]$, so that $r = \dim(X) - \dim(Y)$. $\qquad\square$

Recall from Section IX.1 that a constructible subset of a variety is a finite union of locally closed subsets, and that it inherits a variety structure in the natural way.

Theorem 4.4. *Let $\sigma: X \to Y$ be a morphism between varieties over an algebraically closed field. If A is a constructible subset of X then $\sigma(A)$ is a constructible subset of Y.*

PROOF. Since a constructible subset of X is a subvariety, it suffices to prove that $\sigma(X)$ is a constructible subset of Y. Moreover, no generality is lost in assuming that X is irreducible. Hence, it suffices to prove that $\sigma(X)$ is constructible in the case where both X and Y are irreducible and σ is dominant.

We assume this and proceed by induction on the dimension of Y. The result is trivial if $\dim(Y) = 0$. Suppose $\dim(Y) > 0$, and that the theorem has been established in the lower cases. By the first part of Theorem 4.3, $\sigma(X)$ contains a non-empty open subset U of Y. Let $W_1, \ldots, W_t$ be the irreducible components of $Y \setminus U$; evidently, we may assume that $U \neq Y$. By Proposition 1.3, each W_i is of strictly smaller dimension than Y. Let $Z_{i1}, \ldots, Z_{ii'}$, be the irreducible components of $\sigma^{-1}(W_i)$. By our inductive hypothesis, each $\sigma(Z_{ij})$ is a constructible subset of W_i, and hence also is a constructible subset of Y. Now $\sigma(X)$ is the union of U and the $\sigma(Z_{ij})$'s, and thus is constructible. $\qquad\square$

Theorem 4.5. *Let $\sigma: X \to Y$ be a dominant morphism between irreducible varieties over an algebraically closed field. Suppose that, for every closed irreducible subset W of Y, each irreducible component of $\sigma^{-1}(W)$ has dimension $\dim(W) + \dim(X) - \dim(Y)$. Then σ is an open map.*

PROOF. If we apply the assumption with W any 1-point subset of Y, we find that σ is surjective. Now let W be any closed irreducible subset of Y, and let $Z_1, \ldots, Z_k$ be the irreducible components of $\sigma^{-1}(W)$. Put

$$r = \dim(X) - \dim(Y).$$

Indicating the closure of a set by $'$, note that Z_i is also an irreducible component of $\sigma^{-1}(\sigma(Z_i)')$, so that the assumption of the theorem gives

$$\dim(\sigma(Z_i)') + r = \dim(Z_i) = \dim(W) + r,$$

whence $\dim(\sigma(Z_i)') = \dim(W)$, and therefore $\sigma(Z_i)' = W$. Thus, the restriction of σ to Z_i is dominant from Z_i to W.

Now let x be a point of X, and let U be an open subset of X containing x. We must show that $\sigma(x)$ lies in the interior of $\sigma(U)$. Suppose this is not the case, so that $\sigma(x) \in (Y \setminus \sigma(U))'$. By Theorem 4.4, $\sigma(U)$ is a constructible subset of Y. Therefore, also $Y \setminus \sigma(U)$ is constructible. Hence, there is a closed set C in Y and an open set E in Y such that $\sigma(x) \in (C \cap E)'$ and

$$C \cap E \subset Y \setminus \sigma(U).$$

Moreover, we may choose C to be irreducible, so that $C \cap E$ is dense in C. We know from the above that the irreducible components of $\sigma^{-1}(C)$ all dominate C, via σ. It follows that $\sigma^{-1}(E)$ meets each of these components, whence $\sigma^{-1}(E) \cap \sigma^{-1}(C)$ is dense in $\sigma^{-1}(C)$. On the other hand,

$$\sigma^{-1}(E) \cap \sigma^{-1}(C) = \sigma^{-1}(E \cap C) \subset \sigma^{-1}(Y \setminus \sigma(U)) \subset X \setminus U.$$

Since $X \setminus U$ is closed, it follows that $\sigma^{-1}(C) \subset X \setminus U$. Since $x \in \sigma^{-1}(C) \cap U$, this is a contradiction. $\qquad\square$

5. We append three purely technical results which will be needed when we return to algebraic groups.

Proposition 5.1. *Let X be an irreducible variety, x a normal point of X. Let f be a rational function on X that is not defined at x. There is a closed subvariety Y of X such that $x \in Y$, $1/f$ is defined at some point of Y and $1/f$ takes the value 0 at every point of Y where it is defined.*

PROOF. Let R denote the local ring of x. Without loss of generality, we assume that X is affine. Then we may identify R with the appropriate ring of fractions of $\mathscr{P}(X)$, and we may regard f as an element of $[\mathscr{P}(X)]$. Working in this field, let I denote the ideal of R consisting of the elements g for which $gf \in R$. Since R is a Noetherian ring, it follows from elementary ideal theory that the set of prime ideals of R that are minimal among the prime ideals containing I is finite (see Note 3 at the end of this chapter). Let $P_1, \ldots, P_t$ be all the prime ideals minimal prime over I. Their intersection is the radical of I, and there is a positive integer n such that $P_1^n \cdots P_t^n \subset I$. Write R_1 for the local ring $R[(R \setminus P_1)^{-1}]$. Then we have $P_i R_1 = R_1$ for each $i > 1$. It follows that $P_1^n R_1 \subset IR_1$. Since $If \subset R$, we have therefore $P_1^n f \subset R_1$. Let k be the smallest non-negative integer m for which $P_1^m f \subset R_1$. If we had $k = 0$, we would have $f \in R_1$, so that there would be an element g in $R \setminus P_1$ with $gf \in R$, giving $g \in I$, which contradicts $I \subset P_1$. Thus, we have $k > 0$.

Now choose an element g from $(P_1^{k-1} f) \setminus R_1$. Since x is a normal point, R is integrally closed in $[R]$, whence also R_1 is integrally closed in $[R]$. Therefore, g is not integral over R_1. It follows that $gP_1 R_1 \not\subset P_1 R_1$. In order to see this, note that $P_1 R_1$ has a finite system $(u_1, \ldots, u_r)$ of R_1-module generators. If $gP_1 R_1 \subset P_1 R_1$ we have

$$gu_i = \sum_{j=1}^{r} c_{ij} u_j,$$

with c_{ij} in R_1. Hence the determinant of the matrix with entries $g\delta_{ij} - c_{ij}$ is equal to 0, showing that g is integral over R_1; a contradiction.

On the other hand, we have $gP_1 \subset P_1^k f \subset R_1$. Since, as we have just seen, $(gP_1)R_1 \not\subset P_1 R_1$, and since $P_1 R_1$ is the maximum ideal of the local ring R_1, it follows that $gP_1 R_1 = R_1$, so that $R_1(1/g) = P_1 R_1$.

Now put $h = f/g^k$. Then $h \in P_1^k f \subset R_1$. Moreover, h is a unit of R_1. Indeed, otherwise $h \in P_1 R_1 = R_1(1/g)$, so that $f/g^{k-1} = gh \in R_1$, whence $fP_1^{k-1} \subset R_1$, contradicting the definition of k. Hence, we have

$$1/f = (1/h)(1/g^k) \in P_1 R_1.$$

Let P be the prime ideal $P_1 \cap \mathscr{P}(X)$ of $\mathscr{P}(X)$, and let Y be the set of zeros of P in X. Then Y is an irreducible closed subvariety of X, and $x \in Y$ (because P_1 is contained in the maximum ideal of the local ring R at x). Now $1/f$ defines a rational function on Y. Since $1/f \in P_1 R_1$, it is clear that $(1/f)(y) = 0$ for every point y of Y at which $1/f$ is defined. The elements of R_1 may be written as fractions u/v, with u and v in $\mathscr{P}(X)$ and $v \notin P$. There are points y in Y with $v(y) \neq 0$, i.e., points y at which u/v is defined. In particular, $1/f$ is defined at some point of Y. $\qquad\square$

Proposition 5.2. *Let $\sigma: X \to Y$ be an injective and dominant morphism between irreducible varieties over an algebraically closed field F. Then the field of rational functions of X is a finite purely inseparable algebraic extension of the image, under the transpose of σ, of the field of rational functions of Y.*

PROOF. Applying Theorem 2.1 with a 1-point subset of Y in the place of W and using that σ is injective, we see that we must have $\dim(X) = \dim(Y)$. Therefore the field of rational functions of X is a finite algebraic extension of the image of the field of rational functions of Y. There is an affine patch U of X such that $\sigma(U)$ is contained in some affine patch V of Y. Let $A = \mathscr{P}(U)$, $B = \mathscr{P}(V) \circ \sigma$. Let a be an element of A, and suppose γ and δ are F-algebra homomorphisms $A \to F$ whose restrictions to B coincide. Then γ and δ are points of U whose images, under σ, in Y coincide. Since σ is injective, we have $\gamma = \delta$. Now we can apply Proposition III.2.4 to conclude that a is purely inseparably algebraic over $[B]$. Since the field of rational functions of X may be identified with $[A]$ and, compatibly, the image of the field of rational functions of Y with $[B]$, this is the required result. $\qquad\square$

Proposition 5.3. *Let $\rho: X \to Y$ be a morphism between irreducible varieties whose transpose is an isomorphism of the field of rational functions of Y onto the field of rational functions of X. Then there is an open non-empty subset U of Y such that ρ induces an isomorphism of varieties from $\rho^{-1}(U)$ to U.*

PROOF. Without loss of generality, we assume that Y is affine. Let V be an affine patch of X, and let W denote the closure of $\rho(X \setminus V)$ in Y. Since the irreducible components of $X \setminus V$ are of dimension $< \dim(X)$, the irreducible components of W are of dimension $< \dim(Y)$. Therefore, there is a non-zero element f in $\mathscr{P}(Y)$ such that $f(W) = (0)$. Now we have

$$\rho^{-1}(Y_f) = X_{f \circ \rho} \subset V.$$

This shows that no generality is lost in assuming that both X and Y are affine.

Then, by assumption, the transpose ρ° of ρ maps $[\mathscr{P}(Y)]$ isomorphically onto $[\mathscr{P}(X)]$. Write $\mathscr{P}(X) = F[f_1, \ldots, f_n]$, where F is the base field and $f_i = (g_i \circ \rho)/(h \circ \rho)$, with g_i and h in $\mathscr{P}(Y)$. Then ρ° maps $\mathscr{P}(Y)[h^{-1}]$ isomorphically onto $\mathscr{P}(X)[(h \circ \rho)^{-1}]$, so that ρ induces an isomorphism from $X_{h \circ \rho}$ to Y_h. $\qquad\square$

Notes

1. The dimension-theoretical analysis of morphisms, as presented here, comes from [11].

2. Theorem 4.5 and Proposition 5.1 are due to C. Chevalley.

3. In proving Proposition 5.1, we have used the fact that, in a Noetherian commutative ring R, the set of prime ideal minimal over a proper ideal, J say, is finite. It suffices to show that the radical of J is the intersection of a *finite* family of prime ideals. Suppose this is false. From the family of ideals that are intersections of families of prime ideals, but not of *finite* such families, choose a maximal one, Q say. There are elements a and b in $R \setminus Q$ such that ab belongs to Q. Let A be the radical of $Q + Ra$, and let B be the radical of $Q + Rb$. It is easy to see that $A \cap B = Q$, and this gives a contradiction.

Chapter XI

Local Theory

The content of this chapter is the dimension theory of local rings, and its application to the investigation of tangent spaces to varieties and local properties of morphisms. The resulting technique enables us to use Lie algebras in dealing with coset varieties arising from algebraic groups later on.

Section 1 deals with general results centering around the notion of length of modules. In Section 2, these results are used for defining the "characteristic degree" of a Noetherian local ring. With this, it is shown that, for such a ring, the Krull dimension is equal to the parametric dimension. This equality connects the maximum length of chains of prime ideals with the minimum cardinality of systems of generators for ideals whose radical is the maximum ideal.

The main result of Section 3 is the equality of the Krull dimension with the degree of transcendence of the field of fractions for finitely generated integral domain algebras over a field.

Section 4 contains the background in local ring theory for the notion of singular point of a variety. The basic results concerning singular points are established in Section 5.

In the varieties arising directly from algebraic groups, all points are non-singular, which has the effect that the local behavior of a morphism reveals its properties with regard to the dimensionalities of its image and its fibers. The general auxiliary results used later on for exploiting this are developed in Section 6.

1. We begin with preparations for the dimension theory of local rings. The first lemma is known as the Artin–Rees Lemma.

Lemma 1.1. *Let R be a commutative Noetherian ring, M a finitely generated R-module, N a sub R-module of M, and J an ideal of R. There is a non-negative integer k such that, for every $n \geq k$,*

$$(J^n \cdot M) \cap N = J^{n-k} \cdot ((J^k \cdot M) \cap N).$$

PROOF. Let $(a_1, \ldots, a_r)$ be a system of R-module generators of J, let t be an auxiliary variable over R, and let S be the sub R-algebra $R[a_1 t, \ldots, a_r t]$ of the polynomial algebra $R[t]$. Since R is Noetherian and S is finitely generated as an R-algebra, S is Noetherian.

We imbed the given R-module M in the $R[t]$-module $M[t] = \sum_{n \geq 0} M t^n$ in the evident fashion, and we consider the sub S-module $S \cdot M$ of $M[t]$. Clearly, every system of R-module generators of M is also a system of S-module generators of $S \cdot M$, so that $S \cdot M$ is finitely generated as an S-module. Since S is a Noetherian ring, $S \cdot M$ is therefore Noetherian as an S-module.

Let $(f_1, \ldots, f_q)$ be a system of S-module generators for $(S \cdot M) \cap N[t]$, and let k be the maximum of the degrees of the f_i's. Suppose x is an element of $(J^n \cdot M) \cap N$, where $n \geq k$. Then $x t^n$ belongs to $(S \cdot M) \cap N[t]$, so that there are elements s_i in S such that

$$x t^n = s_1 \cdot f_1 + \cdots + s_q \cdot f_q.$$

The coefficient of t^n in $s_i \cdot f_i$ is

$$\sum_{e=0}^{k} s_i(n-e) f_i(e),$$

where $s_i(n-e)$ is the coefficient of t^{n-e} in s_i and $f_i(e)$ is the coefficient of t^e in f_i. This shows that x belongs to $J^{n-k} \cdot ((J^k \cdot M) \cap N)$, so that

$$(J^n \cdot M) \cap N \subset J^{n-k} \cdot ((J^k \cdot M) \cap N).$$

The reversed inclusion relation is evident. $\square$

Proposition 1.2. *Let R be a Noetherian commutative ring, M a finitely generated R-module, J an ideal of R. Suppose that, for every a in J and every non-zero element x of M, we have $(1 + a) \cdot x \neq 0$. Then $\bigcap_{n>0} J^n \cdot M = (0)$.*

PROOF. Put $N = \bigcap_{n>0} J^n \cdot M$. Then we have $(J^m \cdot M) \cap N = N$ for every non-negative integer m, so that Lemma 1.1 gives $N = J^{n-k} \cdot N$ for all $n \geq k$. Thus $N = J \cdot N$.

Now let $(x_1, \ldots, x_q)$ be a system of R-module generators of N. Then we have

$$x_i = \sum_{j=1}^{q} c_{ij} x_j,$$

with c_{ij} in J. This gives $D \cdot x_j = 0$ for each j, where D is the determinant of the matrix with entries $\delta_{ij} - c_{ij}$. Since D is of the form $1 + a$, with a in J, our assumption implies that each $x_j = 0$, so that $N = (0)$. $\square$

Let R be an arbitrary ring, and let M be an R-module having a finite composition series

$$(0) = M_0 \subset \cdots \subset M_n = M.$$

By the Jordan-Hoelder Theorem, the *length n* of such a series is determined by M. We call it the *R-length* of M and denote it by $L_R(M)$.

A ring R is called an *Artin ring* if it satisfies the minimal condition for left ideals. This implies that R also satisfies the maximal condition for left ideals. We shall not use this result, but we shall *define* an Artin ring as a ring satisfying both the minimal and the maximal condition for left ideals. Familiar elementary arguments show that *R is an Artin ring if and only if every finitely generated R-module has a finite composition series.*

Let A be an Artin ring, and consider the polynomial ring $A[x_1, \ldots, x_q]$, where the x_i's are variables, central in $A[x_1, \ldots, x_q]$. We regard this as a graded ring, the elements of A being of degree 0, while each x_i is of degree 1.

Theorem 1.3. *Let $A[x_1, \ldots, x_q]$ be as described just above, and let M be a finitely generated graded $A[x_1, \ldots, x_q]$-module. There is a polynomial P_M of degree strictly less than q, with rational coefficients, such that, for all sufficiently large n's, $L_A(M_n) = P_M(n)$, where M_n is the component of degree n of M.*

PROOF. If $q = 0$ then, since M is finitely generated, we have $M_n = (0)$ for all sufficiently large n's, and P_M is the zero polynomial. Now we suppose that $q > 0$, and that the theorem has been established in the cases of fewer than q variables.

The endomorphism of M corresponding to x_q yields a morphism of A-modules from M_n to M_{n+1}, whose kernel is K_n, where K is the kernel of the endomorphism corresponding to x_q. This gives an exact sequence of morphisms of A-modules

$$(0) \to K_n \to M_n \to M_{n+1} \to M_{n+1}/x_q \cdot M_n \to (0).$$

The alternating sum of the lengths of the terms of an exact sequence must be 0, whence we obtain

$$L_A(M_{n+1}) - L_A(M_n) = L_A(M_{n+1}/x_q \cdot M_n) - L_A(K_n).$$

Now we can apply our inductive hypothesis to the graded $A[x_1, \ldots, x_{q-1}]$-modules $M/(x_q \cdot M)$ and K. This gives the existence of polynomials $P_{M\,(x_q \cdot M)}$ and P_K of degree strictly less than $q - 1$ such that, for all sufficiently large n's,

$$L_A(M_{n+1}) - L_A(M_n) = P_{M/(x_q \cdot M)}(n + 1) - P_K(n).$$

As a function of n, the expression on the right is a polynomial of degree strictly less than $q - 1$. It follows that, for all sufficiently large n's, $L_A(M_n)$ is given by a polynomial of degree strictly less than q. $\qquad\square$

Theorem 1.4. *Let B be a Noetherian local ring, $m(B)$ its maximum ideal, J an ideal such that $m(B)^e \subset J \subset m(B)$ for some positive e. Let q denote the minimum number of B-module generators of J, and let E be a finitely generated B-module. There is a polynomial $P_{E,J}$ of degree no greater than q, with rational coefficients, such that, for all sufficiently large n's, $L_B(E/J^n \cdot E) = P_{E,J}(n)$.*

PROOF. Consider the filtration of B by the powers of J, and let $G(B)$ denote the associated graded ring $\sum_{n \geq 0} J^n/J^{n+1}$. Similarly, let $G(E)$ denote the graded $G(B)$-module $\sum_{n \geq 0} (J^n \cdot E)/(J^{n+1} \cdot E)$. Since E is finitely generated as a B-module, $G(E)$ is finitely generated as a $G(B)$-module.

Now choose a system $(b_1, \ldots, b_q)$ of B-module generators of J. There is a surjective morphism of graded rings from the polynomial ring

$$(B/J)[x_1, \ldots, x_q]$$

to $G(B)$ sending each x_i onto the element $b_i + J^2$ of J/J^2 and coinciding with the injection on B/J. Using this, we obtain the structure of a graded $(B/J)[x_1, \ldots, x_q]$-module on $G(E)$, for which $G(E)$ is clearly still finitely generated.

We show that B/J is an Artin ring, so that we may then apply Theorem 1.3. Since B/J is a homomorphic image of $B/m(B)^e$, it will suffice to show that $B/m(B)^t$ is an Artin ring for every positive exponent t. Evidently, this is true for $t = 1$, because $B/m(B)$ is a field. Suppose we have already shown that $B/m(B)^t$ is an Artin ring for some t. Consider the canonical exact sequence of morphisms of B-modules

$$(0) \to m(B)^t/m(B)^{t+1} \to B/m(B)^{t+1} \to B/m(B)^t \to (0).$$

The sub B-modules of $m(B)^t/m(B)^{t+1}$ are the subspaces for the vector space structure over the field $B/m(B)$. Since $m(B)^t$ is finitely generated as a B-module, $m(B)^t/m(B)^{t+1}$ is of finite dimension over $B/m(B)$, and so has a finite composition series as a B-module. By inductive hypothesis, $B/m(B)^t$ has a finite composition series as a B-module. The exact sequence shows that therefore the same holds for $B/m(B)^{t+1}$, so that $B/m(B)^{t+1}$ is an Artin ring.

Now we can apply Theorem 1.3 to the $(B/J)[x_1, \ldots, x_q]$-module $G(E)$ and conclude that there is a polynomial P of degree strictly less than q such that $L_{B/J}(J^n \cdot E/J^{n+1} \cdot E) = P(n)$ for all sufficiently large n's. We consider the canonical exact sequence of morphisms of B-modules

$$(0) \to J^n \cdot E/J^{n+1} \cdot E \to E/J^{n+1} \cdot E \to E/J^n \cdot E \to (0).$$

For $n = 0$, we have $E/J^n \cdot E = (0)$. Suppose we have already shown that $E/J^n \cdot E$ has finite B-length for some n. Clearly, $J^n \cdot E/J^{n+1} \cdot E$ has the finite B-length $L_B(J^n \cdot E/J^{n+1} \cdot E) = L_{B/J}(J^n \cdot E/J^{n+1} \cdot E)$. The exact sequence shows that $E/J^{n+1} \cdot E$ again has finite B-length. Thus, $E/J^n \cdot E$ has finite B-length for all n, and the exact sequence shows, moreover, that

$$L_B(E/J^{n+1} \cdot E) - L_B(E/J^n \cdot E) = P(n)$$

for all sufficiently large n's. It follows that, for large n, $L_B(E/J^n \cdot E)$ is given by a polynomial of degree no greater than q. $\square$

2. Consider the result of Theorem 1.4 in the case where $E = B$. This says that there is a polynomial P_J of degree no greater than q, with rational coefficients, such that $L_B(B/J^n) = P_J(n)$ for all sufficiently large n's. We claim that the degree, d_J say, of P_J is the same for all J's. In order to see this, suppose that J' is another ideal like J. There is an exponent $e > 0$ such that $J^e \subset J'$. Now $J^{en} \subset (J')^n$ for all n, so that $B/(J')^n$ is a homomorphic image of B/J^{en}. Hence, for all sufficiently large n's, we have $P_{J'}(n) \le P_J(en)$, which clearly implies that $d_{J'} \le d_J$. By symmetry, it follows that $d_{J'} = d_J$.

We shall denote this degree by $d(B)$ and call it the *characteristic degree* of B. Note that $d(B)$ is at most equal to the minimum number of generators for an ideal with radical $m(B)$. This minimum number will be called the *parametric dimension* of B, and denoted by $r(B)$. Thus, $d(B) \le r(B)$.

On the other hand, we consider chains of prime ideals of B,

$$P_0 \subset \cdots \subset P_n \subset B,$$

where the inclusions are proper. The index n is called the length of the chain. The largest such length, or ∞, if there is no largest, is called the *Krull dimension* of B and denoted $k(B)$. The main result concerning these invariants is as follows.

Theorem 2.1. *If B is a Noetherian local ring then $d(B) = r(B) = k(B)$.*

PROOF. As long as we can, let us choose elements $b_1, b_2, \ldots$ from $m(B)$, as follows: b_1 is any element of $m(B)$ not belonging to any minimal prime ideal of B; if $b_1, \ldots, b_s$ have been chosen, choose b_{s+1} from $m(B)$ not belonging to any prime ideal minimal among those containing $Bb_1 + \cdots + Bb_s$. Since B is Noetherian, this process must come to a halt, i.e., we must reach an index q ($q = 0$ being possible) such that the union of the family of prime ideals minimal among those containing $Bb_1 + \cdots + Bb_q$ ($= (0)$ in the case where $q = 0$) is $m(B)$. Then we see from Lemma X.1.8 that $m(B)$ is the only prime ideal of B containing $Bb_1 + \cdots + Bb_q$.

Now we need the elementary result that *if S is a commutative ring, Q a prime ideal of S and I an ideal contained in Q, then there is a prime ideal Q_0 such that $I \subset Q_0 \subset Q$ and Q_0 is minimal among the prime ideals containing I.* In order to prove this one applies Zorn's Lemma to the family of all prime ideals between I and Q, endowed with the reversal of the inclusion order. The applicability of Zorn's Lemma comes from the nearly evident fact that the intersection of a totally ordered family of prime ideals is a prime ideal.

Thus, $m(B)$ contains a prime ideal P_{q-1} minimal over $Bb_1 + \cdots + Bb_{q-1}$. Next, P_{q-1} contains a prime ideal P_{q-2} minimal over $Bb_1 + \cdots + Bb_{q-2}$, etc. In this way, we obtain a chain of prime ideals

$$P_0 \subset \cdots \subset P_q = m(B),$$

where the inclusions are proper, by the choice of the b_i's. Since $m(B)$ is the only prime ideal containing $Bb_1 + \cdots + Bb_q$, it is the radical of this ideal. It follows that, for some $e > 0$,

$$m(B)^e \subset Bb_1 + \cdots + Bb_q \subset m(B).$$

By the definition of the parametric dimension, this shows that $r(B) \leq q$. The above chain of prime ideals shows that $q \leq k(B)$. Thus, we have $r(B) \leq k(B)$. We have already seen that $d(B) \leq r(B)$. Therefore, it will suffice to prove that $k(B) \leq d(B)$.

If $d(B) = 0$ then $L_B(B/m(B)^n)$ becomes constant as n is made large, and it is easy to see that therefore $m(B)^{n+1} = m(B)^n$ for all sufficiently large n's. An evident application of Proposition 1.2 (noting that $1 + a$ is a unit of B for every element a of $m(B)$) shows that $\bigcap_{n \geq 0} m(B)^n = (0)$. Hence, we must have $m(B)^n = (0)$ for some n, which implies that $m(B)$ is the only prime ideal of B, so that $k(B) = 0$.

Now suppose that $d(B) > 0$ and that the theorem has been established for all local rings whose characteristic degree is strictly smaller than $d(B)$. Consider a chain of prime ideals

$$P_0 \subset \cdots \subset P_t = m(B),$$

with proper inclusions and $t > 0$. Put $B' = B/P_0$. Then B' is a Noetherian local ring, with $m(B') = m(B)/P_0$. Now $B'/m(B')^n$ is isomorphic with $B/(m(B)^n + P_0)$, and $L_{B'}(B'/m(B')^n) = L_B(B/(m(B)^n + P_0)) \leq L_B(B/m(B)^n)$. Hence, $d(B') \leq d(B)$.

In order to proceed, we need the general result that *if A is a Noetherian local ring, and a is a non-zero element of A that is not a zero divisor, then $d(A/Aa) < d(A)$.* This is proved as follows. By Lemma 1.1, there is a non-negative integer k such that, for all $n \geq k$,

$$Aa \cap m(A)^n = (Aa \cap m(A)^k)m(A)^{n-k} \subset Aam(A)^{n-k}.$$

Using this and the canonical A-module isomorphism between the modules $(Aa + m(A)^n)/m(A)^n$ and $Aa/(Aa \cap m(A)^n)$, we obtain

$$L_A((Aa + m(A)^n)/m(A)^n) \geq L_A(Aa/Aam(A)^{n-k}).$$

Since a is not a zero divisor, the multiplication by a is an isomorphism of A-modules from A to Aa, so that the length on the right above coincides with $L_A(A/m(A)^{n-k})$. Using this, we get

$$\begin{aligned}
L_{A/Aa}(A/(Aa + m(A)^n)) &= L_A(A/(Aa + m(A)^n)) \\
&= L_A(A/m(A)^n) - L_A((Aa + m(A)^n)/m(A)^n) \\
&\leq L_A(A/m(A)^n) - L_A(A/m(A)^{n-k}).
\end{aligned}$$

This shows that, for all sufficiently large n's, we have

$$L_{A/Aa}(A/(Aa + m(A)^n)) \leq P_{m(A)}(n) - P_{m(A)}(n - k).$$

The module figuring on the left is isomorphic with $(A/Aa)/m(A/Aa)^n$. Hence, for large n, the left side is equal to $P_{m(A/Aa)}(n)$. As a function of n, the expression on the right is a polynomial of degree strictly smaller than $d(A)$. Hence our inequality shows that $d(A/Aa)$ is strictly smaller than $d(A)$.

Now choose a non-zero element a from P_1/P_0, and put $B^\circ = B'/B'a$. By what we have just proved, we have $d(B^\circ) < d(B')$, and so $d(B^\circ) < d(B)$. Hence, our inductive hypothesis gives $k(B^\circ) \leq d(B^\circ)$, so that $k(B^\circ) < d(B)$.

On the other hand, the chain

$$(P_1/P_0)/B'a \subset \cdots \subset (P_t/P_0)/B'a$$

is a chain of prime ideals of B° with proper inclusions, so that $k(B^\circ) \geq t - 1$. With the last inequality above, this gives $t \leq d(B)$. Thus, we conclude that $k(B)$ is finite and at most equal to $d(B)$. $\qquad\square$

The following generalization of Krull's principal ideal theorem is an almost immediate consequence of Theorem 2.1.

Theorem 2.2. *Let R be a Noetherian commutative ring. Let P be a prime ideal of R that is minimal prime over an ideal J generated by n elements. Then every properly increasing chain of prime ideals ending at P has length at most n.*

PROOF. Let $P_0 \subset \cdots \subset P_k = P$ be such a chain. We must prove that $k \leq n$. Replacing R with R/P_0, if necessary, we reduce the problem to the situation where R is an integral domain. Consider the local ring $R_P = R[(R \setminus P)^{-1}]$. The only prime ideal of R_P containing JR_P is PR_P. Thus, the radical of JR_P is PR_P. Since JR_P is generated by n elements, we have $r(R_P) \leq n$. By Theorem 2.1, this implies that $k(R_P) \leq n$. But

$$P_0 R_P \subset \cdots \subset P_k R_P = PR_P$$

is a properly increasing chain of prime ideals of R_P. Therefore, we must have $k \leq n$. $\qquad\square$

3. Lemma 3.1. *Let R be a commutative ring, x a variable over R. Suppose that $P_0 \subset P_1 \subset P_2$ is a chain of prime ideals of $R[x]$, with proper inclusions. Then $P_0 \cap R \neq P_2 \cap R$.*

PROOF. Put $P = (P_0 \cap R)R[x]$. This is clearly a prime ideal of $R[x]$ that is contained in P_0. We may identify $R[x]/P$ with $(R/(P_0 \cap R))[x]$, and P_1/P, P_2/P are prime ideals of this polynomial ring such that, with proper inclusions,

$$(0) \subset P_1/P \subset P_2/P.$$

Now let K denote the field of fractions of the integral domain $R/(P_0 \cap R)$, and consider the ideals KP_1/P and KP_2/P of $K[x]$. First, suppose that $KP_2/P = K[x]$. Then P_2/P contains a non-zero element μ of $R/(P_0 \cap R)$, and if u is a representative of μ in R we have $u \in (P_2 \cap R) \setminus (P_0 \cap R)$, so that $P_0 \cap R \neq P_2 \cap R$.

If $KP_2/P \neq K[x]$ then both KP_2/P and KP_1/P are non-zero prime ideals of $K[x]$. Since every non-zero prime ideal of $K[x]$ is maximal, we must therefore have $KP_1/P = KP_2/P$. Choose an element p from $P_2 \setminus P_1$. By the last equality, there is a non-zero element μ in $R/(P_0 \cap R)$ such that $\mu(p + P) \in P_1/P$. Let u be a representative of μ in R. Then $up \in P_1$, whence $u \in P_1 \cap R$, while $u \notin P_0 \cap R$. Thus, $P_0 \cap R \neq P_1 \cap R$. $\qquad\square$

Theorem 3.2. *Let R be a commutative ring. If R satisfies the maximal condition for prime ideals, so does the polynomial ring $R[x]$. If R has finite Krull dimension then the same holds for $R[x]$, and*

$$k(R) + 1 \leq k(R[x]) \leq 2k(R) + 1.$$

PROOF. Let $P_0 \subset P_1 \subset \cdots$ be a properly ascending chain of prime ideals of $R[x]$. By Lemma 3.1, the chain $P_0 \cap R \subset P_2 \cap R \subset \cdots$ formed with the even indices is properly ascending. Evidently, this gives the first part of Theorem 3.2.

Now suppose that $k(R) = n$. Then it is clear that the second chain above must stop at $P_{2n} \cap R$ or before, whence the original chain must stop at P_{2n+1} or before. Thus, $k(R[x]) \leq 2n + 1$.

On the other hand, let $Q_0 \subset \cdots \subset Q_n$ be a properly ascending chain of prime ideals of R. Then

$$Q_0 R[x] \subset \cdots \subset Q_n R[x] \subset Q_n R[x] + x R[x]$$

is a properly ascending chain of prime ideals of $R[x]$, showing that

$$k(R[x]) \geq n + 1. \qquad\square$$

Lemma 3.3. *Let P be a minimal non-zero prime ideal of the Noetherian commutative ring R. Then $PR[x]$ is a minimal non-zero prime ideal of $R[x]$.*

PROOF. Let p be a non-zero element of P. We show that $PR[x]$ is minimal prime over $pR[x]$. Suppose that Q is a prime ideal of $R[x]$ containing p and contained in $PR[x]$. Then $Q \cap R$ is a non-zero prime ideal of R contained in P. By assumption on P, we have therefore $Q \cap R = P$, so that $Q = PR[x]$.

Now let J be any prime ideal of $R[x]$ that is contained in $PR[x]$. Then either $J \cap R = P$ or $J \cap R = (0)$. In the first case, $J = PR[x]$. In the second case, (0) is a prime ideal of R, and hence also a prime ideal of $R[x]$. We have $(0) \subset J \subset PR[x]$. Since $PR[x]$ is minimal prime over the principal ideal $pR[x]$, we have from Theorem 2.2 that the inclusions here cannot both be proper. Thus, either $J = (0)$ or $J = PR[x]$. $\qquad\square$

Theorem 3.4. *Let R be a Noetherian commutative ring, and assume that R has finite Krull dimension $k(R)$. Then $k(R[x]) = k(R) + 1$.*

PROOF. If $k(R) = 0$ then Theorem 3.2 gives $k(R[x]) = 1$. Now assume that $k(R) = n > 0$, and that the theorem has been established in the lower cases. Let $P_0 \subset \cdots \subset P_m$ be a properly ascending chain of prime ideals of $R[x]$. In view of Theorem 3.2, it suffices to show that $m \le n + 1$. Write Q_i for $P_i \cap R$ and $Q_i[x]$ for $Q_i R[x]$.

First, suppose that $Q_1 \ne Q_0$. Then $k(R/Q_1) < n$, and we have the following properly ascending chain of prime ideals of $(R/Q_1)[x]$:

$$(0) \subset P_2/Q_1[x] \subset \cdots \subset P_m/Q_1[x].$$

By inductive hypothesis, this gives $m - 1 \le n$, i.e., $m \le n + 1$.

Now consider the case $Q_1 = Q_0$. Then $P_1 \ne Q_1[x]$, because otherwise $P_1 \subset P_0$. Hence we have the proper inclusions

$$(0) \subset P_1/Q_1[x] \subset P_2/Q_1[x]$$

showing that $P_2/Q_1[x]$ is not minimal among the non-zero prime ideals of $(R/Q_1)[x]$.

If $P_2 = Q_2[x]$ then $P_2/Q_1[x] = (Q_2/Q_1)[x]$, and it follows from Lemma 3.3 that Q_2/Q_1 is not a *minimal* non-zero prime ideal of R/Q_1. Hence we have

$$k(R/Q_2) \le k(R/Q_1) - 2 \le n - 2.$$

Therefore, the properly ascending chain

$$(0) \subset P_3/Q_2[x] \subset \cdots \subset P_m/Q_2[x]$$

of prime ideals of $(R/Q_2)[x]$ and our inductive hypothesis give

$$m - 2 \le n - 1,$$

i.e., $m \le n + 1$.

Finally, if $P_2 \ne Q_2[x]$, we have the properly ascending chain

$$(0) \subset P_2/Q_2[x] \subset \cdots \subset P_m/Q_2[x]$$

of prime ideals in $(R/Q_2)[x]$. By Lemma 3.1, $Q_2 \ne Q_0$, so that $k(R/Q_2) < n$. Therefore, the chain and the inductive hypothesis give $m - 1 \le n$, i.e., $m \le n + 1$. $\qquad \square$

Corollary 3.5. *Let R be a Noetherian integral domain, S an integral domain containing R and algebraic over $[R]$. Then $k(S) \le k(R)$.*

PROOF. We suppose, without loss of generality, that $k(R)$ is finite; say $k(R) = n$. Assume the corollary is false, and consider a properly ascending chain $(0) \subset Q_1 \subset \cdots \subset Q_{n+1}$ of prime ideals of S. Choose an element q_i from $Q_i \setminus Q_{i-1}$ (where $Q_0 = (0)$), and put $P_i = Q_i \cap R[q_1, \ldots, q_{n+1}]$. Then the P_i's are prime ideals of $R[q_1, \ldots, q_{n+1}]$, and we have the proper inclusions $(0) \subset P_1 \subset \cdots \subset P_{n+1}$. Now $R[q_1, \ldots, q_i]$ is isomorphic with

$$R[q_1, \ldots, q_{i-1}][x]/T_i,$$

where T_i is a non-zero prime ideal. Hence,

$$k(R[q_1, \ldots, q_i]) \leq k(R[q_1, \ldots, q_{i-1}][x]) - 1.$$

By Theorem 3.4, the right side is equal to $k(R[q_1, \ldots, q_{i-1}])$. Hence, we have $k(R[q_1, \ldots, q_{n+1}]) \leq k(R) = n$, contradicting the existence of the above chain of P_i's. $\qquad\qquad\square$

Theorem 3.6. *Let R and S be integral domains such that $R \subset S$ and S is integral over R. Let Q be a prime ideal of S, and let P_1 be a prime ideal of R containing $Q \cap R$. There is a prime ideal Q_1 of S such that $Q_1 \cap R = P_1$ and $Q \subset Q_1$.*

PROOF. Let γ denote the canonical homomorphism $R/(Q \cap R) \to R/P_1$. The inclusion map $R \to S$ clearly induces an injective ring homomorphism $R/(Q \cap R) \to S/Q$, by means of which we identify $R/(Q \cap Q)$ with a subring of S/Q. Evidently, S/Q is integral over $R/(Q \cap R)$. By Proposition II.3.2, γ extends to a ring homomorphism from a valuation subring T of $[S/Q]$, containing $R/(Q \cap R)$, to an algebraic closure, F say, of $[R/P_1]$. Since S/Q is integral over $R/(Q \cap R)$, we have $S/Q \subset T$. Thus, γ extends to a ring homomorphism $\gamma' : S/Q \to F$. The kernel of γ' is of the form Q_1/Q, where Q_1 is a prime ideal of S containing Q. The kernel of the restriction of γ' to $R/(Q \cap R)$ is the kernel $P_1/(Q \cap R)$ of γ. Hence, $Q_1 \cap R = P_1$. $\qquad\square$

Corollary 3.7. *Let R and S be as in Theorem 3.6. Then S has finite Krull dimension if and only if R has finite Krull dimension, and $k(S) = k(R)$.*

PROOF. Let $P_0 \subset P_1 \subset \cdots$ be a chain of prime ideals of R. It follows at once from Theorem 3.6 that there is a chain $Q_0 \subset Q_1 \subset \cdots$ of prime ideals of S such that $Q_i \cap R = P_i$ for each i. Therefore, if $k(S)$ is finite, so is $k(R)$, and $k(S) \geq k(R)$. By Corollary 3.5, if $k(R)$ is finite, so is $k(S)$, and $k(S) \leq k(R)$. $\qquad\square$

Theorem 3.8. *Let F be a field, and let R be a finitely generated integral domain F-algebra. Then R has finite Krull dimension, and $k(R)$ is equal to the degree of transcendence of $[R]$ over F.*

PROOF. Let r denote the degree of transcendence of $[R]$ over F. By Theorem X.1.2, there is a transcendence basis $(z_1, \ldots, z_r)$ for $[R]$ over F such that R is integral over the polynomial F-algebra $F[z_1, \ldots, z_r]$. By Corollary 3.7, we have $k(R) = k(F[z_1, \ldots, z_r])$. By Theorem 3.4, we have therefore $k(R) = r$.

4. Let R be a Noetherian local ring, $m(R)$ the maximum ideal of R. It is clear from Theorem 2.1 that the cardinality of every system of R-module generators of $m(R)$ is no smaller than the Krull dimension $k(R)$. One calls R

regular if $m(R)$ has a system of *R-module* generators of cardinality $k(R)$. We let $G(R)$ denote the graded ring derived from the filtration of R by the powers of $m(R)$, so that $G(R)_n = m(R)^n/m(R)^{n+1}$. Evidently, $G(R)$ is an algebra over the field $R/m(R)$ in the natural way.

Theorem 4.1. *In the notation introduced above, let $(\gamma_1, \ldots, \gamma_q)$ be an $R/m(R)$-basis of $m(R)/m(R)^2$, and let x_i be a representative of γ_i in $m(R)$. Then $(x_1, \ldots, x_q)$ is a system of R-module generators of $m(R)$. If R is regular, then $G(R)$, as a graded $R/m(R)$-algebra, is isomorphic with the polynomial algebra $(R/m(R))[t_1, \ldots, t_d]$, where the t_i's are independent variables, and $d = k(R)$. Conversely, if $G(R)$ is isomorphic with such a polynomial algebra then R is regular and $d = k(R)$.*

PROOF. Put $P = m(R)$ and $Q = Rx_1 + \cdots + Rx_q$. Then $P = Q + P^2$, and it follows inductively that $P = Q + P^n$ for every positive exponent n. Now R/Q is a Noetherian local ring, with $m(R/Q) = P/Q$. From Proposition 1.2, we see that $\bigcap_{n>0} (P/Q)^n = (0)$. Here, we have

$$P/Q = (Q + P^n)/Q = (P/Q)^n,$$

so that we must have $P/Q = (0)$, i.e., $P = Q$.

Now suppose that R is regular, and put $d = k(R)$. Let $(x_1, \ldots, x_d)$ be a system of R-module generators of P, and let $t_1, \ldots, t_d$ be independent variables over R/P. Consider the surjective morphism of graded R/P-algebras

$$\eta: (R/P)[t_1, \ldots, t_d] \to G(R),$$

where

$$\eta(t_i) = x_i + P^2 \in G(R)_1.$$

The kernel of η is a homogeneous ideal whose component of degree 0 is (0). We shall derive a contradiction from the assumption that η is not injective, in which case the kernel of η contains a non-zero homogeneous element u of strictly positive degree, n say. If $m \geq n$ then the kernel of the restriction of η to the homogeneous component of degree m contains

$$u(R/P)[t_1, \ldots, t_d]_{m-n}.$$

The dimension of the component $(R/P)[t_1, \ldots, t_d]_q$ of degree q is the number of monomials of total degree q in d variables, which is the binomial coefficient $C(q + d - 1, d - 1)$. Thus, the above part of the kernel of η is of dimension $C(m - n + d - 1, d - 1)$. Therefore, the dimension of $G(R)_m$ is no greater than $C(m + d - 1, d - 1) - C(m - n + d - 1, d - 1)$. With n and d fixed, this last integer is of the form $f(m)$, where f is a polynomial of degree strictly less than $d - 1$, and if L_R denotes R-module length we have $L_R(P^m/P^{m+1}) \leq f(m)$. Applying this to the exact sequences

$$(0) \to P^m/P^{m+1} \to R/P^{m+1} \to R/P^m \to (0),$$

we find that, for $m > n$,

$$L_R(R/P^m) \leq L_R(R/P^n) + \sum_{q=n}^{m-1} f(q).$$

Here, the expression on the right is a polynomial in m, of degree strictly less than d.

On the other hand, we know from Theorem 2.1 that the characteristic degree $d(R)$ is equal to $k(R)$, i.e., $d(R) = d$. Hence, for all sufficiently large m's, we have $L_R(R/P^m) = g(m)$, where g is a polynomial of degree d. This contradicts the above. The conclusion is that η is an isomorphism.

Now suppose that $G(R)$ is isomorphic with a polynomial algebra $(R/P)[t_1, \ldots, t_q]$. Then the dimension of $G(R)_n$ is the number of monomials of degree n in q variables, i.e., $L_R(P^n/P^{n+1}) = C(n + q - 1, q - 1)$. It follows that, for $m > 0$,

$$L_R(R/P^m) = \sum_{n=0}^{m-1} C(n + q - 1, q - 1).$$

The sum on the right is the number of monomials in q variables of degree $< m$, and so is equal to $C(m + q - 1, q)$. Therefore, the above expression for $L_R(R/P^m)$ shows that $d(R) = q$, whence also $k(R) = q$. By the first part of Theorem 4.1, P has a system of R-module generators of cardinality q, so that R is regular. $\qquad\square$

Corollary 4.2. *If R is a regular Noetherian local ring then R is an integral domain, and integrally closed in $[R]$.*

PROOF. By Theorem 4.1, $G(R)$ is an integral domain. Clearly, this implies that R is an integral domain. In order to proceed, we introduce the following notation. For a non-zero element x of R, let $\mu(x)$ be the largest exponent n such that x belongs to P^n, where $P = m(R)$. Let x' denote the element $x + P^{\mu(x)+1}$ of $G(R)_{\mu(x)}$. If x and y are non-zero elements of R, since $G(R)$ is an integral domain, we have $x'y' = (xy)'$.

Now let x be a non-zero element of $[R]$, and suppose that x is integral over R. Write $x = a/b$, with a and b in R. There is a non-zero element d in R such that dx^q lies in R for every positive exponent q, i.e., $da^q \in Rb^q$. We show that, for all non-negative exponents n, we have $a \in Rb + P^n$. Evidently, this holds for $n = 0$. Suppose it holds for some n; say $a = cb + e$, with c in R and e in P^n. Then, for all q,

$$de^q = d(a - cb)^q \in Rb^q,$$

whence $d'(e')^q$ belongs to $G(R)(b')^q$ for all q. Since $G(R)$ is a unique factorization domain, this implies that e' belongs to $G(R)b'$. Since e belongs to P^n, this shows that there is an element c_1 in R such that $e - c_1 b$ belongs to P^{n+1}, whence $a \in Rb + P^{n+1}$.

Thus, $a \in Rb + P^n$ for all n. If b does not lie in P then x is evidently in R, because b is then a unit. Now suppose that b lies in P, and consider the local ring R/Rb, with $m(R/Rb) = P/Rb$. If a° is the canonical image of a in R/Rb, then $a^\circ \in \bigcap_{n > 0} m(R/Rb)^n = (0)$. Thus $a^\circ = 0$, which means that $a \in Rb$, so that $x \in R$. $\qquad\qquad\square$

5. Let X be a variety, p a point of X, and $\mathscr{F}_X(p)$ the local ring at p. Since this is the direct limit of the system formed with the algebras $\mathscr{F}_X(U)$ of regular functions on open neighborhoods of p, the evaluations at p define an F-algebra homomorphism $\mathscr{F}_X(p) \to F$, where F is the base field. We indicate this by $f \mapsto f(p)$. A *tangent to X at p* is a differentiation $\tau \colon \mathscr{F}_X(p) \to F$, based on the evaluation $f \mapsto f(p)$. These tangents constitute an F-space, which we call the *tangent space at p*, and which we denote by X_p.

Let U be an affine patch of X containing p. We may evidently identify X_p with U_p. Therefore, it suffices to examine X_p in the case where X is an irreducible affine variety. Assuming this, write A for $\mathscr{P}(X)$, and M for the annihilator of p in A. Then $\mathscr{F}_X(p)$ may be identified with $A[(A \setminus M)^{-1}]$, which is a Noetherian local integral domain with maximum ideal $MA[(A \setminus M)^{-1}]$. Let us write R for this local ring, and P for its maximum ideal. Since R/P is isomorphic with F, we have $R = F + P$, and clearly $F \cap P = (0)$. Let τ be an element of X_p. Then τ annihilates P^2 and hence induces an F-linear map $\tau' \colon P/P^2 \to F$. Conversely, if σ is any F-linear map from P/P^2 to F, then we can define an F-linear map σ^* from R to F by $\sigma^*(\alpha + u) = \sigma(u + P^2)$, where α is an element of F and u is an element of P. It is verified directly that σ^* belongs to X_p. This shows that the F-space X_p is isomorphic with the dual space $(P/P^2)^\circ$ of P/P^2. Since P is finitely generated as an R-module, it follows that X_p is finite-dimensional.

Write d for the F-dimension of P/P^2, and choose elements $u_1, \ldots, u_d$ from P such that $(u_1 + P^2, \ldots, u_d + P^2)$ is an F-basis of P/P^2. Then $P = Ru_1 + \cdots + Ru_d + P^2$, and if we write S for the R-module

$$P/(Ru_1 + \cdots + Ru_d)$$

we have $S = P \cdot S$. Now we see from an evident application of Proposition 1.2 that $S = (0)$, i.e., that $P = Ru_1 + \cdots + Ru_d$. Thus, $\dim(X_p)$ is equal to the (minimum) cardinality of a system of R-module generators for P, so that $\dim(X_p)$ is no smaller than the parametric dimension $r(R)$ of R. By Theorem 2.1, this means that $\dim(X_p) \geq k(R)$. By Theorem 3.8, $k(R)$ is equal to the degree of transcendence of $[R]$ over F, i.e., the degree of transcendence of the field of rational functions of X over F. Our conclusion is that $\dim(X_p) \geq \dim(X)$. We call p a *singular point* of X if $\dim(X_p) > \dim(X)$.

Next, we show that the singular points constitute a closed subset of X. Let us write A as a homomorphic image of an ordinary polynomial F-algebra, by $\gamma \colon F[x_1, \ldots, x_n] \to A$. Let I denote the kernel of γ, and I_p the kernel of the homomorphism $p \circ \gamma$ from $F[x_1, \ldots, x_n]$ to F. Clearly, I_p is generated by the elements $x_i - \gamma(x_i)(p)$, and $I \subset I_p$. For an element f of $F[x_1, \ldots, x_n]$, let f'_i denote the formal derivative of f with respect to x_i. Now define

$$\delta_i(f) = \gamma(f'_i)(p),$$

and $\delta: F[x_1, \ldots, x_n] \to F^n$ by $\delta(f) = (\delta_1(f), \ldots, \delta_n(f))$. Then the images $\delta(x_i - \gamma(x_i)(p))$ are the canonical basis elements of F^n, while $\delta(I_p^2) = (0)$. It is clear from this that δ induces an F-linear isomorphism δ° from I_p/I_p^2 to F^n.

Now let $g_1, \ldots, g_t$ be a set of ideal generators of I, and consider the Jacobian matrix J_p whose rows are the $\delta(g_i)$'s. Evidently, the rank of J_p is equal to the F-dimension of $\delta(I)$. We have $\delta(I) = \delta^\circ((I + I_p^2)/I_p^2)$. Since δ° is an isomorphism, it follows that the rank of J_p is equal to the F-dimension of $(I + I_p^2)/I_p^2$.

Now the annihilator M of p in A is isomorphic with I_p/I, whence M/M^2 is isomorphic with $I_p/(I + I_p^2)$. On the other hand, M/M^2 is isomorphic with P/P^2. Hence, we have $\dim(I_p/(I + I_p^2)) = \dim(X_p)$, so that

$$\dim(X_p) + \operatorname{rank}(J_p) = \dim(I_p/I_p^2) = n.$$

Thus, $\dim(X_p)$ is strictly larger than $\dim(X)$ if and only if $\operatorname{rank}(J_p)$ is strictly smaller than $n - \dim(X)$. If $\dim(X) = d$, this means that the singular points of X are precisely the common zeros in X of the determinants of the $n - d$ by $n - d$ submatrices of the matrix with rows $(\gamma((g_i)'_1), \ldots, \gamma((g_i)'_n))$ (whose value at p is J_p). In particular, the set of singular points is therefore closed in X.

Now we assume that our base field F is algebraically closed, and we show that then not every point of X is singular. Recall from Section III.1 that, since F is algebraically closed (and hence perfect), the field $[A]$ of rational functions on our variety X is a *separable* extension of F. Therefore, by Theorem III.2.1, there is a transcendence base $(t_1, \ldots, t_m)$ of $[A]$ relative to F such that $[A]$ is a finite *separable* algebraic extension of $F(t_1, \ldots, t_m)$. Thus, we may write $[A] = F(t_1, \ldots, t_m)[c]$, where c is separably algebraic over $F(t_1, \ldots, t_m)$. Let f denote the minimum polynomial for c relative to $F(t_1, \ldots, t_m)$, with denominators cleared so that the coefficients of f are polynomials in $F[t_1, \ldots, t_m]$. Now let us view f as an element of $F[t_0, \ldots, t_m]$, where t_0 is an auxiliary variable, and let Y be the set of zeros of f in F^{m+1}. Then Y is clearly an irreducible (because f is an irreducible polynomial) affine variety whose field of rational functions is isomorphic with $[A]$. Using an F-algebra isomorphism, π say, from the field of rational functions of Y to $[A]$, it is easy to construct a morphism from an appropriate irreducible open subvariety of X to Y whose transpose is π. By Proposition X.5.3, it follows that there is a non-empty open subvariety ,U say, of X that is isomorphic with an open subvariety, V say, of Y. This shows that it will suffice to prove that not every point of Y is singular. Indeed, since we have already shown that the set of non-singular points is open, we can then conclude that not every point of V is singular, so that not every point of U is singular.

Now suppose that, contrary to what we have to show, every point of Y is singular. Let us apply our above Jacobian criterion, viewing $\mathscr{P}(Y)$ as a homomorphic image of $F[t_0, \ldots, t_m]$, the kernel of the homomorphism being the ideal generated by f. In the present case, the dimension of our variety Y is m, and the number of auxiliary variables (denoted n earlier) is $m + 1$. The assumption that every point is singular therefore is equivalent to the assumption that the Jacobian matrix be of rank 0. This means that all the partial derivatives of f vanish on Y, and so belong to the ideal generated by f, and therefore are equal to 0. As a polynomial in t_0, our f is irreducible and separable, so that the derivative of f with respect to t_0 is different from 0, giving the desired contradiction.

We summarize our results as follows.

Theorem 5.1. *For every point p of the irreducible variety X, we have* $\dim(X_p) \geq \dim(X)$. *If the base field is algebraically closed then the set of points where the equality holds, i.e., the set of non-singular points, is non-empty and open in X.*

It is clear from the definitions that a point p of X is non-singular if and only if the local ring at p is regular. From Corollary 4.2, we have that *every non-singular point is a normal point.*

6. Proposition 6.1. *Let X be an irreducible variety over a perfect field F. Suppose that p is a non-singular point of X. There is an $F(X)$-basis $(\delta_1, \ldots, \delta_m)$ of the $F(X)$-space of F-linear derivations from $F(X)$ to $F(X)$ satisfying the following requirements:*

(1)   *each δ_i stabilizes the local ring $\mathscr{F}_X(p)$;*
(2)   *if $\delta_{ip}: \mathscr{F}_X(p) \to F$ is the tangent at p defined by $\delta_{ip}(f) = \delta_i(f)(p)$ then $(\delta_{1p}, \ldots, \delta_{mp})$ is an F-basis of X_p;*
(3)   *there are elements $f_1, \ldots, f_m$ in $\mathscr{F}_X(p)$ such that $\delta_i(f_j) = \delta_{ij}$.*

PROOF. Without loss of generality, we assume that X is affine, and we write A for $\mathscr{P}(X)$. Then $F(X) = [A]$, and

$$\mathscr{F}_X(p) = A[(A \setminus M)^{-1}],$$

where M is the annihilator of p in A.

Now let us use the notation and the Jacobian criterion of Section 5. If $m = \dim(X)$ then the rank of the Jacobian matrix J_p is $n - m$ exactly, while the rank of J_q is at most $n - m$ for every point q of X. Write a_{ij} for the element $\gamma((g_i)'_j)$ of A, and let J be matrix with rows $(a_{i1}, \ldots, a_{in})$, so that $J(q) = J_q$. Then the rank of the matrix J is $n - m$, and we choose the indexing of the ideal generators g_i and the variables x_j such that the determinant, D say, of the matrix with entries a_{ij}, where $1 \leq i \leq n - m$ and $1 \leq j \leq n - m$, does not vanish at p.

For $k = 1, \ldots, m$, define elements $u_{k1}, \ldots, u_{kn}$ of $[A]$ as follows: for $j > n - m$, put $u_{kj} = 0$, except when $j = n - m + k$, in which case put $u_{kj} = 1$; since the rank of J is $n - m$ and since $D \neq 0$, there is one and only one choice of elements u_{kj} of $[A]$, with $j \leq n - m$, such that, for all i and k,

$$\sum_{j=1}^{n} a_{ij} u_{kj} = 0.$$

In fact, each u_{kj} belongs to $A[D^{-1}]$.

Now let τ_k be the F-linear derivation from $F[x_1, \ldots, x_n]$ to $[A]$ (viewed as $F[x_1, \ldots, x_n]$-module via γ) sending each x_j onto u_{kj}. From the defining relations for the u_{kj}'s, we have then $\tau_k(I) = (0)$, so that τ_k induces a derivation from A to $[A]$, and so defines an F-linear derivation δ_k from $[A]$ to $[A]$. Since $\tau_k(A) \subset A[D^{-1}]$ and $D(p) \neq 0$, it is clear that δ_k stabilizes $\mathscr{F}_X(p)$. If $f_j = \gamma(x_{n-m+j})$ for $j = 1, \ldots, m$ then we have $\delta_i(f_j) = \delta_{ij}$, by the choice of the u_{kj}'s with $j > n - m$. It is clear from this that the δ_i's are linearly independent over $[A]$. Since F is perfect, $[A]$ is a separable extension of F. Hence, by Theorem III.2.1, the $[A]$-dimension of the space of F-linear derivations of $[A]$ is equal to m. Therefore, $(\delta_1, \ldots, \delta_m)$ is an $[A]$-basis of this space. From $\delta_i(f_j) = \delta_{ij}$, we see also that the tangents δ_{ip} are linearly independent $(i = 1, \ldots, m)$. Hence, they constitute a basis for the m-dimensional F-space X_p. Thus, $(\delta_1, \ldots, \delta_m)$ satisfies all the requirements of the Proposition. $\qquad\square$

Let $\rho: X \to Y$ be a morphism of varieties, and let p be a point of X. Then ρ defines an algebra homomorphism from the local ring at $\rho(p)$ to the local ring at p, and this transposes to a linear map $\rho'_p: X_p \to Y_{\rho(p)}$, called the *differential of ρ at p*.

Let $\rho: X \to Y$ be a dominant morphism between irreducible varieties. One says that ρ is a *separable morphism* if $F(X)$ is separable over $F(Y) \circ \rho$.

Proposition 6.2. *Let ρ be a dominant morphism from an irreducible variety X to an irreducible variety Y over a perfect field F. Suppose that p is a nonsingular point of X such that $\rho(p)$ is non-singular and $\rho'_p: X_p \to Y_{\rho(p)}$ is surjective. Then ρ is a separable morphism.*

PROOF. For a commutative algebra R over our base field F, and an R-module M, we denote the R-module of F-linear derivations from R to M by $\mathrm{Der}_F(R, M)$. In particular, we consider the $F(X)$-space $\mathrm{Der}_F(F(X), F(X))$, and the $F(Y)$-space $\mathrm{Der}_F(F(Y), F(X))$, where $F(X)$ is viewed as an $F(Y)$-module via the injection $g \to g \circ \rho$ from $F(Y)$ to $F(X)$.

In the evident way, the second of these may be viewed as an $F(X)$-space, and we consider the $F(X)$-linear map

$$\rho': \mathrm{Der}_F(F(X), F(X)) \to \mathrm{Der}_F(F(Y), F(X)),$$

where $\rho'(\delta)(g) = \delta(g \circ \rho)$. From Section III.1, it is clear that $F(X)$ is separable over $F(Y) \circ \rho$ if (and only if) ρ' is surjective.

By Theorem III.2.1, the $F(Y)$-dimension of $\mathrm{Der}_F(F(Y), F(Y))$ is equal to $\dim(Y)$, for which we shall write n. Choose an $F(Y)$-basis of $\mathrm{Der}_F(F(Y), F(Y))$ satisfying the requirements of Proposition 6.1 with respect to the non-singular point $\rho(p)$ of Y. Via the injective map $g \mapsto g \circ \rho$ from $F(Y)$ to $F(X)$, this yields a system $(\eta_1, \ldots, \eta_n)$ of elements of $\mathrm{Der}_F(F(Y), F(X))$. Evidently, this system is linearly independent over $F(X)$. On the other hand, by choosing a transcendence basis $(t_1, \ldots, t_n)$ for $F(Y)$ over F such that $F(Y)$ is separably algebraic over $F(t_1, \ldots, t_n)$, we see that the $F(X)$-dimension of

$$\mathrm{Der}_F(F(Y), F(X)),$$

cannot exceed n. Therefore, $(\eta_1, \ldots, \eta_n)$ is an $F(X)$-basis of $\mathrm{Der}_F(F(Y), F(X))$.

On the other hand, let $(\delta_1, \ldots, \delta_m)$ be an $F(X)$-basis of $\mathrm{Der}_F(F(X), F(X))$ satisfying the requirements of Proposition 6.1 with respect to the non-singular point p of X, and write

$$\rho'(\delta_i) = \sum_{j=1}^{n} f_{ji}\eta_j,$$

with each f_{ji} in $F(X)$.

Now there are elements $g_1, \ldots, g_n$ in $\mathcal{F}_Y(\rho(p))$ such that $\eta_i(g_j) = \delta_{ij}$, so that $f_{ji} = \rho'(\delta_i)(g_j) = \delta_i(g_j \circ \rho)$. But $g_j \circ \rho$ belongs to $\mathcal{F}_X(p)$, which is stabilized by δ_i. Hence, we have $f_{ji} \in \mathcal{F}_X(p)$ for all i and j. If τ is any element of $\mathrm{Der}_F(F(Y), F(X))$ that sends $\mathcal{F}_Y(\rho(p))$ into $\mathcal{F}_X(p)$, let $\tau_{\rho(p)}$ denote the tangent to Y at $\rho(p)$ that is obtained by following up the restriction to $\mathcal{F}_Y(\rho(p))$ of τ with the evaluation at p. Then we have

$$\rho'(\delta_i)_{\rho(p)} = \sum_{j=1}^{n} f_{ji}(p)\eta_{j\rho(p)}.$$

The expression on the left coincides with $\rho_p'(\delta_{ip})$. Since ρ_p' is surjective, and $(\delta_{1p}, \ldots, \delta_{mp})$ is an F-basis of X_p, while $(\eta_{1\rho(p)}, \ldots, \eta_{n\rho(p)})$ is an F-basis of $Y_{\rho(p)}$, it follows that the rank of the matrix with entries $f_{ij}(p)$ must be equal to n, whence the rank of the matrix with entries f_{ij} must also be equal to n, so that ρ' is surjective. $\qquad\square$

Proposition 6.3. *Let $\rho: X \to Y$ be a dominant morphism between irreducible varieties over an algebraically closed field. Suppose that ρ is separable. Then there is a non-empty open subset U of X such that, for every p in U, the points p and $\rho(p)$ are non-singular and ρ_p' is surjective.*

PROOF. Since ρ is separable, the map ρ' from $\mathrm{Der}_F(F(X), F(X))$ to $\mathrm{Der}_F(F(Y), F(X))$ used in the proof of Proposition 6.2 is surjective. It is clear from Theorem 5.1, applied first to Y and then to X, that the set, S say, of points p of X such that both p and $\rho(p)$ are non-singular is non-empty and open in X. With reference to any one point p in S, choose bases

$(\delta_1, \ldots, \delta_m)$ of $\mathrm{Der}_F(F(X), F(X))$ and $(\eta_1, \ldots, \eta_n)$ of $\mathrm{Der}_F(F(Y), F(X))$ as in the proof of Proposition 6.2, and, as done there, write

$$\rho'(\delta_i) = \sum_{j=1}^{n} f_{ji}\eta_j.$$

Then the matrix with entries f_{ji} has rank n, because ρ' is surjective. Hence the set, T say, of points p of X such that each f_{ji} belongs to $\mathscr{F}_X(p)$ and the matrix with entries $f_{ji}(p)$ is of rank n is *non-empty* and open in X. Finally, the set, L say, of the points p such that each δ_i stabilizes $\mathscr{F}_X(p)$, each η_j sends $\mathscr{F}_Y(\rho(p))$ into $\mathscr{F}_X(p)$ and the δ_{ip}'s and $\eta_{j\rho(p)}$'s are linearly independent is non-empty and open in X. If we set $U = S \cap T \cap L$ then it is clear from what we saw in proving Proposition 6.2 that U satisfies the requirements of Proposition 6.3. $\qquad\square$

Notes

1. The development of Sections 1 and 2, where it goes beyond the classical theory in generality, efficiency and elegance, is due to P. Samuel and J-P. Serre.

2. A standard example of a non-regular local integral domain is as follows. Let F be a field, x and y independent variables over F. Let R be the factor ring of $F[x, y]$ mod the ideal generated by $x^3 - y^2$. Let I be the ideal of R that is generated by the cosets of x and y. Note that I is a prime ideal, and consider the corresponding local ring $S = R_I$ whose maximum ideal is $J = IR_I$. It is easy to see from Theorem 3.8 that the Krull dimension $k(S)$ is equal to 1. One can verify directly that J is not a principal ideal, showing that S is not regular.

Chapter XII

Coset Varieties

At this point, we have accumulated enough from algebraic geometry for dealing with coset varieties. Suppose that G is an irreducible algebraic group over an algebraically closed field F, and H is an algebraic subgroup of G. The main task for this chapter is the construction of an appropriate variety structure on G/H. In Section 1, it appears that $[\mathscr{P}(G)]^H$ is a suitable candidate for the field $F(G/H)$ of rational functions. Starting with this field, Section 2 provides an imbedding of G/H as an open irreducible subset of a projective variety, and shows that the resulting variety structure of G/H has all of the desirable properties.

Section 3 begins with the presentation of the Grassmann variety as a coset variety. The rest of this section is devoted to an examination of conditions under which a coset variety is quasi-affine, i.e., an open subvariety of an affine variety.

In Section 4 it is shown that if G is as above and solvable then G/H is affine for every algebraic subgroup H. Section 5 is concerned with the representation-theoretical significance of the condition that a coset variety be quasi-affine. It is shown that this condition is equivalent to the condition that every polynomial representation of the subgroup extend to one of the whole group, allowing enlargement of the representation space.

1. Proposition 1.1. *Let G be an algebraic group, H an algebraic subgroup of G. Let N be a finite-dimensional polynomial H-module. There is a polynomial character g of H and a finite-dimensional polynomial G-module M, with module structure $(x, m) \mapsto x(m)$, such that N is a sub vector space of M and $x(n) = g(x)x \cdot n$ for every element x of H and every element n of N, where the $\cdot$ indicates the given H-module structure of N.*

PROOF. Let S denote the space of representative functions on H that is associated with N. Then, via its comodule structure, N may be identified with a sub H-module of the direct sum of a finite family of copies of S. Since every polynomial function on H is the restriction to H of a polynomial function on G, it follows that N may be written in the form U/V, where U is a sub H-module of some finite-dimensional G-module, T say, and V is a sub H-module of U.

Let $n = \dim(V)$, and consider the homogeneous component

$$M = \bigwedge^{n+1}(T)$$

of the exterior algebra built on T. This contains $U \bigwedge^n(V)$ as a sub H-module, which is H-module isomorphic with $N \otimes \bigwedge^n(V)$, owing to the fact that $V \bigwedge^n(V) = (0)$. Now $\bigwedge^n(V)$ is 1-dimensional, and the action of an element x of H on $\bigwedge^n(V)$ is the scalar multiplication by the determinant, $g(x)$ say, of the linear automorphism of V that corresponds to x. If u is any fixed non-zero element of $\bigwedge^n(V)$, then the composite of the map from N to $N \otimes \bigwedge^n(V)$ sending each n onto $n \otimes u$ with the injection

$$N \otimes \bigwedge^n(V) \to \bigwedge^{n+1}(T)$$

coming from $N = U/V$ identifies N with a sub vector space of M so as to satisfy the requirements of the proposition. $\square$

Proposition 1.2. *Let G be an irreducible algebraic group, H an algebraic subgroup of G. Let f be an element of $[\mathscr{P}(G)]^H$. There is a polynomial character g of H and elements s and t of $\mathscr{P}(G)$ such that $f = s/t$ and, for every x in H, $x \cdot s = g(x)s$ and $x \cdot t = g(x)t$. With respect to the action of G on $[\mathscr{P}(G)]$ from the left and the induced action of $\mathscr{L}(G)$ by derivations, the element-wise fixer of $[\mathscr{P}(G)]^H$ in G coincides with H, and the annihilator of $[\mathscr{P}(G)]^H$ in $\mathscr{L}(G)$ coincides with $\mathscr{L}(H)$.*

PROOF. We assume, without loss of generality, that $f \neq 0$, and we consider the polynomial H-module $(\mathscr{P}(G)f) \cap \mathscr{P}(G)$. This contains a simple sub H-module $V \neq (0)$. Let V° denote the dual H-module, and let g and M be as in Proposition 1.1, with V° in the place of N. Evidently, V° is a simple H-module also with respect to the H-module structure of M coming from its G-module structure. Like every finite-dimensional polynomial G-module, M is isomorphic with a sub G-module of the direct sum of a finite set of copies of $\mathscr{P}(G)$. Therefore, the simplicity of V° implies that there is an *injective* morphism, ρ say, of H-modules from V°, viewed as a sub H-module of M, to $\mathscr{P}(G)$. With the original H-module structure of V°, we have, for every x in H and every α in V°,

$$x \cdot \rho(\alpha) = g(x)\rho(x \cdot \alpha).$$

Now let $(v_1, \ldots, v_n)$ be an F-basis of V such that $v_1(1_G) \neq 0$, while $v_i(1_G) = 0$ for every $i > 1$. Let $(\alpha_1, \ldots, \alpha_n)$ be the dual basis of V°, so that

$\alpha_i(v_j) = \delta_{ij}$. Put $k_i = \rho(\alpha_i)$, so that each k_i is a non-zero element of $\mathscr{P}(G)$. Choose an element y from G such that $k_1(y) \neq 0$, and put

$$s = \sum_{i=1}^{n} (k_i \cdot y)v_i.$$

For x in H, write

$$x \cdot v_i = \sum_{i=1}^{n} h_{ji}(x)v_j.$$

Then we have

$$x \cdot k_i = g(x)\rho(x \cdot \alpha_i) = g(x)\sum_{j=1}^{n} h_{ij}(x^{-1})k_j.$$

By direct substitution, we find that $x \cdot s = g(x)s$ for every x in H. Now s belongs to $(\mathscr{P}(G)f) \cap \mathscr{P}(G)$, so that we have $s = tf$, with t in $\mathscr{P}(G)$. Since $s(1_G) = k_1(y)v_1(1_G) \neq 0$, we have $s \neq 0$, whence also $t \neq 0$. From the fact that f is fixed under the action of H, we see that $x \cdot t = g(x)t$ for every x in H. Thus, $f = s/t$, with s and t as required.

Now let E be a finite set of H-semi-invariants in $\mathscr{P}(G)$ such as is given by Theorem II.2.1. Let U denote the sub G-module of $\mathscr{P}(G)$ that is generated by the elements of E, and let V be the direct sum of n copies of U, where n is the number of elements of $E = (e_1, \ldots, e_n)$. Let v stand for the point $(e_1, \ldots, e_n)$ of V. Then, by Theorem II.2.1, an element x of G belongs to H if and only if $x \cdot v$ is a scalar multiple of v. Clearly, this implies that, for every element τ of $\mathscr{L}(H)$, the transform $\tau \cdot v$ is also a scalar multiple of v. Conversely, it follows from Proposition IV.5.1 that every element τ of $\mathscr{L}(G)$ satisfying this condition belongs to $\mathscr{L}(H)$.

Suppose that x is an element of G that fixes every element of $[\mathscr{P}(G)]^H$. We wish to show that x belongs to H. Suppose this is false. Then the elements v and $x \cdot v$ of V are linearly independent. Therefore, there are elements α and β in V° such that

$$\alpha(v) = 1, \qquad \alpha(x \cdot v) = 0, \qquad \beta(v) = 1, \qquad \beta(x \cdot v) = 1.$$

Now the rational function $(\alpha/v)/(\beta/v)$ belongs to $[\mathscr{P}(G)]^H$, by the choice of v. It is defined at 1_G and at x, and takes the values 1 and 0 there. Hence, the transform of this rational function under the action of x on $[\mathscr{P}(G)]$ is defined at 1_G and takes the value 0 there. This contradicts our assumption that x fixes the elements of $[\mathscr{P}(G)]^H$. Our conclusion is that the element-wise fixer of $[\mathscr{P}(G)]^H$ in G coincides with H.

Finally, let τ be an element of $\mathscr{L}(G)$ such that the corresponding derivation of $[\mathscr{P}(G)]$ annihilates $[\mathscr{P}(G)]^H$. We wish to show that τ belongs to $\mathscr{L}(H)$. Suppose this is false. Then the elements v and $\tau \cdot v$ of V are linearly independent. Therefore, there are elements α and β in V° such that

$$\alpha(v) = 0, \qquad \alpha(\tau \cdot v) = 1, \qquad \beta(v) = 1.$$

We have

$$\tau \cdot [(\alpha/v)/(\beta/v)] = ((\beta/v)\tau \cdot (\alpha/v) - (\alpha/v)\tau \cdot (\beta/v))/(\beta/v)^2.$$

This is defined at 1_G, and its value there is

$$\tau(\alpha/v) - \alpha(v)\tau(\beta/v) = \tau(\alpha/v) = \alpha(\tau \cdot v) = 1,$$

contradicting our assumption that the derivation corresponding to τ annihilates $[\mathscr{P}(G)]^H$. Our conclusion is that the annihilator of $[\mathscr{P}(G)]^H$ in $\mathscr{L}(G)$, with respect to the action by derivations, coincides with $\mathscr{L}(H)$. $\qquad\square$

2. Let F be a field, G an irreducible affine algebraic F-group, H an algebraic subgroup of G. Let $(f_1, \ldots, f_n)$ be a system of field generators for $[\mathscr{P}(G)]^H$ over F. We know from Proposition 1.2 that each f_i can be written as a fraction of H-semi-invariants. An evident adjustment of numerators and denominators yields H-semi-invariants $u_0, \ldots, u_n$ in $\mathscr{P}(G)$, all of the same weight, such that $f_i = u_i/u_0$ for $i = 1, \ldots, n$. Let S denote the sub G-module of $\mathscr{P}(G)$ that is generated by all the u_i's, and let T be the direct sum of $n + 1$ copies of S. Let t denote the point $(u_0, \ldots, u_n)$ of T. We consider the projective variety T^* whose points are the 1-dimensional subspaces of T, and we let t^* be the point of T^* corresponding to t. The action of G by linear automorphisms on T induces an action of G by algebraic variety automorphisms on T^*. We know from Proposition 1.2 that the element-wise fixer of $[\mathscr{P}(G)]^H$ in G coincides with H. It is clear from this that the fixer of t^* in G is precisely H.

The map α from G to T^* defined by $\alpha(x) = x \cdot t^*$ is evidently a morphism of algebraic varieties. *We assume that our base field is algebraically closed.* Then we have from Theorem X.4.4 that $\alpha(G)$ is a constructible subset of T^*, so that it inherits a variety structure from T^*. Moreover, since G is irreducible, so is $\alpha(G)$. The closure of $\alpha(G)$ in T^* is therefore an irreducible projective variety. By Theorem X.4.3, $\alpha(G)$ contains a non-empty open subset of its closure in T^*. Evidently, this closure is stable under the action of G, and G acts transitively on $\alpha(G)$. It follows that $\alpha(G)$ is open in its closure. Thus, $\alpha(G)$ is an open subvariety of a projective variety. Such varieties are called *quasi-projective* varieties.

Clearly, α induces a bijective map α^H from the set G/H of cosets to $\alpha(G)$. We endow G/H with the variety structure coming from that of $\alpha(G)$ via α^H. If π is the canonical map from G to G/H then $\pi = (\alpha^H)^{-1} \circ \alpha$, so that π is now a morphism of varieties. It is clear from the construction of α that the field of rational functions of the variety G/H is mapped isomorphically onto $[\mathscr{P}(G)]^H$ by the transpose of π. By Theorem III.2.3, $[\mathscr{P}(G)]$ is separable over $[\mathscr{P}(G)]^H$. Thus, the canonical map $\pi: G \to G/H$ is a separable morphism of varieties.

By Proposition XI.6.3, there is a non-singular point x in G such that $\pi(x)$ is non-singular and the differential π'_x is surjective from G_x to $(G/H)_{xH}$.

Since π is compatible with the actions of G on G and G/H, it follows from the transitivity of the action of G on G that all the points of G and G/H are non-singular and that π'_x is surjective for every x in G. Taking x to be the neutral element 1 of G and identifying $\mathscr{L}(G)$ with the tangent space to G at 1, we have that π'_1 maps $\mathscr{L}(G)$ surjectively onto the tangent space to G/H at H. From the last part of Proposition 1.2, we see that the kernel of π'_1 is precisely $\mathscr{L}(H)$. Therefore, π'_1 induces a linear isomorphism from $\mathscr{L}(G)/\mathscr{L}(H)$ to the tangent space to G/H at H. Since H, like every point, is a non-singular point of G/H, this shows that

$$\dim(G) = \dim(H) + \dim(G/H)$$

By Theorem X.4.3, there is a non-empty open subset U of G/H having the following property: if W is a closed irreducible subset of G/H meeting U, and Z is an irreducible component of $\pi^{-1}(W)$ meeting $\pi^{-1}(U)$, then

$$\dim(Z) = \dim(W) + \dim(G) - \dim(G/H).$$

Now the G-translates of U cover G/H, and their inverse images in G are the corresponding G-translates of $\pi^{-1}(U)$. It follows that the above dimension relation holds for *every* closed irreducible subset W of G/H and *every* irreducible component Z of $\pi^{-1}(W)$. Therefore, we can apply Theorem X.4.5 and conclude that π *is an open map.*

In order to analyze π further, we need the following technical result concerning regular functions.

Lemma 2.1. *Let* $\sigma: X \to Y$ *be an open surjective and separable morphism between irreducible varieties over an algebraically closed field. Assume that Y is a normal variety, and that f is a regular function on X that is constant on each* $\sigma^{-1}(y)$ *with* $y \in Y$. *Then* $f = h \circ \sigma$, *where h is a regular function on Y.*

PROOF. Let F denote the base field, and define a map

$$\rho: X \to Y \times F,$$

by $\rho(x) = (\sigma(x), f(x))$. Let Z denote the closure of $\rho(X)$ in $Y \times F$. Then Z is an irreducible subvariety of $Y \times F$. By Theorem X.4.3, $\rho(X)$ contains an irreducible open subset V of Z. Let $\eta: Z \to Y$ denote the restriction to Z of the canonical projection morphism from $Y \times F$ to Y. It is clear from the constancy property of f that the restriction of η to V is injective. Indicating restrictions by subscripts, we have $\eta_V \circ \rho = \sigma_{\rho^{-1}(V)}$. Since $\rho^{-1}(V)$ is open in X and since σ is an open map, $\sigma(\rho^{-1}(V))$ is open in Y, and therefore dense in Y. Thus, η_V is dominant from V to Y.

Consider the field map $q \mapsto q \circ \rho$ from $F(V)$ to $F(\rho^{-1}(V)) = F(X)$. This sends the subfield $F(Y) \circ \eta_V$ of $F(V)$ onto the subfield $F(Y) \circ \sigma_{\rho^{-1}(V)}$ of $F(\rho^{-1}(V))$. Since σ is a separable morphism, $F(X)$ is separable over $F(Y) \circ \sigma$, i.e., $F(\rho^{-1}(V))$ is separable over $F(Y) \circ \sigma_{\rho^{-1}(V)}$. Therefore, also the subfield $F(V) \circ \rho$ is separable over $F(Y) \circ \sigma_{\rho^{-1}(V)}$. Thus, $F(V) \circ \rho$ is separable over $F(Y) \circ \eta_V \circ \rho$. Since the transpose of ρ is injective, this implies that $F(V)$

is separable over $F(Y) \circ \eta_V$. Our conclusion is that η_V is a separable morphism. Now we know that η_V is injective, separable and dominant. By Proposition X.5.2, we must therefore have $F(Y) \circ \eta_V = F(V)$. Thus, the transpose of the morphism $\eta: Z \to Y$ is an isomorphism of $F(Y)$ onto $F(Z)$.

Let g denote the restriction to Z of the canonical projection $Y \times F \to F$. Clearly, g is a regular function on Z. By what we have just seen concerning η, we have $g = h \circ \eta$, where h is an element of $F(Y)$. Now

$$f = g \circ \rho = h \circ \eta \circ \rho = h \circ \sigma.$$

It remains only to show that h is actually a regular function on Y. Suppose this is not the case. Since Y is a normal variety, we can apply Proposition X.5.1 and conclude that $1/h$ is defined at some point y of Y and takes the value 0 at y. On the other hand, we have $((1/h) \circ \sigma)f = 1$. Evaluating this at a point x of X such that $\sigma(x) = y$, we get a contradiction. Thus, h is indeed regular on Y. $\qquad\square$

Essentially, the above establishes the following main result on coset varieties.

Theorem 2.2. *Let F be an algebraically closed field, G an irreducible affine algebraic F-group, H an algebraic subgroup of G. Then G/H can be endowed with the structure of an F-variety, actually, a quasi projective variety, such that the following requirements are fulfilled:*

(1) $\dim(G/H) + \dim(H) = \dim(G)$;
(2)   *the canonical map $\pi: G \to G/H$ is a separable open morphism of varieties;*
(3) $F(G/H) = [\mathscr{P}(G)]^H$;
(4)   *for every morphism $\alpha: G \to V$ of algebraic F-varieties that is constant on the cosets xH, the induced map α^H from G/H to V is a morphism of algebraic F-varieties;*
(5)   *the canonical map from $G \times (G/H)$ to G/H coming from the composition of G is a morphism of varieties.*

PROOF. We have already established (1), (2) and (3). Clearly, (5) follows from our above construction of the variety structure on G/H. It remains only to verify (4).

Let U be an open subset of V. Then $\alpha^{-1}(U)$ is an open subset of G. Since π is an open map, $\pi(\alpha^{-1}(U))$ is open in G/H. Since $\pi(\alpha^{-1}(U)) = (\alpha^H)^{-1}(U)$, this shows that α^H is continuous.

Now let f be an element of $\mathscr{F}_V(U)$. Then $f \circ \alpha$ is an element of $\mathscr{F}_G(\alpha^{-1}(U))$ that is constant on the intersections of $\alpha^{-1}(U)$ with the cosets xH. We know that every point of G/H is non-singular. By the remark at the end of Section XI.5, this implies that G/H is a normal variety. Therefore, we can apply Lemma 2.1 in the evident way and conclude that $f \circ \alpha = h \circ \pi$, where h is an element of $\mathscr{F}_{G/H}((\alpha^H)^{-1}(U))$. Since $f \circ \alpha = f \circ \alpha^H \circ \pi$, it follows that

$f \circ \alpha^H = h$, showing that α^H satisfies the sheaf condition for a morphism of varieties. $\square$

Note that (4), *together with the crude part of* (2) *that π is a morphism of varieties, determines the algebraic variety structure of G/H.* This remedies the defect that our above construction of the variety structure of G/H was not technically natural.

The appropriate extension of Theorem 2.2 to the case where G is not irreducible is as follows. Let G_1 denote the irreducible component of the neutral element in G. By writing G as a union of cosets xG_1H, we see that G/H is a disjoint union of a finite family of subsets that are G-translates of $(G_1H)/H$, which may be identified with $G_1/(H \cap G_1)$. Transporting the variety structure of $G_1/(H \cap G_1)$ to each of these G-translates, we obtain a variety structure on G/H satisfying all the requirements of Theorem 2.2, with the evident modification of (3).

Let F, G and H be as in Theorem 2.2. In the case where H is normal in G. Theorem II.4.4 establishes the structure of an affine algebraic F-group on G/H. We claim that, in this case, *the variety structure of G/H given by Theorem 2.2 is the same as that obtained from Theorem II.4.4.*

In proving this, let us denote the factor group with its variety structure as obtained from Theorem II.4.4 by $G//H$. We have $\mathscr{P}(G//H) = \mathscr{P}(G)^H$, so that $F(G//H) = [\mathscr{P}(G)^H]$. Since the element-wise fixer of $\mathscr{P}(G)^H$ in G coincides with H, we know from Lemma IV.5.3 that $[\mathscr{P}(G)^H] = [\mathscr{P}(G)]^H$. Now let us apply part (4) of Theorem 2.2 to the canonical morphism $G \to G//H$. This gives the conclusion that the identity map $\gamma: G/H \to G//H$ is a morphism of varieties. The equality $[\mathscr{P}(G)^H] = [\mathscr{P}(G)]^H$ says that the transpose of γ is an isomorphism of $F(H//G)$ onto $F(G/H)$. By Proposition X.5.3, this implies that there is a non-empty open subset U of $G//H$ such that the restriction of γ to $\gamma^{-1}(U)$ is an isomorphism of varieties. Here, this means simply that U is open also for the topology of G/H and that the variety structures of U induced by those of G/H and $G//H$ coincide. Clearly, the same is therefore true for every G-translate xU of U. Since $G//H$ is the union of a finite family of these translates, we can apply Proposition IX.1.3 and conclude that γ^{-1} is a morphism of varieties, thus establishing our above claim.

In particular, it is now clear that the universal mapping property described in Theorem II.4.4 extends from the category of affine algebraic groups to the category of algebraic varieties, where it is part (4) of Theorem 2.2.

3. Let F be an algebraically closed field, V an n-dimensional F-space, d an integer with $0 < d \leq n$. We consider the Grassmann variety $\mathscr{G}_d(V)$ whose points are the d-dimensional subspaces of V. Let G be the group of all linear automorphisms of V, viewed as an irreducible affine algebraic F-group in the canonical fashion. Let s denote a point of $\mathscr{G}_d(V)$, to be kept fixed in our discussion. We consider the morphism of varieties $\sigma: G \to \mathscr{G}_d(V)$, where

$\sigma(x) = x(s)$. Let H denote the stabilizer of s in G. Then σ is constant on the cosets xH and induces the bijective map $\sigma^H \colon G/H \to \mathscr{G}_d(V)$ in the canonical fashion. Let π denote the canonical morphism from G to G/H, so that $\sigma^H \circ \pi = \sigma$. We know from Theorem 2.2 that σ^H is a morphism of varieties. By Proposition X.5.2, the field $F(G/H)$ is (purely inseparably) algebraic over $F(\mathscr{G}_d(V)) \circ \sigma^H$. Hence, we have $\dim(\mathscr{G}_d(V)) = \dim(G/H)$.

In order to determine this dimension, choose an F-basis of V whose initial section of length d is an F-basis of s. Let f_{ij} denote the corresponding matrix entry functions on G. These are algebraically independent and, together with the reciprocal of the determinant function, generate $\mathscr{P}(G)$ as an F-algebra. The annihilator of H in $\mathscr{P}(G)$ is generated as an ideal by the functions f_{ij} with $j \le d$ and $i > d$. This shows that $\dim(H) = n^2 - d(n - d)$, so that $\dim(G/H) = d(n - d)$.

We identify $\mathscr{L}(G)$ with $\mathrm{End}_F(V)$ in the canonical fashion, and we denote the matrix entry functions on $\mathscr{L}(G)$ that correspond to our basis of V by g_{ij}, so that each f_{ij} is the restriction of g_{ij} to G. The Lie algebra $\mathscr{L}(H)$ is evidently contained in the stabilizer of s in $\mathrm{End}_F(V)$. The annihilator in $\mathrm{End}_F(V)^\circ$ of this stabilizer is spanned over F by the functions g_{ij} with $j \le d$ and $i > d$. This shows that the dimension of the stabilizer of s in $\mathrm{End}_F(V)$ is equal to $n^2 - d(n - d)$, which we know to be the dimension of $\mathscr{L}(H)$. Therefore, *$\mathscr{L}(H)$ coincides with the stabilizer of s in $\mathrm{End}_F(V)$.*

We shall prove that *σ^H is actually an isomorphism of varieties, so that $\mathscr{G}_d(V)$ may be identified with G/H.* This will follow from arguments we have used before once we have shown that σ^H is a *separable* morphism of varieties.

Let us view the differential σ'_1 of σ at the neutral element 1 of G as a linear map from $\mathscr{L}(G)$ to $\mathscr{G}_d(V)_s$. From the fact that σ is constant on H, it is clear that $\mathscr{L}(H)$ is contained in the kernel of σ'_1. Conversely, let τ be an element of the kernel of σ'_1. Let a and b be elements of $\bigwedge^d(V^\circ)$, choosing b so that it does not vanish on the line in $\bigwedge^d(V)$ that is determined by s. Then a/b is an element of the local ring of $\mathscr{G}_d(V)$ at s. The action of G by algebra automorphisms on $\bigwedge(V^\circ)$ induces an action of $\mathscr{L}(G)$ by derivations, which we indicate by a dot. Then $\sigma'_1(\tau)(a/b)$ is the value at s of the rational function represented by $((\tau \cdot a)b - (\tau \cdot b)a)/b^2$. Let s^* denote an F-space generator of the canonical image of s in $\bigwedge^d(V)$. Then, since $\sigma'_1(\tau) = 0$, we have

$$(\tau \cdot a)(s^*)b(s^*) - (\tau \cdot b)(s^*)a(s^*) = 0.$$

We choose s^* so that $b(s^*) = 1$. Then the above gives

$$a(\tau \cdot s^*) = -(\tau \cdot a)(s^*) = -(\tau \cdot b)(s^*)a(s^*),$$

where $\tau \cdot s^*$ is the transform of s^* by the derivation of $\bigwedge(V)$ corresponding to τ. Keeping b fixed and letting a range over $\bigwedge^d(V^\circ)$, we see from this that $\tau \cdot s^* \in Fs^*$, which means that τ stabilizes s. From the above, we know that this means that τ belongs to $\mathscr{L}(H)$. Our conclusion is that the kernel of σ'_1 coincides with $\mathscr{L}(H)$.

Since G acts transitively on $\mathcal{G}_d(V)$ by variety automorphisms, every point of $\mathcal{G}_d(V)$ is non-singular. In particular, the dimension of the tangent space $\mathcal{G}_d(V)_s$ is equal to the dimension of $\mathcal{G}_d(V)$, which we have shown to be equal to $\dim(\mathcal{L}(G)/\mathcal{L}(H))$. This shows that σ'_1 is surjective. Since $\sigma'_1 = (\sigma^H)'_H \circ \pi'_1$, this implies that $(\sigma^H)'_H$ is surjective. By Proposition XI.6.2, it follows that σ^H is a *separable* morphism of varieties. Since σ^H is bijective, this implies that the transpose of σ^H is an isomorphism from $F(\mathcal{G}_d(V))$ to $F(G/H)$. By the same argument we made at the end of Section 2, we see from this that σ^H is an isomorphism of varieties.

We know from Theorem IX.5.3 that $\mathcal{G}_d(V)$ is a complete variety. Hence, G/H is a complete variety. The elements of $\mathcal{P}(G)^H$ may be regarded as everywhere regular functions on the variety G/H. By (3) and (4) of Proposition IX.5.2, the only everywhere regular functions on an irreducible complete variety are the constants. Thus, we have $\mathcal{P}(G)^H = F$.

This phenomenon, representing an extreme opposite to the situation of a *normal* subgroup H, should be viewed in the light of Theorem 3.1 below. There, and subsequently, a variety is called *quasi-affine* if it is isomorphic with an open subvariety of an affine variety.

Theorem 3.1. *Let G be an irreducible algebraic group over an algebraically closed field, H an algebraic subgroup of G. The following conditions are mutually equivalent*: (1) *the element-wise fixer of $\mathcal{P}(G)^H$ in G coincides with H*; (2) $[P(G)^H] = [P(G)]^H$; (3) *the variety G/H is quasi-affine*.

PROOF. Lemma IV.5.3 says that (1) implies (2). Now suppose that (2) holds, and let K be the element-wise fixer of $\mathcal{P}(G)^H$ in G. Evidently, it follows from (2) that the elements of $[\mathcal{P}(G)]^H$ are fixed under action of K. By Proposition 1.2, we must therefore have $K = H$. Thus, (1) and (2) are equivalent.

Next, we show that (2) implies (3). Let F denote the base field. By Proposition II.3.6, $[\mathcal{P}(G)]^H$ is finitely generated as a field over F. Therefore, it follows from (2) that there is a finitely generated sub F-algebra A of $\mathcal{P}(G)^H$ such that $[A] = [\mathcal{P}(G)]^H$. Since $\mathcal{P}(G)^H$ is locally finite as a right G-module, we may choose A so that it is stable under the action of G from the right. Let $\mathcal{S}(A)$ denote the irreducible affine algebraic F-variety whose points are the F-algebra homomorphisms from A to F. Clearly, G acts by variety automorphisms on $\mathcal{S}(A)$ via the right G-module structure of A. Let $\rho: G \to \mathcal{S}(A)$ be the restriction morphism. Evidently, ρ is compatible with the actions of G on G from the left and on $\mathcal{S}(A)$. It is clear from Theorem II.3.3 that $\rho(G)$ contains a non-empty open subset of $\mathcal{S}(A)$. Since G acts transitively on $\rho(G)$, and since ρ is a G-morphism, it follows that $\rho(G)$ is open in $\mathcal{S}(A)$, so that it is a quasi-affine variety.

Clearly, ρ is constant on the cosets xH, and we know from Theorem 2.2 that the induced map $\rho^H: G/H \to \mathcal{S}(A)$ is a morphism of varieties. Since A separates the points of G/H, this morphism is injective. Thus, ρ^H defines a bijective morphism of varieties $\sigma: G/H \to \rho(G)$. Since $\rho(G)$ is open in $\mathcal{S}(A)$

and since $[A] = F(G/H)$, the transpose of σ is an isomorphism of $F(\rho(G))$ onto $F(G/H)$. Now the argument of the end of Section 2 can be applied again and shows that σ is an isomorphism of varieties. Thus, (2) implies (3).

Finally, we show that (3) implies (2). Assuming that (3) holds, we show that every element of $[\mathscr{P}(G)]^H$ belongs to $[\mathscr{P}(G)^H]$. Write such an element, f say, in the form s/t, where s and t are H-semi-invariants in $\mathscr{P}(G)$ of the same weight, ρ say. We know from Proposition 1.2 that this is possible. It suffices to show that, for some x in G, we have $f \cdot x \in [\mathscr{P}(G)^H]$. Therefore, we assume without loss of generality that t does not vanish at the neutral element 1 of G. Let V denote the set of zeros of t in G. Since t is an H-semi-invariant, we have $VH = V$. Let π denote the canonical morphism from G to G/H. Using our last remark and the fact that π is an open map, we see that $\pi(V)$ is closed in G/H. Since $t(1) \neq 0$, the point H of G/H does not belong to $\pi(V)$. By assumption, G/H is an open subvariety of some affine variety, S say. Let T denote the closure of $\pi(V)$ in S. Since $\pi(V)$ is closed in the open subset G/H of S, we have $T \cap (G/H) = \pi(V)$. Therefore, the point H of G/H does not belong to T, so that there is an element of $\mathscr{P}(S)$ that vanishes on T but not at H. The restriction of this function to G/H is an everywhere regular function g on G/H such that $g(\pi(V)) = (0)$, but $g(H) \neq 0$. Put $h = g \circ \pi$. By Proposition IX.1.2, h belongs to $\mathscr{P}(G)^H$. Clearly, $h(V) = (0)$ and $h(1) \neq 0$. By Theorem II.3.5, the vanishing of h on the set V of zeros of t implies that there is some positive exponent e such that h^e belongs to $\mathscr{P}(G)t$. Accordingly, let us write $h^e = ut$, with $u \in \mathscr{P}(G)$. Since h^e is fixed under the action of H, while t is a semi-invariant of weight ρ, it is clear that u is a semi-invariant of weight $1/\rho$. Hence, both us and ut are elements of $\mathscr{P}(G)^H$. Since $f = (us)/(ut)$, this shows that f belongs to $[\mathscr{P}(G)^H]$. $\square$

4. Let G be an algebraic group, V an algebraic variety. Suppose there is given a group homomorphism from G to the group of variety automorphisms of V such that the corresponding map from $G \times V$ to V is a morphism of varieties. Then we say that V is a *strict G-variety*.

Lemma 4.1. *Let G be a unipotent algebraic group over an algebraically closed field, and let V be an affine strict G-variety. Then every G-orbit in V is closed.*

PROOF. Without loss of generality, we assume that G is irreducible. Let T be a G-orbit in V. In proving that T is closed, we assume without loss of generality that T is dense in V. Then we know from Theorem X.4.3 that T contains a non-empty open subset, U say, of V. Now suppose that, contrary to what we wish to prove, we have $T \neq V$. Then $U \neq V$, so that there is a non-zero element f in $\mathscr{P}(V)$ such that $f(V \setminus U) = (0)$, and hence

$$f(V \setminus T) = (0).$$

The action of G on V defines the structure of a right polynomial G-module on $\mathscr{P}(V)$, in the evident way. Let M denote the right sub G-module of $\mathscr{P}(V)$ that is generated by f. Since G is unipotent, there is a non-zero element m in M such that $m \cdot x = m$ for every element x of G. This implies that m is constant on T. Since T is dense in V, it follows that m is a constant. Since f vanishes on $V \setminus T$, so does every element of M. This gives the contradiction $m = 0$. $\qquad\square$

Proposition 4.2. *Let G be an irreducible algebraic group over an algebraically closed field, H an algebraic subgroup of G. If the variety G/H_1 is affine, so is the variety G/H.*

PROOF. It is clear from Proposition IX.1.2 that the transpose of the canonical morphism from G to G/H_1 is an isomorphism of $\mathscr{P}(G/H_1)$ onto $\mathscr{P}(G)^{H_1}$. Let us write A for $\mathscr{P}(G)$ and identify A^{H_1} with $\mathscr{P}(G/H_1)$. We know that the element-wise fixer of $[A]^H$ in G coincides with H. Hence, $[A]^{H_1}$ is a finite Galois extension of $[A]^H$ whose Galois group is isomorphic with H/H_1 in the evident way.

Let τ denote the trace map from $[A]^{H_1}$ to $[A]^H$. Then τ is not the zero map and, if $x_1, \ldots, x_n$ is a system of representatives in H for the elements of H/H_1, we have

$$\tau(f) = \sum_{i=1}^{n} x_i \cdot f.$$

Now let f be an element of $[A]^H$. Since $[A]^{H_1}$ coincides with $[A^{H_1}]$, by Theorem 3.1, we can write $f = u/v$, where u and v are elements of A^{H_1}. There is an element g in $[A]^{H_1}$ such that $\tau(vg) \neq 0$. Write g in the form a/b with a and b in A^{H_1}, and put

$$h = g \prod_{i=1}^{n} x_i \cdot b.$$

Then h belongs to A^{H_1} and we still have $\tau(vh) \neq 0$. Since f is fixed under the action of H, and since $f = (uh)/(vh)$, we have $f = \tau(uh)/\tau(vh)$, which shows that f belongs to $[A^H]$. Thus, $[A]^H = [A^H]$, which implies that the element-wise fixer of A^H in G coincides with H.

Now A^{H_1} is finitely generated as an algebra. Since H/H_1 is finite, it follows that A^{H_1} is finitely generated also as an A^H-module. By Proposition II.3.7, it follows that A^H is finitely generated as an algebra. Let ρ denote the restriction map from G to the affine algebraic variety $\mathscr{S}(A^H)$. By Theorem 2.2, the induced map ρ^H from G/H to $\mathscr{S}(A^H)$ is a morphism of varieties. Since the element-wise fixer of A^H in G coincides with H, this map ρ^H is injective.

Now let F denote the base field, and let $u: A^H \to F$ be an element of $\mathscr{S}(A^H)$. Since A^{H_1} is integral over A^H, it follows from Proposition II.3.2 that u extends to an F-algebra homomorphism $v: A^{H_1} \to F$. By assumption on H_1, this point v of $\mathscr{S}(A^{H_1})$ is a point of G/H_1, and therefore is the canonical image of an element x of G. Thus, ρ is surjective, so that ρ^H is bijective.

Since $F(G/H) = [A]^H = [A^H]$, the transpose of ρ^H is an isomorphism of $F(\mathscr{S}(A^H))$ onto $F(G/H)$. By an argument used already several times in this chapter, we can now conclude that ρ^H is an isomorphism of varieties. $\quad\square$

Theorem 4.3. *Let G be an irreducible solvable algebraic group over an algebraically closed field, H an algebraic subgroup of G. Then the variety G/H is affine.*

PROOF. Proposition 4.2 reduces the theorem to the case where H is irreducible. Then, by Theorem VI.3.2, we can write $H = H_u \rtimes T$, where T is a toroid, and $G = G_u \rtimes S$, where S is a toroid containing T. It follows immediately from Theorem V.5.3 that S is a direct product $T \times T'$. By Corollary VI.1.2, every unipotent element of G belongs to G_u. Hence, we have $H_u \subset G_u$. Now G is the semidirect product $(G_u \rtimes T) \rtimes T'$, and H is an algebraic subgroup of $G_u \rtimes T$. This shows that it suffices to deal with the case where $S = T$, so that $G = G_u \rtimes T$ and $H = H_u \rtimes T$.

In this case, we consider the variety morphism ρ from G to G_u/H_u that is defined as the composite of the projection $G \to G_u$ of the semidirect product presentation with the canonical morphism $G_u \to G_u/H_u$. This is constant on the cosets xH, and we know from Theorem 2.2 that the induced map ρ^H from G/H to G_u/H_u is a morphism of varieties. On the other hand, let τ be the morphism from G_u to G/H coming from the injection $G_u \to G$ and the canonical morphism from G to G/H. This is constant on the cosets xH_u and induces the morphism τ^{H_u} from the variety G_u/H_u to the variety G/H. Evidently, ρ^H and τ^{H_u} are mutually inverse. Thus, the variety G/H is isomorphic with the variety G_u/H_u.

Now consider Proposition 1.2 in the case where H is unipotent. In this case, the weight of the semi-invariants s and t is necessarily the constant 1, so that the first part of Proposition 1.2 says that $[\mathscr{P}(G)]^H$ coincides with $[\mathscr{P}(G)^H]$. By Theorem 3.1, this implies that G/H is quasi-affine. Moreover, the proof of Theorem 3.1 has shown that G/H is isomorphic with a G-orbit in an affine strict G-variety. If G is unipotent, it follows therefore from Lemma 4.1 that the variety G/H is affine. Thus, the above G_u/H_u is affine, whence also the original G/H is affine. $\qquad\square$

5. Theorem 3.1 has a representation-theoretical aspect involving the following notion. An algebraic subgroup H of an algebraic group G is said to be *observable* if every finite-dimensional polynomial H-module is a sub H-module of a polynomial G-module.

Proposition 5.1. *Let G be an algebraic group, H an algebraic subgroup of G. Suppose that, for every 1-dimensional polynomial H-module that is a sub H-module of a polynomial G-module, the dual H-module is also a sub H-module of a polynomial G-module. Then H is an observable subgroup of G.*

PROOF. Let N be a finite-dimensional polynomial H-module, and let M and g be as in Proposition 1.1. Let R be the 1-dimensional polynomial H-module determined by the polynomial character g, so that the automorphism of R corresponding to an element x of H is the scalar multiplication by $g(x)$.

In our proof of Proposition 1.1, R appeared as a sub H-module of the polynomial G-module $\bigwedge^n(T)$. Therefore, it follows from our present assumption that there is a polynomial G-module, S say, such that the dual of R, whose associated character is $1/g$, is a sub H-module of S. Now it is clear from Proposition 1.1 that N may be identified with a sub H-module of $M \otimes S$. $\square$

Theorem 5.2. *Let G be an irreducible algebraic group, H an algebraic subgroup of G. Then H is observable in G if and only if $[\mathscr{P}(G)]^H = [\mathscr{P}(G)^H]$.*

PROOF. First, suppose that H is observable in G, and let f be a non-zero element of $[\mathscr{P}(G)]^H$. We must show that f belongs to $[\mathscr{P}(G)^H]$. For this, it evidently suffices to show that the H-fixed part of the polynomial G-module $(\mathscr{P}(G)f) \cap \mathscr{P}(G)$ is not (0). We do this by copying the construction of the first part of the proof of Proposition 1.2. Thus, we start with any non-zero simple sub H-module V of this module. By the present assumption, V° is a sub H-module of some finite-dimensional polynomial G-module W. Since W is isomorphic with a sub G-module of a direct sum of a finite family of copies of $\mathscr{P}(G)$ and since V° is simple, it follows that there is an injective morphism of H-modules $\gamma: V^\circ \to \mathscr{P}(G)$. Choose a vector space basis $(v_1, \ldots, v_n)$ of V such that $v_1(1_G) \neq 0$ and $v_i(1_G) = 0$ for every $i > 1$. Let $(\mu_1, \ldots, \mu_n)$ be the dual basis of V°, and put $g_i = \gamma(\mu_i)$. Then $\sum_{i=1}^{n} (g_i \cdot x)v_i$ is an H-fixed element of $(\mathscr{P}(G)f) \cap \mathscr{P}(G)$ for every element x of G. Choose x so that $g_1(x) \neq 0$. The value at 1_G of our element is $g_1(x) \neq 0$, so that our element is not 0. This proves the necessity of the condition of the theorem.

In order to prove the sufficiency, we show first that it suffices to deal with the case where the base field, F say, is algebraically closed. Let F' be an algebraic closure of F, and let G' and H' be the affine algebraic F'-groups obtained from G and H by canonical base field extension. Thus, we have $\mathscr{P}(G') = \mathscr{P}(G) \otimes F'$, and when we identify G with its canonical image in G' then G is dense in G', and H' is the closure of H in G'. It follows that

$$[\mathscr{P}(G')^{H'}] = [\mathscr{P}(G)^H] \otimes F' \quad \text{and} \quad [\mathscr{P}(G')]^{H'} = [\mathscr{P}(G)]^H \otimes F'.$$

It is clear from this that if the condition of the theorem holds for the pair (G, H) then it holds also for the pair (G', H'). A similar but easier consideration shows that if H' is observable in G' then H is observable in G.

It remains only to establish the sufficiency of the condition in the case where F is algebraically closed. We show that if F is algebraically closed and $[\mathscr{P}(G)]^H = [\mathscr{P}(G)^H]$ then the condition of Proposition 5.1 is satisfied, so that H is then observable in G.

Suppose that V is a 1-dimensional polynomial H-module that is contained in a polynomial G-module. Let v be an F-space generator of V, and let f be the element σ/v of $\mathscr{P}(G)$, where σ is the element of V° such that $\sigma(v) = 1$. For every element y of H, we have $y \cdot f = f(y)f$. Moreover, for every element x of G, we have $y \cdot (f \cdot x) = f(y)f \cdot x$ whenever y belongs to H.

Hence, each fraction $(f \cdot x)/f$ belongs to $[\mathscr{P}(G)]^H$. By our present assumption, there is a non-zero element g_x in $\mathscr{P}(G)^H$ such that $(f \cdot x)g_x$ belongs to $\mathscr{P}(G)^H f$. Let Z denote the set of zeros of f in G, and write Z' for the complement of Z in G. Clearly, g_x must vanish on $Z \cap x^{-1}Z'$. Since the family of open sets $x^{-1}Z'$ covers the closed set Z and since G is a Noetherian space, there is a finite set $(x_1, \ldots, x_n)$ in G such that Z is the union of the family of sets $Z \cap x_i^{-1}Z'$. Put $g = g_{x_1} \cdots g_{x_n}$. Then g is a non-zero element of $\mathscr{P}(G)^H$ that vanishes on Z. Since F is algebraically closed, we can apply Theorem II.3.5 and conclude that there is an exponent $e > 0$ such that $g^e = hf$ with h in $\mathscr{P}(G)$. Since g is fixed under the action of H, we must have $y \cdot h = f(y)^{-1}h$ for every element y of H. This shows that the sub H-module of $\mathscr{P}(G)$ that is spanned by h may be identified with V°, so that V° is contained in a polynomial G-module. $\square$

Notes

1. The simplest example of a coset variety that is quasi-affine but not affine is as follows. Let G be the multiplicative group of matrices $\begin{pmatrix} a & b \\ c & d \end{pmatrix}$ with $ad - bc = 1$, over an algebraically closed field F. Let H be the subgroup of matrices $\begin{pmatrix} 1 & b \\ 0 & 1 \end{pmatrix}$. Let α, β, γ, δ be the matrix entry functions on G with values a, b, c, d, respectively. We regard G as an affine algebraic F-group with $\mathscr{P}(G) = F[\alpha, \beta, \gamma, \delta]$, where $\alpha\delta - \beta\gamma = 1$. Let A be the sub F-algebra $F[\alpha, \gamma]$. One verifies directly that A is stable under the action of G from the right, and that the element-wise fixer of A with respect to the action of G from the left coincides with H. With a little more computation, one sees that, actually, $A = \mathscr{P}(G)^H$. From the proof of Theorem 3.1, it is now clear that the variety G/H is isomorphic with the restriction image of G in $\mathscr{S}(A)$. Clearly, $\mathscr{S}(A)$ is the 2-dimensional F-space, and the image of G is the complement of the point $(0, 0)$. From Note IX.1, we know that this quasi-affine variety is not affine.

2. The following corollaries are worth noting: if G is an irreducible solvable algebraic group over an algebraically closed field then every algebraic subgroup of G is observable in G; if G is any irreducible algebraic group over an algebraically closed field, and H is an algebraic subgroup of G that is either unipotent or coincides with its commutator subgroup, then G/H is quasi-affine.

3. The following procedure shows that, *if G is an irreducible affine algebraic group over the algebraically closed field F, and if H is a linearly reductive algebraic subgroup of G, then the variety G/H is affine.*

First, observe that, since H is linearly reductive, the module V in the proof of Proposition 1.1 may be taken to be (0), whence the polynomial character g of Proposition 1.1 may be taken to be the constant 1. Therefore, the polynomial character g of Proposition 1.2 may also be taken to be the constant 1, whence $[\mathscr{P}(G)]^H$ coincides with $[\mathscr{P}(G)^H]$.

By Corollary V.3.2, $\mathscr{P}(G)^H$ is finitely generated as an F-algebra, so that we have the irreducible affine F-variety $\mathscr{S}(\mathscr{P}(G)^H)$. By the above, the canonical map from G/H to this variety is injective. We have the canonical H-module decomposition $\mathscr{P}(G) = \mathscr{P}(G)^H + \mathscr{P}(G)_H$. Now let σ be an element of $\mathscr{S}(\mathscr{P}(G)^H)$, and let J be its kernel in $\mathscr{P}(G)^H$. Our decomposition of $\mathscr{P}(G)$ shows that $J\mathscr{P}(G) \subset J + \mathscr{P}(G)_H$, whence we see that $J\mathscr{P}(G)$ does not coincide with $\mathscr{P}(G)$. If M is a maximal ideal of $\mathscr{P}(G)$ containing $J\mathscr{P}(G)$ then $M \cap \mathscr{P}(G)^H = J$, and it follows that σ extends to an F-algebra homomorphism from $\mathscr{P}(G)$ to F. Thus, the canonical map from G/H to $\mathscr{S}(\mathscr{P}(G)^H)$ is actually bijective. Now it follows as in Section 2 (for the case where H is normal in G) that this map is an isomorphism of varieties.

4. The discussion preceding Lemma 2.1 of the differential of the canonical morphism π from G to G/H appealed to Proposition 1.2 for the conclusion that the kernel of π'_1 coincides with $\mathscr{L}(H)$. This is justified by noting that if τ is an element of the kernel of π'_1 in $\mathscr{L}(G)$ then the derivation of $[\mathscr{P}(G)]$ effected by τ annihilates $[\mathscr{P}(G)]^H$. Indeed, if f is an element of $[\mathscr{P}(G)]^H$, so is $f \cdot x$ for every element x of G, and this is defined at 1 for all elements x of some non-empty open subset S_f of G. Indicating the action of τ as a derivation of $[\mathscr{P}(G)]$ by a dot, we have

$$(\tau \cdot f)(x) = ((\tau \cdot f) \cdot x)(1) = (\tau \cdot (f \cdot x))(1) = \tau'_1(f \cdot x) = 0$$

for every x in S_f, whence $\tau \cdot f = 0$.

Chapter XIII

Borel Subgroups

This chapter contains the basic ingredients for the detailed structure theory of algebraic groups that leads to the classification of the simple groups. This theory is based on certain families of subgroups, such as toroids and *Borel subgroups*, i.e., irreducible maximal solvable subgroups. To some extent, the consideration of Borel subgroups reduces the structure theory to that of solvable groups.

The principal result of Section 1 is Borel's fixed point theorem concerning actions of solvable groups on complete varieties. This is the key for the study of groups via their Borel subgroups.

Let G be an irreducible algebraic group over an algebraically closed field. Section 2 establishes the conjugacy of the Borel subgroups of G and characterizes the subgroups P of G that contain a Borel subgroup by the property that G/P is a complete variety. Section 3 shows that G is the union of the family of its Borel subgroups. The main result of Section 4 is that the centralizer of a toroid S in G is irreducible and that its Borel subgroups are its intersections with the Borel subgroups of G that contain S. Section 5 shows that the algebraic subgroups of G containing a Borel subgroup are irreducible and coincide with their normalizers. Finally, it introduces the Weyl group of G, deriving the simplest basic property with regard to a maximal toroid of G.

1. Proposition 1.1. *Let G be an irreducible algebraic group over an algebraically closed field, and let V be a strict G-variety. Each G-orbit of minimal dimension is closed in V.*

PROOF. By Theorem X.4.4, each orbit $G \cdot v$ is a constructible subset of V, and thus a subvariety. Since G is irreducible, so is $G \cdot v$. Without loss of generality, we replace V with the closure of $G \cdot v$ in V, and so assume that $G \cdot v$ is dense in V. Then we know from Theorem X.4.3 that $G \cdot v$ contains a non-empty open subset of V. From the transitivity of the action of G on $G \cdot v$, it follows that $G \cdot v$ is open in V. Now $V \setminus G \cdot v$ is stable under the action of G, and every irreducible component of $V \setminus G \cdot v$ is of strictly smaller dimension than V, i.e., of strictly smaller dimension than $G \cdot v$. If $G \cdot v$ is of minimal dimension among the G-orbits, $V \setminus G \cdot v$ must therefore be empty. $\square$

Lemma 1.2. *Let F be an algebraically closed field, and let $\rho: X \to Y$ be a bijective morphism between irreducible algebraic F-varieties. Suppose that an arbitrary group G acts transitively by variety automorphisms on X and on Y, and that ρ commutes with the action of G. Then ρ is a closed map.*

PROOF. By Proposition X.5.2, $F(X)$ is a finite algebraic extension of $F(Y) \circ \rho$. As in the proof of Theorem X.4.3, with $r = 0$, we see from this that there are affine patches U of X and V of Y such that ρ restricts to a *finite* morphism from U to V. By Proposition X.2.2, this finite morphism is a closed map. The same is therefore true for the restriction of ρ to $g \cdot U$ for every element g of G. Now let C be a closed subset of X. Then, for every g in G, the image $\rho(C \cap g \cdot U)$ is closed in $g \cdot V$, i.e., $\rho(C) \cap g \cdot V$ is closed in $g \cdot V$. Since Y is the union of a finite family of such affine patches $g \cdot V$, it follows that $\rho(C)$ is closed in Y. $\square$

The next theorem is a fundamental tool theorem for the structure theory of algebraic groups. It is known as *Borel's fixed point theorem.*

Theorem 1.3. *Let G be an irreducible solvable algebraic group over an algebraically closed field, and let X be a complete strict G-variety. Then the set X^G of G-fixed points of X is not empty.*

PROOF. Making an induction on the dimension of G, we suppose that the theorem has been established in the lower cases. Then we know that $X^{[G, G]}$ is not empty. Being closed in X, this is a complete variety, and it is evidently stable under the action of G. This action factors through $G/[G, G]$, and we see from Theorem XII.2.2 that $X^{[G, G]}$ thereby becomes a strict $G/[G, G]$-variety. This reduces the theorem to the case where G is commutative.

In this case, choose a point x from X such that the orbit $G \cdot x$ is of minimal dimension. We know from Proposition 1.1 that $G \cdot x$ is closed in X, so that it is a complete variety. Let G^x denote the fixer of x in G. Then the map from G to $G \cdot x$ sending each element g of G onto $g \cdot x$ induces a bijective morphism of varieties

$$\eta: G/G^x \to G \cdot x.$$

Since G is commutative, it is clear that G/G^x is an *affine* variety (note that we could get to this point without making an induction by using Theorem XII.4.3).

On the other hand, from the fact that $G \cdot x$ is complete, we find that G/G^x is complete, as follows. Let V be an arbitrary variety, and consider the projection morphism

$$\pi \colon (G/G^x) \times V \to V.$$

Using the above η, we may factor this as shown:

$$(G/G^x) \times V \quad \xrightarrow{\ \eta \times i_V\ } \quad (G \cdot x) \times V \quad \xrightarrow{\ \pi'\ } \quad V.$$

Since $G \cdot x$ is complete, the projection π' is a closed map. Applying Lemma 1.2, we see that $\eta \times i_V$ also is a closed map. Therefore, the composite π is a closed map, showing that G/G^x is complete.

Since G/G^x is also affine, we must therefore have $G = G^x$, which means that x belongs to X^G. $\qquad\square$

2. Let G be an irreducible algebraic group. A *Borel subgroup* of G is an irreducible solvable algebraic subgroup of G that is maximal in the family of all such subgroups. We shall see later on that a Borel subgroup is actually maximal in the family of all solvable subgroups of G.

Let V be a vector space of finite dimension $n > 0$. A *full flag* in V is an n-tuple $(S_1, \ldots, S_n)$ of subspaces of V such that S_i is of dimension i and $S_i \subset S_{i+1}$. These flags may be regarded as points of the projective variety $\mathscr{G}_1(V) \times \cdots \times \mathscr{G}_n(V)$. It is easy to see from the definition of the variety structure of $\mathscr{G}_d(V)$ as given in Section IX.4 that the full flags constitute a closed subset of $\mathscr{G}_1(V) \times \cdots \times \mathscr{G}_n(V)$. Now let us suppose that the base field is algebraically closed, and let G be the irreducible algebraic group of all linear automorphisms of V. In the canonical fashion, G acts by variety automorphisms on $\mathscr{G}_1(V) \times \cdots \times \mathscr{G}_n(V)$, and the set of full flags is stable under the action of G. Moreover, it is clear that G acts transitively on this set. It follows that the full flags constitute an *irreducible* closed subvariety of $\mathscr{G}_1(V) \times \cdots \times \mathscr{G}_n(V)$. In particular, *the variety of full flags is an irreducible projective variety.*

Theorem 2.1. *Let G be an irreducible algebraic group over an algebraically closed field. For every Borel subgroup B of G, the algebraic variety G/B is a projective variety, and every Borel subgroup of G is a conjugate of B.*

PROOF. Let C be a Borel subgroup of the largest possible dimension. By Theorem II.2.1, there is an injective polynomial representation of G on a finite-dimensional vector space V having a 1-dimensional subspace S_1 whose stabilizer in G coincides with C. Consider the induced representation of C on V/S_1. By Theorem IV.1.1, C stabilizes a full flag in V/S_1. Hence, there is a full flag $(S_1, \ldots, S_n)$ in V whose stabilizer in G is precisely C. We regard this flag as a point, p say, of the irreducible projective variety of all full flags of V. Let

$$\eta\colon G/C \to G \cdot p,$$

be the canonical bijective morphism. Clearly,

$$\dim(G \cdot p) = \dim(G) - \dim(C).$$

For every full flag q, the identity component of the fixer of q in G is an irreducible solvable algebraic subgroup T, and

$$\dim(G \cdot q) = \dim(G) - \dim(T) \geq \dim(G) - \dim(C).$$

Thus, the orbit $G \cdot p$ is an orbit of minimal dimension. By Proposition 1.1, this implies that $G \cdot p$ is closed in the flag variety, so that it is a projective variety. By Lemma 1.2, the map η is a closed map, and the argument of the end of Section 1 shows that G/C is therefore a complete variety. We know from Theorem XII.2.2 that G/C is a subvariety of a projective variety, W say. Since G/C is complete, it must be closed in W, so that it is a projective variety.

Now consider the canonical action of an arbitrary Borel subgroup B on the projective variety G/C. By Theorem 1.3, there is a point xC of G/C that is fixed under the action of B. This means that $BxC = xC$, whence $x^{-1}Bx \subset C$. Since $x^{-1}Bx$ is a Borel subgroup, it follows that $x^{-1}Bx = C$. $\qquad\square$

Theorem 2.2. *Let G be an irreducible algebraic group over an algebraically closed field. An algebraic subgroup P of G contains a Borel subgroup of G if and only if G/P is complete.*

PROOF. If P contains a Borel subgroup B of G, then the canonical morphism π from G to G/P is constant on the cosets xB and therefore induces a morphism π^B from G/B to G/P. Since G/B is complete and π^B is surjective, it follows from Proposition IX.5.2 that G/P is complete.

Now suppose that G/P is complete, and let B be any Borel subgroup of G. By Theorem 1.3, there is a point xP of G/P that is fixed under the action of B. Thus, $BxP = xP$, showing that P contains the Borel subgroup $x^{-1}Bx$ of G. $\qquad\square$

Proposition 2.3. *Let G be an irreducible algebraic group over an algebraically closed field. Let α be an automorphism of G leaving the elements of some Borel subgroup B fixed. Then α is the identity automorphism.*

PROOF. Consider the map δ from G to G where $\delta(x) = \alpha(x)x^{-1}$. This is a morphism of varieties that is constant on each coset xB. Therefore, δ defines a morphism of varieties δ^B from G/B to G. Since G/B is complete, we know from Proposition IX.5.2 that $\delta^B(G/B)$ is closed in G and complete. Since G is affine, this implies that $\delta^B(G/B)$ consists of a single point, which means that α is the identity map. $\qquad\square$

Proposition 2.4. *Let G be as above. If a Borel subgroup of G is nilpotent then it coincides with G.*

PROOF. We make an induction on the dimension of G, and suppose that the proposition has been established in the lower cases. Let B be a nilpotent Borel subgroup of G. If B is trivial then it follows from Theorem 2.1 that G is trivial. Therefore, we suppose that B is non-trivial. For any group K, let $\mathscr{C}(K)$ denote the center of K. By Proposition 2.3, we have $\mathscr{C}(B) \subset \mathscr{C}(G)$, and it is clear from Theorem 2.1 that $\mathscr{C}(G)_1 \subset B$. Hence, we have $\mathscr{C}(G)_1 = \mathscr{C}(B)_1$. The nilpotency of B implies that $\mathscr{C}(B)_1$ is non-trivial. Now $B/\mathscr{C}(G)_1$ is a nilpotent Borel subgroup of $G/\mathscr{C}(G)_1$, and our inductive hypothesis gives $B = G$. $\qquad\square$

We denote the normalizer of a subset K of G by $\mathscr{N}_G(K)$, and the centralizer by $\mathscr{C}_G(K)$.

Proposition 2.5. *Let G be as above, and let T be a maximal toroid in G. Let $C = \mathscr{C}_G(T)_1$. Then C is nilpotent and coincides with $\mathscr{N}_G(C)_1$.*

PROOF. There is a Borel subgroup S of C such that $T \subset S$. Evidently, T is a maximal toroid in S. By Theorem VI.3.2, we have $S = S_u \rtimes T$. Since T is central in S, this means that S is the direct product of S_u and T, so that S is nilpotent. By Proposition 2.4, C is therefore nilpotent.

Now consider the map from $\mathscr{N}_G(C)_1 \times T$ to C that sends each (x, t) onto xtx^{-1}. Since T is stable under every automorphism of C, this map defines a morphism of varieties

$$\delta: \mathscr{N}_G(C)_1 \times T \to T.$$

For every positive integer e, let $T(e)$ denote the subgroup of T consisting of the elements t such that $t^e = 1_T$. If t belongs to $T(e)$ then $\delta(\mathscr{N}_G(C)_1 \times (t))$ is evidently contained in $T(e)$. Since $T(e)$ is finite, while $\delta(\mathscr{N}_G(C)_1 \times (t))$ is irreducible, it follows that t belongs to the center of $\mathscr{N}_G(C)_1$. Thus, every $T(e)$ is contained in the center of $\mathscr{N}_G(C)_1$. Since the union of the family of subgroups $T(e)$ is dense in T, it follows that T lies in the center of $\mathscr{N}_G(C)_1$, whence $\mathscr{N}_G(C)_1 = C$. $\qquad\square$

3. Lemma 3.1. *Let G be an irreducible algebraic group over an algebraically closed field, and let H be an irreducible algebraic subgroup of G. If G/H is complete then $\cup_{x \in G} xHx^{-1}$ is closed in G. If there is an element of H whose set of fixed points in G/H is finite then $\cup_{x \in G} xHx^{-1}$ contains a non-empty open subset of G.*

PROOF. Let τ denote the map from $G \times G$ to $G \times G$ sending each (x, y) onto (x, xyx^{-1}). Evidently, τ is an automorphism of the variety $G \times G$. Let π denote the canonical morphism from G to G/H, and consider the map $(\pi \times i_G) \circ \tau$ from $G \times G$ to $(G/H) \times G$. Let S denote the image of $G \times H$ under this map. We claim that S is closed in $(G/H) \times G$. Since $\pi \times i_G$ is an open map, it suffices to show that $(\pi \times i_G)^{-1}(S)$ is closed in $G \times G$. This set is easily seen to coincide with $\tau(G \times H)$, which is closed in $G \times G$. Thus, S is closed in $(G/H) \times G$.

Now suppose that G/H is complete. Then the canonical projection from $(G/H) \times G$ to G is a closed map, so that the projection image of S in G is closed. This projection image is precisely $\cup_{x \in G} xHx^{-1}$, so that the first part of the lemma is established.

For each point xH of G/H, the inverse image of xH in S with respect to the projection from $(G/H) \times G$ to G/H is isomorphic with xHx^{-1} as a variety, and thus has always the same dimension $\dim(H)$. By Theorem X.4.3, this implies that $\dim(H) = \dim(S) - \dim(G/H)$, whence $\dim(S) = \dim(G)$.

Now suppose that there is an element h in H whose set of fixed points in G/H is finite. This means that the inverse image of (h) in S, with respect to the projection from S to G, is finite. By Theorem X.2.1, it follows that the dimension of the projection image of S in G is at least equal to $\dim(S)$, i.e., by the above, at least equal to $\dim(G)$. Thus, the projection from S to G is a *dominant* morphism. By Theorem X.4.3, the projection image of S in G therefore contains a non-empty open subset of G. This establishes the second part of the lemma. $\qquad\square$

Proposition 3.2. *Let G be as above, and let S be any toroid in G. There is an element s in S such that every element of G that commutes with s belongs to $\mathscr{C}_G(S)$.*

PROOF. Let V be a finite-dimensional polynomial G-module such that the representation of G on V is injective. We may write V as a direct sum of S-stable subspaces V_i corresponding to mutually distinct polynomial characters f_i such that every element s of S acts as the scalar multiplication by $f_i(s)$ on V_i. For each index pair (i, j), let S_{ij} be the set of all elements s of S for which $f_i(s) = f_j(s)$. For $i \neq j$, S_{ij} is closed in S, but does not coincide with S. Since S is irreducible, there is an element s in S not belonging to any one of these S_{ij}'s. Now, if x is an element of G that commutes with s then x stabilizes each V_i, whence it commutes with every element of S. $\qquad\square$

Theorem 3.3. *Let G be an irreducible algebraic group over an algebraically closed field, and let B be a Borel subgroup of G. Then $\cup_{x \in G} xBx^{-1}$ coincides with G.*

PROOF. Choose a maximal toroid T in G, and write C for $\mathscr{C}_G(T)_1$. By Proposition 2.5, C is nilpotent, so that it follows from Theorem VI.3.1 and the maximality of T that $C = C_u \times T$. By Proposition 3.2, there is an element t in (T)

such that $\mathscr{C}_G(T)_1 = \mathscr{C}_G(t)_1$. We show that the fixed point set for t in G/C is finite, and then apply Lemma 3.1.

Let x be an element of G such that $txC = xC$. Then $x^{-1}tx$ is a semisimple element of C, and therefore belongs to T. Hence, every element of T commutes with $x^{-1}tx$ or, equivalently, every element of xTx^{-1} commutes with t. By the choice of t, this implies that $xTx^{-1} \subset C$, whence $xTx^{-1} = T$. Thus, the conjugation effected by x on G stabilizes T, so that it must also stabilize C, which means that x belongs to $\mathscr{N}_G(C)$. We know from Proposition 2.5 that $\mathscr{N}_G(C)_1 = C$. Therefore, the fixed point set for t in G/C is in bijective correspondence with a subset of the finite set $\mathscr{N}_G(C)/\mathscr{N}_G(C)_1$.

Now we have from Lemma 3.1 that $\cup_{x \in G} xCx^{-1}$ contains a non-empty open subset of G. The nilpotent irreducible algebraic subgroup C of G is contained in some Borel subgroup, B say, of G, and $\cup_{x \in G} xBx^{-1}$ contains a non-empty open set of G. On the other hand, by the first part of Lemma 3.1, $\cup_{x \in G} xBx^{-1}$ is closed in G. Therefore, it must coincide with G. $\qquad\square$

We shall see later on that $B = \mathscr{N}_G(B)$. At this point, we record the following weaker result.

Corollary 3.4. *If G and B are as in Theorem 3.3 then $B = \mathscr{N}_G(B)_1$.*

PROOF. Evidently, B is a Borel subgroup of $\mathscr{N}_G(B)_1$. Since B is normal in $\mathscr{N}_G(B)_1$, the corollary follows at once from Theorem 3.3, with $\mathscr{N}_G(B)_1$ in the place of G.

4. Lemma 4.1. *Let G be an algebraic group over an algebraically closed field, and let U be an irreducible unipotent algebraic subgroup of G. Suppose that s is a semisimple element of G that normalizes U. Then the centralizer of s in U is irreducible.*

PROOF. First, we deal with the case where U is commutative. Let U^s denote the centralizer of s in U, and let U_s be the subgroup of U consisting of the elements $sus^{-1}u^{-1}$ with u in U. We claim that $U^s \cap U_s = (1)$. In order to see this, consider an element $v = sus^{-1}u^{-1}$ of this intersection. Let S be the closure in G of the group generated by s. Since s is semisimple, S is linearly reductive. We have $sus^{-1} = vu$, whence $s^e us^{-e} = v^e u$ for every integer e. Hence $xux^{-1}u^{-1}$ belongs to U^s for every element x of the group generated by s, and therefore also for every element x of S. Consider the map δ from S to U^s that sends each x onto $xux^{-1}u^{-1}$. One sees directly that δ is a morphism of algebraic groups. It follows that $\delta(S)$ is a reductive subgroup of U^s. Since U^s is unipotent, $\delta(S)$ is therefore trivial. In particular, $v = \delta(s) = 1$. Thus, $U^s \cap U_s = (1)$.

Now consider the map γ from U to U_s that sends each u onto $sus^{-1}u^{-1}$. This is evidently a surjective morphism of algebraic groups, whence U_s is irreducible. The kernel of γ is U^s, whence

$$\dim(U) = \dim(U^s) + \dim(U_s).$$

Since $U^s \cap U_s = (1)$, the morphism of algebraic groups from $U^s \times U_s$ to U coming from the composition of U is injective. This implies that
$$\dim((U^s)_1 U_s) = \dim((U^s)_1 \times U_s) = \dim(U),$$
whence $(U^s)_1 U_s = U$. Hence, the images of $U^s \times U_s$ and $(U^s)_1 \times U_s$ in U coincide, so that we must have $U^s = (U^s)_1$. This establishes the lemma in the case where U is commutative.

Now we proceed by induction on the dimension of U, and so suppose that $U \neq (1)$ and that the lemma has been established in the lower cases. Let Z denote the irreducible component of the neutral element in the center of U. Since U is unipotent and non-trivial, we have $Z \neq (1)$. Hence, our inductive hypothesis implies that $(U/Z)^s$ is irreducible.

Let uZ be an element of $(U/Z)^s$, so that $sus^{-1}u^{-1}$ belongs to Z. It follows that $xux^{-1}u^{-1}$ belongs to Z for every element x of S. Now consider the map δ from S to Z that sends each x onto $xux^{-1}u^{-1}$. One verifies directly that δ is a polynomial cocycle for S in Z with respect to the conjugation action of S on Z. By Proposition VI.2.2, this cocycle is a coboundary, i.e., there is an element z in Z such that $xux^{-1}u^{-1} = xzx^{-1}z^{-1}$ for every element x of S. This shows that $z^{-1}u$ belongs to U^s.

Our conclusion is that $(U/Z)^s$ coincides with the canonical image of U^s. The kernel of the canonical morphism from U^s to $(U/Z)^s$ is Z^s, which we know to be irreducible. Since this morphism is surjective to the irreducible group $(U/Z)^s$, it follows that U^s is irreducible. $\qquad\square$

Theorem 4.2. *Let G be an irreducible algebraic group over an algebraically closed field, S a toroidal algebraic subgroup of G. Then $\mathscr{C}_G(S)$ is irreducible.*

PROOF. Let x be any element of $\mathscr{C}_G(S)$, and let B be a Borel subgroup of G. By Theorem 3.3, x belongs to some conjugate of B. This means that the fixed point set, P say, for x in G/B is non-empty. Being closed in the complete variety G/B, this set P is a complete variety. Since the elements of S commute with x, it is clear that P is stable under the action of S on G/B. Therefore, we know from Theorem 1.3 that there is a fixed point for S in P. If this is zB then zBz^{-1} contains S as well as x. Now x belongs to the centralizer of S in zBz^{-1}. If we show that this centralizer is irreducible we may conclude that x belongs to $\mathscr{C}_G(S)_1$. Thus, it suffices to establish the theorem in the case where G is solvable.

In that case, we have $G = G_u \rtimes T$, where T is a maximal toroid in G containing S. Clearly, $\mathscr{C}_G(S)$ is the semidirect product $(\mathscr{C}_G(S) \cap G_u) \rtimes T$. By Proposition 3.2, there is an element s in S such that $\mathscr{C}_G(S) = G^s$. Now we have $\mathscr{C}_G(S) = G_u^s \rtimes T$. By Lemma 4.1, G_u^s is irreducible, so that $\mathscr{C}_G(S)$ is irreducible. $\qquad\square$

Lemma 4.3. *Let G and S be as in Theorem 4.2, and let B be a Borel subgroup of G containing S. Let P be the fixed point set for S in G/B, and let p be a fixed point for B in G/B. Then $\mathscr{C}_G(S) \cdot p$ coincides with the irreducible component of p in P.*

PROOF. Let Q denote the irreducible component of p in P, and let π be the canonical map from G to G/B. First, we note that $\pi^{-1}(Q)$ is irreducible. In order to see this, let H be an irreducible component of $\pi^{-1}(Q)$. Since B is irreducible, so is the product set HB, whence $HB = H$. It follows that $Q\setminus\pi(H)$ coincides with $\pi(\pi^{-1}(Q)\setminus H)$. Since π is an open map, this shows that $\pi(H)$ is closed in Q. Since Q is irreducible, we must therefore have $\pi(H) = Q$ for at least one of the irreducible components H of $\pi^{-1}(Q)$. Since $HB = H$, it follows that $H = \pi^{-1}(Q)$.

Let x be an element of $\pi^{-1}(Q)$. Then $S \cdot \pi(x)$ is the singleton $(\pi(x))$, which means that $x^{-1}Sx$ is contained in B. Let δ be the variety morphism from $\pi^{-1}(Q) \times S$ to B/B_u that sends each (x, s) onto $x^{-1}sxB_u$. If s is an element of S that is of finite order, then the image of $\pi^{-1}(Q) \times (s)$ under δ lies in a finite subgroup of the toroid B/B_u. Since it is irreducible, this image is therefore the singleton $(t^{-1}stB_u)$, where t is an element of G such that $p = tB$. Thus, we have

$$(t^{-1}s^{-1}t)(x^{-1}sx) \in B_u,$$

for every element x of $\pi^{-1}(Q)$ and every torsion element s of S. Since the torsion subgroup of S is dense in S, it follows that the above holds for *every* element s of S. Hence, we have $x^{-1}Sx \subset t^{-1}StB_u$ for every element x of $\pi^{-1}(Q)$. From the fact that p is fixed under the action of B, it follows that t normalizes B, and hence also B_u. Therefore, $t^{-1}StB_u$ is actually a subgroup of G. Clearly, it is therefore an irreducible solvable algebraic subgroup of G, and its maximum unipotent normal subgroup coincides with B_u. Each of the groups $t^{-1}St$ and $x^{-1}Sx$ is a maximal toroid in $t^{-1}StB_u$. Therefore, there is an element b in B_u such that $b^{-1}x^{-1}Sxb = t^{-1}St$, so that xbt^{-1} belongs to $\mathscr{N}_G(S)$. Thus, we have $\pi^{-1}(Q) \subset \mathscr{N}_G(S)tB$.

Evidently, $\mathscr{C}_G(S)tB$ is contained in $\pi^{-1}(Q)$, so that we have the inclusions

$$\mathscr{C}_G(S)tB \subset \pi^{-1}(Q) \subset \mathscr{N}_G(S)tB_u.$$

The set $\mathscr{N}_G(S)tB_u$ is the union of a finite family of translates of the closed irreducible subset $\mathscr{N}_G(S)_1tB_u$. By considering the conjugation action of $\mathscr{N}_G(S)_1$ on the torsion subgroup of S, we see that $\mathscr{N}_G(S)_1$ is contained in $\mathscr{C}_G(S)$. By Theorem 4.2, $\mathscr{C}_G(S)$ is irreducible, and therefore $\mathscr{N}_G(S)_1 = \mathscr{C}_G(S)$. Thus, the irreducible components of $\mathscr{N}_G(S)tB_u$ are translates of $\mathscr{C}_G(S)tB_u$. Since $\pi^{-1}(Q)$ is irreducible, it must be contained in one of these translates. Since it contains $\mathscr{C}_G(S)tB_u$, we must therefore have $\pi^{-1}(Q) = \mathscr{C}_G(S)tB_u$. Applying π, we find $Q = \mathscr{C}_G(S) \cdot p$.　　　　　$\square$

Theorem 4.4. *Let G be an irreducible algebraic group over an algebraically closed field. Let S be a toroidal subgroup of G, and let B be a Borel subgroup of G containing S. Then $\mathscr{C}_G(S) \cap B$ is a Borel subgroup of $\mathscr{C}_G(S)$.*

PROOF. Applying Lemma 4.3, with the point B of G/B taking the place of p, we conclude that the canonical image of $\mathscr{C}_G(S)$ in G/B is closed in G/B, and therefore complete. The canonical map induces a bijective morphism from

$\mathscr{C}_G(S)/(\mathscr{C}_G(S) \cap B)$ to the canonical image of $\mathscr{C}_G(S)$ in G/B. Therefore, we can apply the argument of the end of Section 1 and conclude that the variety $\mathscr{C}_G(S)/(\mathscr{C}_G(S) \cap B)$ is complete. By Theorem 2.2, this implies that $\mathscr{C}_G(S) \cap B$ contains a Borel subgroup of $\mathscr{C}_G(S)$. We have $\mathscr{C}_G(S) \cap B = \mathscr{C}_B(S)$, so that we know from Theorem 4.2 that this group is irreducible. Since it is solvable, it must therefore coincide with the Borel subgroup it contains. $\qquad\square$

5. Theorem 5.1. *Let G be an irreducible algebraic group over an algebraically closed field, and let B be a Borel subgroup of G. Then $\mathscr{N}_G(B) = B$.*

PROOF. Let x be an element of $\mathscr{N}_G(B)$, and let T be a maximal toroid in G that is contained in B. Then xTx^{-1} is also a maximal toroid in G that is contained in B. Because the maximal toroids in B are conjugates in B, there is an element b in B such that xb normalizes T. In order to conclude that x belongs to B, it suffices to show that xb belongs to B. Thus, we assume without loss of generality that x normalizes T, as well as B. We proceed by induction on the dimension of G, and so assume that the theorem has been established in the lower cases.

Let $S = \mathscr{C}_T(x)_1$, and let us first deal with the case where S is non-trivial. We know from Theorem 4.2 that $\mathscr{C}_G(S)$ is irreducible, and from Theorem 4.4 that $\mathscr{C}_G(S) \cap B$ is a Borel subgroup of $\mathscr{C}_G(S)$. From the inductive hypothesis, we have that every Borel subgroup of $\mathscr{C}_G(S)/S$ coincides with its normalizer. Clearly, S must be contained in every Borel subgroup of $\mathscr{C}_G(S)$. Hence, it follows that every Borel subgroup of $\mathscr{C}_G(S)$ coincides with its normalizer in $\mathscr{C}_G(S)$. In particular, this applies to $\mathscr{C}_G(S) \cap B$. Since x belongs to $\mathscr{C}_G(S)$ and normalizes B, we have therefore $x \in B$.

It remains to deal with the case where S is trivial. We know from Corollary 3.4 that $B = \mathscr{N}_G(B)_1$. It follows that some power of x belongs to B, so that the subgroup, H say, of G that is generated by x and B is an *algebraic* subgroup of G, with $H_1 = B$. Let δ be the algebraic group endomorphism of T that sends each t onto $xtx^{-1}t^{-1}$. Our assumption that S is trivial means that the kernel $\mathscr{C}_T(x)$ of δ is finite. This implies that $\dim(\delta(T)) = \dim(T)$, so that δ is surjective. In particular, this shows that T is contained in $[H, H]$. Since $B = B_u T$, it follows that B is contained in $B_u[H, H]$.

By Theorem II.2.1, there is a finite-dimensional polynomial G-module V such that H coincides with the stabilizer in G of some 1-dimensional subspace L of V. Let μ be the corresponding polynomial character of H, so that every element h of H acts on L as the scalar multiplication by $\mu(h)$. Then, since $B \subset B_u[H, H]$, it is clear that B is contained in the kernel of μ.

Now let v be a non-zero element of L, and consider the map ρ from G to V that sends each g onto $g \cdot v$. Since B lies in the kernel of μ, the map ρ is constant on the cosets gB, and so induces the morphism ρ^B from G/B to V. Since G/B is complete and V is affine, the image of G/B under ρ^B reduces to the single point v. By the definition of L, this implies that $G = H$. Since G is irreducible, it follows that $G = H_1 = B$, so that $x \in B$. $\qquad\square$

Corollary 5.2. *With G as in Theorem 5.1, let P be an algebraic subgroup of G containing a Borel subgroup of G. Then P is irreducible, and* $P = \mathcal{N}_G(P)$.

PROOF. Let B be a Borel subgroup of G that is contained in P, and let x be an element of $\mathcal{N}_G(P)$. Then B and xBx^{-1} are Borel subgroups of P_1. Hence, there is an element p in P_1 such that px normalizes B. By Theorem 5.1, px belongs to B, so that x belongs to $P_1 B = P_1$. Thus $\mathcal{N}_G(P) = P_1 = P$. $\square$

It is an immediate consequence of this corollary that *every Borel subgroup of G is maximal in the family of all solvable subgroups of G.*

Corollary 5.3. *Let G be as above, B a Borel subgroup of G. Then* $B = \mathcal{N}_G(B_u)$.

PROOF. Write P for $\mathcal{N}_G(B_u)$. Since P contains B, we know from Corollary 5.2 that P is irreducible. From the conjugacy of Borel subgroups, it follows that B_u is maximal in the family of irreducible unipotent subgroups of G. Hence, P/B_u has no non-trivial irreducible unipotent subgroups. Therefore, if C is a Borel subgroup of P/B_u, then C is a toroid. Now it follows from Proposition 2.4 that $P/B_u = C$. Thus, P/B_u is solvable, whence P is solvable, so that $P = B$. $\square$

Let G and B be as above, and let $\mathscr{B}$ denote the set of all Borel subgroups of G. We define a map $\gamma \colon G/B \to \mathscr{B}$ by $\gamma(xB) = xBx^{-1}$. By Theorems 2.1 and 5.1, γ is bijective. If L is any subset of G, we denote the fixed point set for L in G/B by $(G/B)^L$. Then γ maps $(G/B)^L$ onto the set $\mathscr{B}(L)$ of all Borel subgroups of G that contain L.

In particular, let T be a maximal toroid in G, and choose B so that $T \subset B$. Clearly, $(G/B)^T$ is stable under the action of $\mathcal{N}_G(T)$. We claim that $\mathcal{N}_G(T)$ acts transitively on $(G/B)^T$. By the above, this is equivalent to saying that $\mathcal{N}_G(T)$ acts transitively on $\mathscr{B}(T)$, by conjugation. Let X and Y be members of $\mathscr{B}(T)$. There is an element g in G such that $gXg^{-1} = Y$. Now T and gTg^{-1} are maximal toroids in Y. Therefore, there is an element y in Y such that yg belongs to $\mathcal{N}_G(T)$. Since $(yg)X(yg)^{-1} = Y$, this proves the transitivity.

Next, we show that $\mathscr{C}_G(T)$ is contained in every member of $\mathscr{B}(T)$. We know from Theorem 4.2 that $\mathscr{C}_G(T)$ is irreducible. Hence, we have from Proposition 2.5 that $\mathscr{C}_G(T)$ is nilpotent and coincides with $\mathcal{N}_G(T)_1$. In particular, it is clear from the irreducibility and nilpotency that $\mathscr{C}_G(T)$ is contained in some member, X say, of $\mathscr{B}(T)$. As above, for every member Y of $\mathscr{B}(T)$ there is an element z in $\mathcal{N}_G(T)$ such that $zXz^{-1} = Y$. Hence

$$\mathscr{C}_G(T) = z\mathscr{C}_G(T)z^{-1} \subset Y.$$

It follows that the action of $\mathcal{N}_G(T)$ on $(G/B)^T$ factors through the finite group $\mathcal{N}_G(T)/\mathcal{N}_G(T)_1 = \mathcal{N}_G(T)/\mathscr{C}_G(T)$. In particular, $\mathscr{B}(T)$ is therefore finite. Finally, suppose that x is an element of $\mathcal{N}_G(T)$ such that $xBx^{-1} = B$.

Then x belongs to B, by Theorem 5.1. Thus x belongs to $\mathcal{N}_B(T)$. Hence, for every element t of T, we have

$$xtx^{-1}t^{-1} \in T \cap [B, B] \subset T \cap B_u = (1),$$

so that x lies in $\mathcal{C}_G(T)$.

The group $\mathcal{N}_G(T)/\mathcal{C}_G(T)$ is called the *Weyl group* of G with respect to T. Since all the maximal toroids T of G are conjugates, the isomorphism class of the Weyl group is determined by G. The above has shown that *the Weyl group acts simply transitively on the finite set of all Borel subgroups of G containing T.*

Notes

1. Let V be a finite-dimensional vector space over an algebraically closed field F. It is easy to see that the Borel subgroups of $\mathrm{Aut}_F(V)$ are precisely the stabilizers of the full flags in V.

2. It is not true that every maximal solvable subgroup of an irreducible algebraic group over an algebraically closed field is a Borel subgroup. For example, let F be an algebraically closed field of characteristic other than 2. Let G be the group of all n by n orthogonal matrices of determinant 1 with entries in F, where $n > 2$. Let D be the subgroup of G consisting of the elements whose entries off the diagonal are all equal to 0. Then D is the direct product of $n - 1$ copies of the group of order 2, and D coincides with its centralizer in G. If D were contained in a Borel subgroup of G, it would follow from Theorem VI.3.2 that D is contained in a toroid of G. Thus, D is not contained in any Borel subgroup of G.

3. Let G be an irreducible algebraic group over an algebraically closed field. Suppose that G has no infinite unipotent subgroup. Use Proposition 2.4 for showing that G is a toroid.

4. For the detailed structure and classification theory, we refer the reader to [6], [1], [16] and [7].

Chapter XIV

Applications of Galois Cohomology

The theme of this chapter is the use of Galois theory for extending the structure theory of algebraic groups. The applicability of Galois theory stems from the fact that solvable algebraic groups are made up from the additive and multiplicative groups of the base field, and Section 1 provides the technical preparations for exploiting this.

The main result of Section 2 is the extension of the basic semidirect product decomposition for solvable groups from the case of an algebraically closed base field to that of a perfect base field.

Section 3 contains the crucial cross-section result that is used in Section 4 for showing that if G is an irreducible algebraic group over an algebraically closed field, and H is a unipotent irreducible algebraic subgroup of G such that G/H is an affine variety then, as a variety, G is the direct product of G/H and H. This cross-section result is of considerable importance for the theory of group extensions, of which we give only a glimpse in Theorem 4.2.

1. Let F be a field, G an affine algebraic F-group. One says that G *is defined over a subfield K of F* if there is given a K-form for the Hopf algebra $\mathscr{P}(G)$, i.e., a K-Hopf algebra A such that $\mathscr{P}(G) = A \otimes F$. In this situation, an algebraic subgroup H of G is said to be *associated with A* if the kernel of the restriction map from $\mathscr{P}(G)$ to $\mathscr{P}(H)$ is generated as an ideal by its intersection with A, so that $\mathscr{P}(H) = A_H \otimes F$, where A_H stands for the restriction image of A. Thus, H is then defined over K, by the K-form A_H of $\mathscr{P}(H)$.

We are interested in the case where K is the fixed part F^S for a group S of automorphisms of F. In this case, if G is defined over K, the group S acts on G by abstract group automorphisms in the following way. Regard G as the group of all K-algebra homomorphisms from A to F. Then the transform of an element x of G by an element σ of S is simply the composite $\sigma \circ x$.

If σ' is the automorphism of G corresponding to the element σ of S, we have

$$f \circ \sigma' = \sigma \circ (i_A \otimes \sigma^{-1})(f),$$

for every element f of $\mathscr{P}(G)$. In particular, this shows that, although σ' is not an automorphism of affine algebraic F-groups, it is a homeomorphism from G to G with respect to the Zariski topology of G. In fact, if J_Q is the annihilator of a subset Q of G in $\mathscr{P}(G)$, then $(i_A \otimes \sigma)(J_Q)$ is the annihilator of $\sigma'(Q)$ in $\mathscr{P}(G)$, and if Q is the annihilator in G of a subset J of $\mathscr{P}(G)$, then $\sigma'(Q)$ is the annihilator in G of $(i_A \otimes \sigma)(J)$.

Proposition 1.1. *Let F be a Galois extension of a field K, with Galois group S. Let G be an affine algebraic F-group that is defined over K by the K-form A of $\mathscr{P}(G)$. Then the S-stable algebraic subgroups of G are precisely the algebraic subgroups that are associated with A.*

PROOF. Let H be an algebraic subgroup of G, and let J be its annihilator in $\mathscr{P}(G)$. First, suppose that H is associated with A, so that J is generated as an ideal by $A \cap J$. Clearly, this implies that $(i_A \otimes \sigma)(J) = J$ for every element σ of S. By the remark just preceding the statement of the proposition, this means that $\sigma'(H) = H$ for every σ in S.

Now suppose that H is S-stable. Then J is S-stable with respect to the S-module structure of $\mathscr{P}(G)$ coming from the action of S on the tensor factor F of $A \otimes F$. Let f be an element of J, and write it in the form $\sum_{i=1}^{n} f_i \otimes c_i$, where the c_i's are K-linearly independent elements of F, and the f_i's belong to A. Let M be the smallest Galois extension of K that is contained in F and contains each c_i. Then M is finite over K, and S induces the full Galois group of M relative to K. Therefore, every K-linear endomorphism of M is an M-linear combination of restrictions to M of elements of S. In particular, this holds for the endomorphisms sending one of the c_i's to 1 and annihilating all the others. Since J is S-stable, it follows that each f_i belongs to J. Hence, we have $J = (J \cap A) \otimes F$, which means that H is associated with A. $\square$

Proposition 1.2. *Let F, K, S, G and A be as in Proposition 1.1. Suppose that H is a properly normal algebraic subgroup of G that is associated with A. Then G/H is defined over K, with A^H as the K-form of $\mathscr{P}(G/H)$.*

PROOF. Let f be an element of $\mathscr{P}(G/H) = (A \otimes F)^H$. For every element y of G, every element x of H and every element σ of S, we have

$$(i_A \otimes \sigma)(f)(\sigma'(y)\sigma'(x)) = \sigma(f(yx)) = \sigma(f(y)) = (i_A \otimes \sigma)(f)(\sigma'(y)),$$

whence

$$\sigma'(x) \cdot ((i_A \otimes \sigma)(f)) = (i_A \otimes \sigma)(f).$$

Since H is S-stable (by Proposition 1.1), this shows that $(A \otimes F)^H$ is S-stable with respect to the S-module structure of $A \otimes F$ coming from the action of S on the tensor factor F. Exactly as in the proof of Proposition 1.1, we see from this that $(A \otimes F)^H = A^H \otimes F$. $\square$

Lemma 1.3. *Let S be a finite group of automorphisms of a field F. Then every multiplicative and every additive cocycle for S in F is a coboundary.*

PROOF. There is an element t in F such that $\sum_{\sigma \in S} \sigma(t) = 1$. Let f be an additive cocycle for S in F, so that

$$f(\sigma\tau) = \sigma(f(\tau)) + f(\sigma),$$

for all elements σ and τ of S. Put $c = \sum_{\tau \in S} \tau(t)f(\tau)$. One verifies directly that $c - \sigma(c) = f(\sigma)$ for every element σ of S, which means that f is a coboundary.

Now let f be a multiplicative cocycle, so that

$$f(\sigma\tau) = \sigma(f(\tau))f(\sigma),$$

for all elements σ and τ of S, and all the values of f are different from 0. Since the elements of S are linearly independent as maps from F to F, there is an element u in F such that $\sum_{\tau \in S} f(\tau)\tau(u) \neq 0$. If e denotes this element of F, one verifies directly that $e\sigma(e)^{-1} = f(\sigma)$ for every element σ of S, which means that f is a coboundary. $\square$

Let F be a Galois extension of a field K. We endow the Galois group S for F relative to K with the *Krull topology*, in which a fundamental system of neighborhoods of the neutral element of S consists of the element-wise fixers of the finite Galois extensions of K that are contained in F. If G is an affine algebraic F-group that is defined over K, we shall, from now on, write $\sigma(x)$ for the transform of an element x of G by the automorphism σ' corresponding to an element σ of S. A *Galois cocycle for S in G* is a map f from S to G that satisfies the identity $f(\sigma\tau) = \sigma(f(\tau))f(\sigma)$ and is continuous with respect to the Krull topology of S and the *discrete* topology of G. For every element x of G, the map g from S to G defined by

$$g(\sigma) = \sigma(x)^{-1}f(\sigma)x,$$

is also a Galois cocycle for S in G. In the case where f is the constant map whose value is the neutral element of G, this shows that the coboundaries are actually Galois cocycles.

Suppose that G has a series of S-stable algebraic subgroups

$$(1) = G_0 \subset \cdots \subset G_n = G,$$

such that each G_i is properly normal in G_{i+1} and G_{i+1}/G_i is isomorphic, as an affine algebraic F-group and S-module, with either the additive group of F or the multiplicative group. Then we say that G is *split solvable* with respect to the S-fixed part K of F.

Proposition 1.4. *Let F be a Galois extension of a field K with Galois group S. Suppose that G is an affine algebraic F-group that is defined over K so as to be split solvable with respect to K. Then every Galois cocycle for S in G is a coboundary.*

PROOF. Let f be a Galois cocycle for S in G, and let the G_i's be as in the above definition of "split solvable." Starting with $i = n$, suppose that we have already found an element u_i of G such that $\sigma(u_i)^{-1} f(\sigma) u_i$ belongs to G_i for every element σ of S. Write this element of G_i as $g(\sigma)$, and let g' be the map from S to G_i/G_{i-1} obtained by composing g with the canonical map from G_i to G_i/G_{i-1}. Then g' is a Galois cocycle for S in G_i/G_{i-1}. By assumption on the series of G_i's, we may regard g' as an additive or multiplicative cocycle for S in F.

By the continuity assumption on Galois cocycles, there is a finite Galois extension M of K that is contained in F and such that, if T is the element-wise fixer of M in S, our cocycle is constant on the cosets of T in S. This, together with the cocycle identity, implies that g' actually maps S into M and defines a cocycle for the Galois group of M relative to K in the additive or multiplicative group of M. By Lemma 1.4, this cocycle is a coboundary. For the original cocycle f, this means that there is an element u in G_i such that $\sigma(u_i)^{-1} f(\sigma) u_i$ belongs to $\sigma(u)^{-1} u G_{i-1}$ for every element σ of S. Put $u_{i-1} = u_i u^{-1}$. Then $\sigma(u_{i-1})^{-1} f(\sigma) u_{i-1}$ belongs to G_{i-1} for every element σ of S. When we reach the index 0, we have the desired conclusion. $\qquad\square$

2. Theorem 2.1. *Let K be a perfect field, F an algebraic closure of K and S the Galois group of F relative to K. Let U be an irreducible unipotent affine algebraic F-group that is defined over K. Then every Galois cocycle for S in U is a coboundary.*

PROOF. Making an induction on the dimension of U, we suppose that the theorem has been established in the lower cases. First, we deal with the case where U is non-abelian. Evidently, $[U, U]$ is S-stable. It follows from Propositions 1.1 and 1.2 that $U/[U, U]$ is defined over K such that, if A is the given K-form for $\mathscr{P}(U)$, then $A^{[U, U]}$ is the K-form for $\mathscr{P}(U/[U, U])$. Let f be a Galois cocycle for S in U. In the canonical fashion, f defines a Galois cocycle for S in $U/[U, U]$, which we denote by f'. Our inductive hypothesis applies to f' and gives the result that there is an element u in U such that $f(\sigma)$ belongs to $\sigma(u)u^{-1}[U, U]$ for every element σ of S. If $g(\sigma) = \sigma(u)^{-1} f(\sigma) u$, then g is a Galois cocycle for S in $[U, U]$ to which we can again apply our inductive hypothesis. It follows that f is a coboundary.

It remains to deal with the case where U is abelian. Let p be the characteristic of F. There is a non-negative integer r such that u^{p^r} is the neutral element for every element u of U. Let e be the smallest such r; in the case where $p = 0$, we agree that $e = 1$. If $e > 1$, let V be the image of the morphism from U to U sending each u onto $u^{p^{e-1}}$. Then V is an irreducible algebraic S-stable subgroup of U, and V is neither trivial nor coincides with U. Therefore, we can apply the inductive hypothesis to V and to U/V, which reduces the problem to the case where $e = 1$.

In that case, we know from Theorem VI.5.4 that, if A is the given K-form of $\mathscr{P}(U)$, the affine algebraic K-group $\mathscr{G}(A)$ is an algebraic vector group whose algebra of polynomial functions is A. Evidently, this implies that U is identifiable, as an affine algebraic F-group and S-module, with the direct product of a finite family of copies of the additive group of F, so that the result follows from Lemma 1.3. $\qquad\square$

Theorem 2.2. *Let K be a perfect field, G an irreducible affine algebraic K-group, U an irreducible unipotent normal algebraic subgroup of G. Then U is properly normal in G.*

PROOF. Let F and S be as in Theorem 2.1, and let G^F and U^F denote the groups obtained from G and U by the canonical base field extension. Then every element h of $\mathcal{G}(\mathcal{P}(G)^U)$ is the restriction to $\mathcal{P}(G)^U$ of an element x of G^F. For every element σ of S, the restriction of $\sigma(x)$ to $\mathcal{P}(G)^U$ also coincides with h. Therefore, $\sigma(x)x^{-1}$ belongs to U^F, so that the map sending each element σ of S onto $\sigma(x)x^{-1}$ is a Galois cocycle for S in U^F. By Theorem 2.1, there is an element y in U^F such that $\sigma(x)x^{-1} = \sigma(y)y^{-1}$ for every element σ of S. This says that the point $y^{-1}x$ of G^F is fixed under the action of S, which means that $y^{-1}x$ belongs to G. Since its restriction to $\mathcal{P}(G)^U$ coincides with h, this shows that the restriction map from G to $\mathcal{G}(\mathcal{P}(G)^U)$ is surjective, i.e., that U is properly normal in G. $\qquad\square$

Theorem 2.3. *Let K be a perfect field, G an irreducible solvable affine algebraic K-group. There is a linearly reductive algebraic subgroup R of G such that $G = G_u \rtimes R$.*

PROOF. Let F be an algebraic closure of K, and let S be the Galois group of F relative to K. Let us write H for the extended group G^F. We have $H = H_u \rtimes T$, where T is a toroid.

By Theorem 2.2, G/G_u is the affine algebraic K-group $\mathcal{G}(\mathcal{P}(G)^{G_u})$. Since it is abelian and since $(G/G_u)_u$ is trivial, we know from Theorem V.5.1 that G/G_u is linearly reductive. Since K is perfect, we can apply Proposition V.1.2 and conclude that $\mathcal{P}(G)^{G_u} \otimes F$ is semisimple as a G-module, and therefore also as an H-module. This shows that H/G_u^F is linearly reductive, whence H_u must coincide with G_u^F. Thus, H_u is S-stable. We shall show that T may be chosen so as to be S-stable.

Let Z be the normalizer of T in H_u. It is seen immediately from the semidirect product decomposition that Z coincides with the centralizer of T in H_u. First, we deal with the case where H_u is abelian. In this case, Z is clearly the intersection of H_u with the center of H, whence Z is S-stable. By Proposition 1.2, the definition of H_u over K yields a definition of H_u/Z over K. By Theorem 2.1, every Galois cocycle for S in H_u/Z is a coboundary.

Let σ be an element of S. We know that $\sigma(T)$ is an *algebraic* subgroup of H, that $H = \sigma(H_u)\sigma(T) = H_u\sigma(T)$, and that $H_u \cap \sigma(T) = (1)$. Thus H is the semidirect product $H_u \rtimes \sigma(T)$, whence $\sigma(T)$ is a maximal toroid of H. Therefore, there is an element h_σ in H_u such that $\sigma(T) = h_\sigma T h_\sigma^{-1}$. Let $f(\sigma)$ denote the canonical image of h_σ in H_u/Z. Then $f(\sigma)$ depends only on σ, and not on the particular choice of h_σ. We see directly from this and the definition of f that $f(\sigma\tau) = \sigma(f(\tau))f(\sigma)$ for all elements σ and τ of S. From the fact that the annihilator of T in $\mathscr{P}(H)$ is finitely generated as an ideal, one sees that the stabilizer of T in S is an open subgroup of S. Hence, f has the continuity property making it a Galois cocycle for S in H_u/Z. From the above, we know that there is therefore an element h in H_u such that h_σ belongs to $\sigma(h)h^{-1}Z$ for every element σ of S, whence we see that $h^{-1}Th$ is S-stable.

Now we deal with the general case by making an induction on the dimension of H_u. Identifying T with its canonical image in $H/[H_u, H_u]$, we write $H/[H_u, H_u] = (H_u/[H_u, H_u]) \rtimes T$. Applying what we have just proved to this situation, we conclude that there is an element h in H_u such that $[H_u, H_u] \rtimes h^{-1}Th$ is S-stable. Clearly, we can apply the inductive hypothesis to this group, thus obtaining an element h_1 of $[H_u, H_u]$ such that $(hh_1)^{-1}T(hh_1)$ is stable under the action of S.

Now we have $H = H_u \rtimes T$, where T is an S-stable toroid. Let x be any element of G, and write $x = yz$, with y in H_u and z in T. Then, for every element σ of S, we have $x = \sigma(x)$, and therefore $\sigma(y)^{-1}y = \sigma(z)z^{-1}$. This element belongs to $H_u \cap T$, so that it must be the neutral element. Thus, y and z are fixed under the action of S. In particular, y belongs to $H_u^S = G_u$. This shows that G is the semidirect product $G_u \rtimes T^S$, and it is clear that T^S is linearly reductive. $\qquad\square$

3. Proposition 3.1. *Let F be an algebraically closed field, G an irreducible affine algebraic F-group. Suppose that H is an irreducible solvable algebraic subgroup of G, and that the variety G/H is affine. Then there is a non-empty open subset U of G/H and a variety morphism ρ from U to G whose composite with the canonical map from G to G/H is the identity map on U.*

Proof. Let π denote the canonical map from G to G/H, and consider the image $\mathscr{P}(G/H) \circ \pi$ of $\mathscr{P}(G/H)$ in $\mathscr{P}(G)$. Evidently, this is contained in $\mathscr{P}(G)^H$. Conversely, if f is an element of $\mathscr{P}(G)^H$, we see immediately from part (4) of Theorem XII.2.2 that f belongs to $\mathscr{P}(G/H) \circ \pi$. Thus, we have

$$\mathscr{P}(G/H) \circ \pi = \mathscr{P}(G)^H.$$

Accordingly, we identify $\mathscr{P}(G/H)$ with $\mathscr{P}(G)^H$.

Let us write K for the field $[\mathscr{P}(G)^H]$ which, here, coincides with the field $[\mathscr{P}(G)]^H$ of rational functions of G/H. Let A denote the sub K-algebra of $[\mathscr{P}(G)]$ that is generated by $\mathscr{P}(G)$. Our aim is to show that there is a K-algebra homomorphism from A to K. In fact, if α is such a K-algebra homomorphism, and if U is the set of points of G/H at which every element of $\alpha(\mathscr{P}(G))$ is defined, then U is a non-empty open subset of G/H, and the restriction of α to $\mathscr{P}(G)$ transposes into a variety morphism ρ from U to G such that $\pi \circ \rho$ is the identity map on U.

We know from Theorem III.2.3 that $[\mathscr{P}(G)]$ is separable over K. By Theorem III.2.1, this implies that there is a transcendence basis $(x_1, \ldots, x_n)$

for $[\mathscr{P}(G)]$ over K such that $[\mathscr{P}(G)]$ is a finite separable algebraic extension of K. Accordingly, we write

$$[\mathscr{P}(G)] = K(x_1,\ldots,x_n)[a],$$

where a is separably algebraic over $K(x_1,\ldots,x_n)$.

Now let $(p_1,\ldots,p_k)$ be a system of K-algebra generators of A, and write p_i in the form $\sum_j p_{ij}a^j$, where each p_{ij} belongs to $K(x_1,\ldots,x_n)$. Since a is separable over $K(x_1,\ldots,x_n)$, we have $f'(a) \neq 0$, where f is the monic minimum polynomial for a relative to $K(x_1,\ldots,x_n)$, and f' is the formal derivative of f. Let $q_0,\ldots,q_s$ be the coefficients of f, and let

$$R = K[p_{10},\ldots,p_{kd},q_0,\ldots,q_s].$$

By Theorem II.3.3, there is a non-zero element r in R such that every K-algebra homomorphism from R to an algebraic closure, L say, of K not annihilating r extends to a K-algebra homomorphism from $R[a]$ to L not annihilating $f'(a)$. Clearly, we can find elements $k_1,\ldots,k_n$ in K such that r, the q_t's and the p_{ij}'s are in the specialization sub K-algebra of $K(x_1,\ldots,x_n)$ for the specialization $x_i \mapsto k_i$, and such that this specialization does not annihilate r. The restriction to R of this specialization extends to a K-algebra homomorphism σ from $R[a]$ to L such that $\sigma(R) \subset K$ and $\sigma(f'(a)) \neq 0$. If $\sigma(f)$ denotes the polynomial whose coefficients are the $\sigma(q_t)$'s, we have $\sigma(f)(\sigma(a)) = 0$, but $\sigma(f)'(\sigma(a)) = \sigma(f'(a)) \neq 0$. This shows that $\sigma(a)$ is separable over K.

Let K' denote the composite in L of the family of all intermediate fields between K and L that are separable over K. Our conclusion is that there is a K-algebra homomorphism from A to K'.

Now let us make the canonical base field extension from F to K' and consider the extended group $G^{K'}$ with the algebraic subgroup $H^{K'}$. Let us denote these groups more simply by G' and H'. We have $\mathscr{P}(G') = \mathscr{P}(G) \otimes K'$ and $\mathscr{P}(G')^{H'} = \mathscr{P}(G)^H \otimes K'$. It is clear that the element-wise fixer of $\mathscr{P}(G)^H$ in G coincides with H. The condition that an element of G' leave the elements of $\mathscr{P}(G)^H$ fixed is equivalent to the condition that this element be a zero of a certain ideal, Q say, of $\mathscr{P}(G)$. Let J be the annihilator of H in $\mathscr{P}(G)$. The remark we just made concerning H implies that the radical of Q coincides with J. It follows that every element of G' leaving the elements of $\mathscr{P}(G)^H$ fixed is a zero of J, which means that it belongs to H'. Thus, the element-wise fixer of $\mathscr{P}(G)^H$ in G' coincides with H', whence also the element-wise fixer of $\mathscr{P}(G')^{H'}$ in G' coincides with H'. It follows that G'/H' may be identified with the affine algebraic K'-variety resulting from G/H by the canonical base field extension, and that the canonical map from G' to G'/H' then becomes the canonical extension π' of π.

The injection $\mathscr{P}(G)^H \to K'$ defines a point of our affine algebraic K'-variety G'/H'. Let x denote this point. As we have shown above, there is a K-algebra homomorphism from A to K'. Let y denote its restriction to $\mathscr{P}(G)$. Then y may be viewed as a point of G', and it is clear that $\pi'(y) = x$.

Let S denote the Galois group of K' relative to K, and consider the natural actions of S on G', H' and G'/H'. Evidently, π' is compatible with the actions of S on G' and G'/H', in the sense that, for every σ in S and every z in G', we have $\pi'(\sigma(z)) = \sigma(\pi'(z))$. Since the point x of G'/H' is fixed under the action of S, it follows that $\pi'(\sigma(y)) = x$. Since $\pi'^{-1}(x)$ is the coset yH', it follows that $y^{-1}\sigma(y)$ belongs to H'.

Define the map f from S to H' by $f(\sigma) = \sigma(y)^{-1}y$. Then it is easy to see that f is a Galois cocycle for S in H'. Since F is algebraically closed and H' comes from H by canonical base field extension, it is clear that H' is split solvable relative to K. Therefore, it follows from Proposition 1.4 that there is an element z in H' such that $f(\sigma) = \sigma(z)z^{-1}$ for every element σ of S. This is equivalent to the statement that yz is fixed under the action of S, so that yz belongs to G^K. Since z belongs to H', we have $g(yz) = g(y)$ for every element g of $\mathscr{P}(G)^H$. Using this and the fact that y is the restriction to $\mathscr{P}(G)$ of a K-algebra homomorphism from A to K', we see that yz extends, by K-linear extension, to a K-algebra homomorphism from A to K. $\square$

4. Theorem 4.1. *Let F be an algebraically closed field, G an irreducible affine algebraic F-group, H an irreducible unipotent algebraic subgroup of G such that G/H is affine. There is a variety morphism from G/H to G whose composite with the canonical map π from G to G/H is the identity map on G/H.*

PROOF. Using the result of Proposition 3.1 and the translation action of G on G/H in the evident way, we obtain a covering of G/H by a finite family of non-empty open sets U_i, with associated morphisms ρ_i to G, as described in Proposition 3.1. Then, for every element u of $U_i \cap U_j$, we have

$$\rho_j(u)^{-1}\rho_i(u) \in H.$$

We shall prove by induction on the dimension of H that the ρ_i's can be so modified that they fit together to make up a morphism σ from G/H to G such that $\pi \circ \sigma$ is the identity map on G/H.

First, suppose that H is 1-dimensional. Then, by Corollary VI.5.5, H is identifiable with the additive group of F. By refining the above covering of G/H, we arrange that each U_i is a principal open set $(G/H)_{q_i}$, where q_i is a non-zero element of $\mathscr{P}(G/H)$. If t is our isomorphism from H to F, then the function on $U_i \cap U_j$ whose value at each point u is $t(\rho_j(u)^{-1}\rho_i(u))$ is an element f_{ij} of $\mathscr{P}(U_i \cap U_j) = \mathscr{P}(G/H)[1/(q_iq_j)]$. There is a positive integer e such that, for each index pair (i, j), we have $q_i^e f_{ij} \in \mathscr{P}(U_j)$. Since the U_i's cover G/H, the q_i's have no common zero in G/H. This implies that there are elements r_i in $\mathscr{P}(G/H)$ such that $\sum_i r_i q_i^e = 1$. Now define the element g_j of $\mathscr{P}(U_j)$ by

$$g_j = \sum_i r_i q_i^e f_{ij}.$$

Let γ_j denote the morphism from U_j to H given by $t \circ \gamma_j = g_j$, and define the morphism σ_j from U_j to G by $\sigma_j(u) = \rho_j(u)\gamma_j(u)$. Identifying H with F

by means of t and writing out $\sigma_j(u)^{-1}\sigma_i(u)$ additively, one sees that the restrictions to $U_i \cap U_j$ of σ_i and σ_j coincide. Thus, the σ_i's fit together to make up the required morphism σ from G/H to G.

Now suppose that the dimension of H is greater than 1, and that the existence of a σ has been established in the lower cases. There is an irreducible normal algebraic subgroup Q of H such that H/Q is of dimension 1. Let the U_i's and ρ_i's be as above, let η denote the canonical map from G to G/Q, and put $\tau_i = \eta \circ \rho_i$. Now observe that, with u in $U_i \cap U_j$, the element $\rho_j(u)^{-1}\rho_i(u)Q$ of H/Q depends only on $\tau_j(u)$ and $\tau_i(u)$, and that the map sending each u in $U_i \cap U_j$ onto this element is a morphism from $U_i \cap U_j$ to G/Q. Clearly, the above proof of the existence of σ in the case of a 1-dimensional H extends without change to the present situation and yields the existence of a morphism τ from G/H to G/Q whose composite with the canonical morphism from G/Q to G/H is the identity map on G/H. The existence of τ shows that the variety G/Q is isomorphic with the direct product of the varieties G/H and H/Q, so that the variety G/Q is affine. Therefore, we can apply our inductive hypothesis to the pair (G, Q) and conclude that there is a morphism γ from G/Q to G such that $\eta \circ \gamma$ is the identity map on G/Q. Clearly, the morphism $\gamma \circ \tau$ from G/H to G satisfies our requirement. $\square$

Theorem 4.1 can be applied to the structure theory of unipotent groups. In particular, it yields the following result.

Theorem 4.2. *Let H be an irreducible unipotent algebraic group over an algebraically closed field F. Then $\mathscr{P}(H) = F[t_1, \ldots, t_n]$, where the t_i's are algebraically independent over F, and $\delta(t_i) - t_i \otimes 1 - 1 \otimes t_i$ belongs to $F[t_1, \ldots, t_{i-1}] \otimes F[t_1, \ldots, t_{i-1}]$ for each $i > 1$ and is 0 for $i = 1$.*

PROOF. Clearly, the result holds when H is of dimension 1. Therefore, we suppose that the dimension of H is greater than 1, and that the theorem has been established in the lower cases. There is a 1-dimensional irreducible central algebraic subgroup Z of H. Let π be the canonical morphism from H to H/Z, and let σ be a morphism of algebraic varieties from H/Z to H as obtained in Theorem 4.1, so that $\pi \circ \sigma$ is the identity map on H/Z. Let γ be the variety morphism from H to Z given by $\gamma(x) = x\sigma(\pi(x))^{-1}$. Then we have $x = \gamma(x)\sigma(\pi(x))$, whence we see that the multiplication map is an isomorphism of F-algebras from $(\mathscr{P}(Z) \circ \gamma) \otimes (\mathscr{P}(H/Z) \circ \pi)$ to $\mathscr{P}(H)$. We know that the transpose of π is an isomorphism of Hopf algebras from $\mathscr{P}(H/Z)$ to $\mathscr{P}(H)^Z$.

By inductive hypothesis, $\mathscr{P}(H/Z) = F[t_1, \ldots, t_{n-1}]$, where the t_i's are as described in the theorem. We regard them as elements of $\mathscr{P}(H)^Z$ by identifying them with the $t_i \circ \pi$'s. On the other hand, we have $\mathscr{P}(Z) = F[t]$, where t is a group homomorphism from Z to F. Put $t_n = t \circ \gamma$. Then it is clear from the above that $\mathscr{P}(H) = F[t_1, \ldots, t_n]$, that the t_i's are algebraically independent over F and that, for $i < n$, the $\delta(t_i)$'s are as required by the theorem.

In order to analyze $\delta(t_n)$, consider two elements x and y of H. We have

$$t_n(x) + t_n(y) = t(x\sigma(\pi(x))^{-1}y\sigma(\pi(y))^{-1})$$

$$= t(xy\sigma(\pi(y))^{-1}\sigma(\pi(x))^{-1})$$

$$= t_n(xy) + t(\sigma(\pi(x)\pi(y))\sigma(\pi(y))^{-1}\sigma(\pi(x))^{-1}).$$

The second summand on the right is clearly of the form $f(\pi(x), \pi(y))$, where f is a polynomial function on $(H/Z) \times (H/Z)$. Hence, it is clear that $\delta(t_n)$ is as required. $\qquad\square$

Notes

1. Proposition 3.1 is a specialized adaptation of a basic cross-section result on orbit varieties of solvable groups due to M. Rosenlicht [12].

2. Essentially, the material of Section 4 is due to J-P. Serre, who analyzed the structure of commutative unipotent groups with a group extension technique based on Theorem 4.1. See Chapter VII of [14] for this.

3. Let F be an algebraically closed field, G an irreducible *solvable* algebraic F-group, H an irreducible algebraic subgroup of G. Recall that, by Theorem XII.4.3, the variety G/H is affine. Using the reduction to the case where G is unipotent that is made in the proof of that theorem, one can apply Theorem 4.1 and show that *there is a variety morphism from G/H to G whose composite with the canonical morphism from G to G/H is the identity map on G/H.*

Chapter XV

Algebraic Automorphism Groups

For an algebraic group G, let $\mathscr{W}(G)$ denote the group of all algebraic group automorphisms of G. In this chapter, we examine the possibility of endowing $\mathscr{W}(G)$ with the structure of an algebraic group in such a way that G becomes a strict $\mathscr{W}(G)$-variety. The example of a toroid of dimension greater than 1 shows that this is not always possible. However, good questions remain concerning suitable subgroups of $\mathscr{W}(G)$ or suitably restricted groups G.

Section 1 gives a general description of the appropriate algebraic group structures for the "algebraic" subgroups of $\mathscr{W}(G)$. Section 2 deals with the passage from $\mathscr{W}(G)$ to $\mathscr{W}(H)$, where H is an algebraic subgroup of G, and with the passage to $\mathscr{W}(G/H)$ in the case where H is normal in G. Also, it contains the general results concerning the canonical map from $\mathscr{W}(G)$ to the group of Lie algebra automorphisms of $\mathscr{L}(G)$.

The principal results are developed in Sections 3 and 4. They depend on the assumption that the base field is algebraically closed and of characteristic 0. In particular, Theorem 4.3 characterizes the groups G for which $\mathscr{W}(G)$ is an algebraic group in the appropriate way.

1. Let F be a field, G an affine algebraic F-group, $\mathscr{W}(G)$ the group of all affine algebraic group automorphisms of G. Let K be a subgroup of $\mathscr{W}(G)$, and suppose that K has been endowed with the structure of an affine algebraic F-group such that G is a strict K-variety, i.e., such that the map from $K \times G$ to G that sends each (α, x) onto $\alpha(x)$ is a morphism of varieties. Then, for every element f of $\mathscr{P}(G)$, the F-valued function on $K \times G$ sending each (α, x) onto $f(\alpha(x))$ is an element of $\mathscr{P}(K) \otimes \mathscr{P}(G)$. This means that there are elements $g_1, \ldots, g_n$ of $\mathscr{P}(G)$ and elements $h_1, \ldots, h_n$ of $\mathscr{P}(K)$ such that

$$ f \circ \alpha = \sum_{i=1}^{n} h_i(\alpha)g_i, $$

for every element α of K. This shows that a necessary condition for the existence of an affine algebraic group structure on K for which G becomes a strict K-variety is that $\mathscr{P}(G)$ be locally finite as a right K-module. Moreover, the above shows that, for every element f of $\mathscr{P}(G)$ and every linear function τ on $\mathscr{P}(G)$, the function τ/f on K that is defined by

$$(\tau/f)(\alpha) = \tau(f \circ \alpha),$$

must then belong to $\mathscr{P}(K)$. We say that a subgroup K of $\mathscr{W}(G)$ is an *algebraic automorphism group* of G if $\mathscr{P}(G)$ is locally finite as a right K-module, and K is endowed with the structure of an affine algebraic F-group such that $\mathscr{P}(K)$ coincides with the smallest sub Hopf algebra of $\mathscr{R}_F(K)$ containing all these functions τ/f.

Let L be any subgroup of $\mathscr{W}(G)$ with the property that $\mathscr{P}(G)$ is locally finite as a right L-module. Clearly, the functions on L of the form τ/f as above are then representative functions on L. Let $\mathscr{R}_G(L)$ denote the smallest Hopf algebra of representative functions on L that contains all these functions τ/f. We show that $\mathscr{R}_G(L)$ *is finitely generated as an F-algebra.*

Evidently, there is a finite-dimensional L-stable sub F-space S of $\mathscr{P}(G)$ that generates $\mathscr{P}(G)$ as an F-algebra. Let T denote the sub F-space of $\mathscr{R}_G(L)$ spanned by the functions τ/f with f in S and τ in $\mathscr{P}(G)^\circ$. If $(g_1, \ldots, g_n)$ is an F-basis for the space spanned by the transforms $f \circ \alpha$ with α in L, and if

$$f \circ \alpha = \sum_{i=1}^{n} h_i(\alpha)g_i,$$

then the space of functions τ/f, with τ ranging over $\mathscr{P}(G)^\circ$, is the space spanned by the functions $h_1, \ldots, h_n$. This shows that the space T is finite-dimensional. One verifies directly that T is stable under the right and left translation actions of L on $\mathscr{R}_G(L)$.

Now let f and g be elements of $\mathscr{P}(G)$, and let τ be an element of $\mathscr{P}(G)^\circ$. The restriction of τ to the L-orbit $(fg) \circ L$ is an F-linear combination of evaluations at the elements of G. Since these evaluations are F-algebra homomorphisms, it follows that $\tau/(fg)$ is contained in the F-algebra generated by the functions ρ/f and σ/g, where ρ and σ range over $\mathscr{P}(G)^\circ$. This shows that the sub algebra of $\mathscr{R}_G(L)$ that is generated by the elements of T contains every function τ/f with f in $\mathscr{P}(G)$ and τ in $\mathscr{P}(G)^\circ$. Since T is stable under the translation actions of L, it follows that $\mathscr{R}_G(L)$ is generated by the elements of T and $\eta(T)$, where η is the antipode of $\mathscr{R}_G(L)$. Thus, $\mathscr{R}_G(L)$ is finitely generated as an F-algebra.

Theorem 1.1. *Let G be an algebraic group over the field F, and let K be a subgroup of $\mathscr{W}(G)$ such that $\mathscr{P}(G)$ is locally finite as a right K-module. Then $\mathscr{G}(\mathscr{R}_G(K))$ may be identified with a subgroup of $\mathscr{W}(G)$ so as to become an algebraic automorphism group of G whose algebra of polynomial functions is $\mathscr{R}_G(K)$. As such, it coincides with the intersection of the family of all algebraic automorphism groups of G that contain K.*

PROOF. Let σ be an element of $\mathcal{G}(\mathcal{R}_G(K))$, and let x be an element of G. Regarding x as an element of $\mathcal{P}(G)^\circ$, we consider the values $\sigma(x/f)$, where f ranges over $\mathcal{P}(G)$. Define the F-valued function σ_x on $\mathcal{P}(G)$ by setting $\sigma_x(f) = \sigma(x/f)$. Using that x and σ are F-algebra homomorphisms, we see immediately that σ_x is an F-algebra homomorphism, i.e., that σ_x belongs to G.

Let δ denote the comultiplication of $\mathcal{P}(G)$, let f be an element of $\mathcal{P}(G)$, and write

$$\delta(f) = \sum_{i=1}^{n} f_{1i} \otimes f_{2i}.$$

Then, for every element α of K and all elements x and y of G, we have

$$((xy)/f)(\alpha) = (xy)(f \circ \alpha) = f(\alpha(x)\alpha(y))$$

$$= \sum_{i=1}^{n} f_{1i}(\alpha(x))f_{2i}(\alpha(y)),$$

which shows that

$$(xy)/f = \sum_{i=1}^{n} (x/f_{1i})(y/f_{2i}).$$

This gives

$$\sigma_{xy}(f) = \sum_{i=1}^{n} \sigma_x(f_{1i})\sigma_y(f_{2i}).$$

Hence, we have

$$\sigma_{xy} = (\sigma_x \otimes \sigma_y) \circ \delta = \sigma_x \sigma_y.$$

Thus, the map sending each element x of G onto σ_x is a group homomorphism $\sigma^*: G \to G$.

For a fixed element f of $\mathcal{P}(G)$, the functions x/f with x in G all lie in some finite-dimensional sub F-space of $\mathcal{R}_G(K)$. In fact, if

$$f \circ \alpha = \sum_{i=1}^{n} h_i(\alpha)g_i,$$

then we have

$$x/f = \sum_{i=1}^{n} g_i(x)h_i.$$

Therefore, the restriction of σ to the set of these functions x/f coincides with a finite F-linear combination of evaluations at elements of K, i.e., there are elements $c_1, \ldots, c_n$ of F and elements $\alpha_1, \ldots, \alpha_n$ of K such that

$$\sigma(x/f) = \sum_{i=1}^{n} c_i x(f \circ \alpha_i)$$

for every element x of G. This means that

$$f \circ \sigma^* = \sum_{i=1}^{n} c_i f \circ \alpha_i.$$

In particular, this shows that $f \circ \sigma^*$ belongs to $\mathscr{P}(G)$, so that σ^* is a morphism of affine algebraic groups from G to G.

Next, we observe that if the element σ of $\mathscr{G}(\mathscr{R}_G(K))$ is the canonical image of an element α of K we have $\sigma^* = \alpha$. Indeed, for every f in $\mathscr{P}(G)$ and every x in G, we have

$$f(\sigma^*(x)) = \sigma(x/f) = (x/f)(\alpha) = f(\alpha(x)).$$

Now let γ denote the comultiplication of $\mathscr{R}_G(K)$, let σ and τ be elements of $\mathscr{G}(\mathscr{R}_G(K))$, and let h be an element of $\mathscr{R}_G(K)$. Writing i for the identity map on $\mathscr{R}_G(K)$, we have

$$(\sigma\tau)(h) = \tau((\sigma \otimes i)(\gamma(h))).$$

If σ is the canonical image, α' say, of an element α of K then $(\sigma \otimes i)(\gamma(h))$ is simply the translate $h \cdot \alpha$, and the above reads

$$(\alpha'\tau)(h) = \tau(h \cdot \alpha).$$

On the other hand,

$$(\alpha'\tau)(h) = \alpha'((i \otimes \tau)(\gamma(h))) = (i \otimes \tau)(\gamma(h))(\alpha).$$

Thus, we have

$$(i \otimes \tau)(\gamma(h))(\alpha) = \tau(h \cdot \alpha).$$

In particular, for $h = x/f$, this gives

$$(i \otimes \tau)(\gamma(x/f))(\alpha) = \tau(x/(f \circ \alpha)) = (f \circ \alpha)(\tau^*(x)) = (\tau^*(x)/f)(\alpha).$$

Hence, we have

$$(i \otimes \tau)(\gamma(x/f)) = \tau^*(x)/f.$$

Applying an arbitrary element σ of $\mathscr{G}(\mathscr{R}_G(K))$ to this last equality, we obtain

$$(\sigma \otimes \tau)(\gamma(x/f)) = f(\sigma^*(\tau^*(x))).$$

The expression on the left is equal to $f((\sigma\tau)^*(x))$. Letting f range over $\mathscr{P}(G)$, we conclude from this that $(\sigma\tau)^*(x)$ coincides with $\sigma^*(\tau^*(x))$. Since this holds for every element x of G, we have $(\sigma\tau)^* = \sigma^* \circ \tau^*$.

Clearly, if e is the neutral element of $\mathscr{G}(\mathscr{R}_G(K))$ then e^* is the identity map on G. Therefore, the last result above shows that every σ^* is in fact an element of $\mathscr{W}(G)$, its inverse being $(\sigma^{-1})^*$. The map sending each σ onto σ^* is therefore a group homomorphism from $\mathscr{G}(\mathscr{R}_G(K))$ to $\mathscr{W}(G)$. Evidently, this homomorphism is injective, and we use it for identifying $\mathscr{G}(\mathscr{R}_G(K))$ with a subgroup of $\mathscr{W}(G)$. This subgroup contains K, because $(\alpha')^* = \alpha$ for every element α of K.

Now it is clear that, by the above identification, $\mathscr{G}(\mathscr{R}_G(K))$ is an algebraic group of automorphisms of G, that K is dense in $\mathscr{G}(\mathscr{R}_G(K))$, and that the restriction map is an isomorphism of Hopf algebras from $\mathscr{P}(\mathscr{G}(\mathscr{R}_G(K)))$ to

$\mathscr{R}_G(K)$. Finally, if L is any algebraic group of automorphisms of G containing K, then the restriction map from $\mathscr{R}_G(L)$ to $\mathscr{R}_G(K)$ is evidently surjective and transposes to an injective morphism of algebraic groups from $\mathscr{G}(\mathscr{R}_G(K))$ to $\mathscr{G}(\mathscr{R}_G(L)) = L$. In $\mathscr{W}(G)$, this becomes the statement that $\mathscr{G}(\mathscr{R}_G(K))$ is contained in L. $\qquad\square$

2. Proposition 2.1. *Let G be an algebraic group over an algebraically closed field, and let K be an algebraic automorphism group of G. Suppose that H is a K-stable algebraic subgroup of G. Then the image K_H of K in $\mathscr{W}(H)$ is an algebraic group of automorphisms of H, and the canonical map $K \to K_H$ is a morphism of algebraic groups. If H is normal in G then the same facts hold for G/H in the place of H.*

PROOF. Let τ denote the canonical map from K to K_H, and let ρ denote the restriction map from $\mathscr{P}(G)$ to $\mathscr{P}(H)$. Evidently, for f in $\mathscr{P}(G)$ and α in K, we have $\rho(f \circ \alpha) = \rho(f) \circ \tau(\alpha)$. This shows that $\mathscr{P}(H)$ is locally finite as a right K_H-module, and also that τ is a morphism of algebraic groups from K to $\mathscr{G}(\mathscr{R}_H(K_H))$. Since the base field is algebraically closed, it follows that K_H is an algebraic subgroup of $\mathscr{G}(\mathscr{R}_H(K_H))$, whence these two groups coincide.

Now suppose that H is normal in G, and let σ denote the canonical map from K to the corresponding subgroup $K_{G/H}$ of $\mathscr{W}(G/H)$. The $K_{G/H}$-module structure of $\mathscr{P}(G/H)$, i.e., of $\mathscr{P}(G)^H$, when lifted to a K-module structure via σ, becomes the restriction to $\mathscr{P}(G)^H$ of the K-module structure of $\mathscr{P}(G)$. It is clear from this that $\mathscr{P}(G/H)$ is locally finite as a $K_{G/H}$-module, and that σ is a morphism of algebraic groups from K to $\mathscr{G}(\mathscr{R}_{G/H}(K_{G/H}))$. This yields the second part of the proposition in the same way by which we obtained the first part. $\qquad\square$

Proposition 2.2. *Let G and K be as in Proposition 2.1. Suppose that L is another algebraic automorphism group of G that is normalized by K. Then LK is an algebraic automorphism group of G.*

PROOF. It is easy to see that $\mathscr{P}(G)$ is locally finite as a right LK-module. Evidently, the restriction maps from $\mathscr{R}_G(LK)$ to $\mathscr{R}_G(L)$ and $\mathscr{R}_G(K)$ are surjective morphisms of Hopf algebras. Their transposes are injective morphisms of affine algebraic groups from L and K to $\mathscr{G}(\mathscr{R}_G(LK))$. Since the base field is algebraically closed, L and K thus become *algebraic* subgroups of $\mathscr{G}(\mathscr{R}_G(LK))$, and it follows that LK is an *algebraic* subgroup of $\mathscr{G}(\mathscr{R}_G(LK))$. Therefore, LK coincides with $\mathscr{G}(\mathscr{R}_G(LK))$. $\qquad\square$

Proposition 2.3. *Let F be a field, G an affine algebraic F-group, K an algebraic automorphism group of G. Then the restriction to K of the canonical group homomorphism from $\mathscr{W}(G)$ to the affine algebraic group of all Lie algebra automorphisms of $\mathscr{L}(G)$ is a morphism of affine algebraic F-groups.*

PROOF. For every element α of $\mathcal{W}(G)$, let α' denote the corresponding Lie algebra automorphism of $\mathcal{L}(G)$. Then the transform by α' of an element τ of $\mathcal{L}(G)$ is given by

$$\alpha'(\tau)(f) = \tau(f \circ \alpha)$$

for every element f of $\mathcal{P}(G)$. Let f° denote the element of $\mathcal{L}(G)^\circ$ given by $f^\circ(\tau) = \tau(f)$. Accordingly, we denote by f°/τ the element of $\mathrm{End}_F(\mathcal{L}(G))^\circ$ defined by

$$(f^\circ/\tau)(e) = f^\circ(e(\tau)) = e(\tau)(f).$$

The algebra of polynomial functions of the group of all Lie algebra automorphisms of $\mathcal{L}(G)$ is generated by the restrictions of the elements of $\mathrm{End}_F(\mathcal{L}(G))^\circ$ and their antipodes. Since $\mathrm{End}_F(\mathcal{L}(G))^\circ$ is spanned over F by the functions f°/τ, it suffices to show that, for each of these, the function on K sending each α onto $(f^\circ/\tau)(\alpha')$ belongs to $\mathcal{P}(K)$, i.e., to $\mathcal{R}_G(K)$. A direct check shows that this function is simply the element τ/f of $\mathcal{R}_G(K)$. $\square$

3. For an algebraic group H, the irreducible component of the neutral element in the center of H will be denoted by $\mathscr{C}_1(H)$.

Theorem 3.1. *Let F be an algebraically closed field of characteristic 0, and let G be an affine algebraic F-group. Let K be a subgroup of $\mathcal{W}(G)$. Then $\mathcal{P}(G)$ is locally finite as a right K-module if and only if the canonical image of K in $\mathcal{W}(\mathscr{C}_1(G_1/G_u))$ is finite.*

PROOF. First, we reduce the theorem to the case where G is irreducible. Let K' denote the canonical image of K in $\mathcal{W}(G_1)$. If $\mathcal{P}(G)$ is locally finite as a K-module, then it is clear that $\mathcal{P}(G_1)$ is locally finite as a K'-module. Therefore, if the theorem holds for G_1, it follows that the image of K' in $\mathcal{W}(\mathscr{C}_1(G_1/G_u))$ is finite. But this coincides with the image of K.

Conversely, suppose that the canonical image of K in $\mathcal{W}(\mathscr{C}_1(G_1/G_u))$ is finite. If the theorem holds for G_1, it follows that $\mathcal{P}(G_1)$ is locally finite as a K'-module. Choose representatives $x_1, \ldots, x_n$ in G for the elements of G/G_1. For every element f of $\mathcal{P}(G)$, let f_i denote the restriction of $x_i \cdot f$ to G_1. In this way, we obtain an injective F-linear map from $\mathcal{P}(G)$ to the direct sum of n copies of $\mathcal{P}(G_1)$, sending each f onto $(f_1, \ldots, f_n)$. If α is any element of K, we have

$$(f \circ \alpha)_i = (x_i \cdot (f \circ \alpha))_{G_1} = ((\alpha(x_i) \cdot f) \circ \alpha)_{G_1} = (\alpha(x_i) \cdot f)_{G_1} \circ \alpha',$$

where α' is the image of α in K'. Using that $\mathcal{P}(G)$ is locally finite as a G-module and $\mathcal{P}(G_1)$ is locally finite as a right K'-module, we see directly from this that all the functions $(f \circ \alpha)_i$, with i and f fixed and α ranging over K, lie in some finite-dimensional space of functions. It follows that the same is true for the functions $f \circ \alpha$, so that $\mathcal{P}(G)$ is locally finite as a right K-module.

Now we assume that G is irreducible, and that $\mathcal{P}(G)$ is locally finite as a right K-module. It is clear from Proposition 2.1 that, consequently,

$\mathscr{P}(\mathscr{C}_1(G/G_u))$ is locally finite with respect to the action of the canonical image of K in $\mathscr{W}(\mathscr{C}_1(G/G_u))$. The group $\mathscr{C}_1(G/G_u)$ is a toroid, T say, and we have

$$\mathscr{P}[T] = F[h_1, \ldots, h_m, h_1^{-1}, \ldots, h_m^{-1}],$$

where the h_i's are algebraically independent polynomial characters of T. For every element α of $\mathscr{W}(T)$, $h_i \circ \alpha$ is again a polynomial character of T, so that there are integers a_{ji} such that $h_i \circ \alpha$ is the product of the $h_j^{a_{ji}}$'s with $j = 1, \ldots, m$. If α ranges over an infinite subset of $\mathscr{W}(T)$, the corresponding set of exponents a_{ji} is unbounded. This shows that, if the canonical image of K in $\mathscr{W}(T)$ is not finite, then $\mathscr{P}(T)$ is not locally finite as a module for this group. Thus, our assumptions on G and K imply that the canonical image of K in $\mathscr{W}(\mathscr{C}_1(G/G_u))$ is finite.

Now suppose that G is irreducible and that the canonical image of K in $\mathscr{W}(\mathscr{C}_1(G/G_u))$ is finite. We must show that then $\mathscr{P}(G)$ is locally finite as a right K-module. Let G' denote the group of inner automorphisms of G. Since the canonical image of $G'K$ in $\mathscr{W}(\mathscr{C}_1(G/G_u))$ coincides with that of K, we may replace K with $G'K$. Therefore, we now suppose that K contains G'. We may write $G = G_u \rtimes P$, where P is linearly reductive. Let L denote the stabilizer of P in K. Since the maximal linearly reductive subgroups of G are conjugate under G', we have $G'L = K$.

Put $T = \mathscr{C}_1(P)$, and let Z denote the element-wise fixer of T in L. Suppose that γ is an element of L whose canonical image in $\mathscr{W}(\mathscr{C}_1(G/G_u))$ is trivial. Then we have $\gamma(t)G_u = tG_u$ for every element t of T. This gives

$$t^{-1}\gamma(t) \in G_u \cap T = (1),$$

so that $\gamma \in Z$. Since the canonical image of L in $\mathscr{W}(\mathscr{C}_1(G/G_u))$ is finite, this shows that Z is of finite index in L. Hence, $G'Z$ is of finite index in $G'L = K$. Evidently, $G'Z$ is normal in K. Therefore, it suffices to show that $\mathscr{P}(G)$ is locally finite as a $G'Z$-module. Since, in any case, $\mathscr{P}(G)$ is locally finite as a G'-module, this simply means showing that $\mathscr{P}(G)$ is locally finite as a Z-module.

The algebra $\mathscr{P}(G)$ is the tensor product $\mathscr{P}(G)^P \otimes \mathscr{P}(G)^{G_u}$. Evidently, $\mathscr{P}(G)^{G_u}$ is stable under the action of $\mathscr{W}(G)$. Since Z stabilizes P, the tensor factor $\mathscr{P}(G)^P$ is stable under the action of Z. Now it suffices to prove that each of $\mathscr{P}(G)^P$ and $\mathscr{P}(G)^{G_u}$ is locally finite as a right Z-module.

First, we deal with $\mathscr{P}(G)^P$. By the restriction map, this is isomorphic with $\mathscr{P}(G_u)$, and it is clear that the restriction map is a morphism of right Z-modules. Therefore, what we wish to show here is an immediate consequence of the general fact that *if U is a unipotent affine algebraic group over a field of characteristic 0 then $\mathscr{P}(U)$ is locally finite as a right $\mathscr{W}(U)$-module.*

In order to see this, recall from Theorem VIII.1.1 that the exponential map is a variety isomorphism from $\mathscr{L}(U)$ to U. In proving this theorem, we showed that if ρ is any polynomial representation of U then

$$\rho \circ \exp = \exp \circ \rho',$$

where ρ' is the extended differential of ρ. This shows that the transpose of the exponential map is an isomorphism of right $\mathscr{W}(U)$-modules from $\mathscr{P}(U)$ to $\mathscr{P}(\mathscr{L}(U))$. Since $\mathscr{W}(U)$ acts by linear automorphisms on $\mathscr{L}(U)$, and since $\mathscr{P}(\mathscr{L}(U))$ is the symmetric algebra built over $\mathscr{L}(U)^{\circ}$, it is clear that $\mathscr{P}(\mathscr{L}(U))$ is locally finite as a $\mathscr{W}(U)$-module, so that the same holds for $\mathscr{P}(U)$.

It remains to be shown only that $\mathscr{P}(G)^{G_u}$ is locally finite as a Z-module. We identify $\mathscr{P}(G)^{G_u}$ with $\mathscr{P}(P)$, noting that this identification is compatible with the actions of Z. It follows from Theorems IV.2.2, VII.3.2 and VII.1.2 that $P = T[P, P]$, and that $\mathscr{L}([P, P])$ is semisimple. The surjective multiplication map from $T \times [P, P]$ to P transposes to an injective F-algebra and Z-module homomorphism from $\mathscr{P}(P)$ to $\mathscr{P}(T) \otimes \mathscr{P}([P, P])$. By definition, Z leaves the elements of $\mathscr{P}(T)$ fixed, and it clearly stabilizes $\mathscr{P}([P, P])$. Therefore, it suffices to prove that *if S is an irreducible affine algebraic group over an algebraically closed field of characteristic 0 such that $\mathscr{L}(S)$ is semisimple then $\mathscr{W}(S)/S'$ is finite.*

In order to see this, consider the canonical image of S' in the group of all Lie algebra automorphisms of $\mathscr{L}(S)$, which we denote by $\mathscr{W}(\mathscr{L}(S))$. This is the image of S under the adjoint representation, and therefore is an algebraic subgroup of $\mathscr{W}(\mathscr{L}(S))$. By Theorem IV.4.1, its Lie algebra is the Lie algebra of all inner derivations of $\mathscr{L}(S)$. Since S is semisimple, we know from Proposition VII.2.6 that this coincides with the Lie algebra of all derivations of $\mathscr{L}(S)$, i.e., with the Lie algebra of $\mathscr{W}(\mathscr{L}(S))$. Therefore, the image of S' coincides with the irreducible component of the neutral element in $\mathscr{W}(\mathscr{L}(S))$, and so is of finite index in $\mathscr{W}(\mathscr{L}(S))$. Since the base field is of characteristic 0, it follows from Corollary IV.3.2 that the canonical map from $\mathscr{W}(S)$ to $\mathscr{W}(\mathscr{L}(S))$ is injective. Hence, we conclude that $\mathscr{W}(S)/S'$ is finite. $\qquad\square$

4. Theorem 4.1. *Let F be an algebraically closed field of characteristic 0, and let G be an affine algebraic F-group. Let Q be the kernel of the canonical homomorphism from $\mathscr{W}(G)$ to $\mathscr{W}(\mathscr{C}_1(G_1/G_u))$. Then Q is an algebraic automorphism group of G, and every irreducible algebraic group of automorphisms of G is an algebraic subgroup of Q.*

PROOF. By Theorem 3.1, $\mathscr{P}(G)$ is locally finite as a right Q-module. Consider the algebraic automorphism group $\mathscr{G}(\mathscr{R}_G(Q))$ of G. We can apply Proposition 2.1 three times in succession and conclude that the canonical map is a morphism of algebraic groups from $\mathscr{G}(\mathscr{R}_G(Q))$ onto an algebraic group of automorphisms of $\mathscr{C}_1(G_1/G_u)$. Since Q is the kernel of this morphism, it is an algebraic subgroup of $\mathscr{G}(\mathscr{R}_G(Q))$, and therefore coincides with it.

Now let R be any irreducible algebraic automorphism group of G. Again, the canonical map is a morphism of algebraic groups from R onto an algebraic group of automorphisms, S say, of $\mathscr{C}_1(G_1/G_u)$. Since R is irreducible, so is S. On the other hand, we know from Theorem 3.1 that S is finite. Therefore, S is trivial, which means that R is contained in Q. Since $\mathscr{R}_G(R)$ is the restriction image of $\mathscr{R}_G(Q)$, it is clear that R is an *algebraic* subgroup of Q. $\qquad\square$

Theorem 4.2. *Let F be an algebraically closed field of characteristic 0, and let G be an irreducible affine algebraic F-group. A subgroup of $\mathscr{W}(G)$ is an algebraic group of automorphisms of G if and only if its canonical image in $\mathscr{W}(\mathscr{L}(G))$ is an algebraic subgroup of $\mathscr{W}(\mathscr{L}(G))$.*

PROOF. It is clear from Proposition 2.3 that the condition of the theorem is necessary. In order to prove the sufficiency, let K be a subgroup of $\mathscr{W}(G)$ whose image in $\mathscr{W}(\mathscr{L}(G))$ is an algebraic subgroup of $\mathscr{W}(\mathscr{L}(G))$. First, we show that the canonical image, Q say, of K in $\mathscr{W}(\mathscr{C}_1(G/G_u))$ is finite. Let us write T for $\mathscr{C}_1(G/G_u)$. The stabilizer of $\mathscr{L}(G_u)$ in $\mathscr{W}(\mathscr{L}(G))$ is evidently an algebraic subgroup of $\mathscr{W}(\mathscr{L}(G))$ containing the canonical image of K as an algebraic subgroup. The canonical map from this stabilizer to $\mathscr{W}(\mathscr{L}(T))$ is clearly a morphism of affine algebraic groups. It follows that the canonical image of Q in $\mathscr{W}(\mathscr{L}(T))$ is an algebraic subgroup of $\mathscr{W}(\mathscr{L}(T))$. Hence we shall know that Q is finite as soon as we have proved that *if H is any algebraic subgroup of $\mathscr{W}(\mathscr{L}(T))$ that lies in the image of $\mathscr{W}(T)$ then H is finite.*

From our discussion of T in the proof of Theorem 3.1, we see that there is an F-basis of $\mathscr{L}(T)$ with respect to which the image of $\mathscr{W}(T)$ appears as the group of all integral matrices of determinant 1 or -1. Let g_{ij} be the polynomial function on H_1 such that, for every element α of H_1, $g_{ij}(\alpha)$ is the (i, j)-entry of the matrix representing α. If H_1 is non-trivial, then at least one of these g_{ij}'s is non-constant, and hence is transcendental over F. Let u be such a g_{ij}. By Theorem II.3.3, there is a non-zero polynomial $p(u)$ in $F[u]$ such that every F-algebra homomorphism $F[u] \to F$ not annihilating $p(u)$ extends to an F-algebra homomorphism $\mathscr{P}(H_1) \to F$, i.e., is the evaluation at some element of H_1. Clearly, this contradicts the fact that u takes only integer values. Thus, H_1 must be trivial, so that H is finite.

Our conclusion is that the canonical image of K in $\mathscr{W}(\mathscr{C}_1(G/G_u))$ is finite. By Theorem 3.1, $\mathscr{P}(G)$ is therefore locally finite as a right K-module, so that we have the algebraic automorphism group $\mathscr{G}(\mathscr{R}_G(K))$. By Proposition 2.3, the canonical homomorphism from this group to $\mathscr{W}(\mathscr{L}(G))$ is a morphism of affine algebraic groups, and we know that this morphism is injective (cf. the end of Section 3). Now K is the inverse image in $\mathscr{G}(\mathscr{R}_G(K))$ of an algebraic subgroup of the image of $\mathscr{G}(\mathscr{R}_G(K))$ in $\mathscr{W}(\mathscr{L}(G))$. Hence, K is an algebraic subgroup of $\mathscr{G}(\mathscr{R}_G(K))$ and therefore coincides with it. $\square$

Theorem 4.3. *Let F be an algebraically closed field of characteristic 0, and let G be an irreducible affine algebraic F-group. Then $\mathscr{W}(G)$ is an algebraic automorphism group of G if and only if one of the following two conditions is satisfied: (1) $\mathscr{C}_1(G)$ is unipotent; (2) the dimension of the center of G/G_u is at most 1.*

PROOF. Clearly, (2) implies that $\mathscr{W}(\mathscr{C}_1(G/G_u))$ is finite. By Theorem 3.1, this implies that $\mathscr{P}(G)$ is locally finite as a $\mathscr{W}(G)$-module, which evidently implies that $\mathscr{W}(G)$ is an algebraic automorphism group of G. Thus, condition (2) is sufficient.

In order to establish the sufficiency of condition (1), let us write, as in the proof of Theorem 3.1, $G = G_u \rtimes P$, and $T = \mathscr{C}_1(P)$. Let L be the stabilizer of P in $\mathscr{W}(G)$. Then $G'L = \mathscr{W}(G)$, so that the canonical image of $\mathscr{W}(G)$ in $\mathscr{W}(\mathscr{C}_1(G/G_u))$ coincides with that of L. In view of Theorem 3.1, it suffices therefore to prove that the canonical image of L in $\mathscr{W}(\mathscr{C}_1(G/G_u))$ is finite, whenever (1) is satisfied. For this, it clearly suffices to show that the image of L in $\mathscr{W}(T)$ is finite.

Consider the adjoint representation of T on $\mathscr{L}(G_u)$. Since T is a toroid, we can decompose $\mathscr{L}(G_u)$ into a direct sum $V_{f_1} + \cdots + V_{f_n}$ of sub T-modules, where the f_i's are mutually distinct morphisms $T \to F^*$, and $t \cdot v = f_i(t)v$ for every element t of T and every element v of V_{f_i}. If α is an element of L, and α_T is its restriction image in $\mathscr{W}(T)$ then the automorphism of $\mathscr{L}(G_u)$ corresponding to α maps each V_{f_i} onto some V_{f_j}, where $f_j = f_i \circ \alpha_T^{-1}$. In this way, we obtain a homomorphism δ from L to the finite group of permutations of the set $(f_1, \ldots, f_n)$. Now let α be an element of the kernel of δ. Then the adjoint action of $\alpha(t)t^{-1}$ on $\mathscr{L}(G_u)$ is trivial for every element t of T. Hence each $\alpha(t)t^{-1}$ centralizes G_u, and therefore lies in the center of G. Let σ be the map from T to the center of G defined by $\sigma(t) = \alpha(t)t^{-1}$. Clearly, σ is a morphism of affine algebraic groups, whence $\sigma(T)$ is an irreducible algebraic subgroup of the center of G, so that

$$\sigma(T) \subset \mathscr{C}_1(G).$$

Since T is linearly reductive, while $\mathscr{C}_1(G)$ is unipotent (condition (1)), it follows that σ is the trivial map, which means that α leaves the elements of T fixed. Thus, δ induces an isomorphism from the image of L in $\mathscr{W}(T)$ to a finite group, so that the image of L in $\mathscr{W}(T)$ is finite. Our conclusion is that condition (1) is sufficient.

It remains to be shown that if neither (1) nor (2) is satisfied then $\mathscr{W}(G)$ is not an algebraic automorphism group of G. By Theorem V.5.3, every subtoroid of a toroid is a direct factor. Hence we may write $T = T_0 \times T_1$, where T_0 is the irreducible component of the neutral element in the intersection of T with the center of G, and T_1 is a complementary subtoroid. First, we consider the case where the dimension of T_0 is greater than 1. Recall that $P = TS$, where $S = [P, P]$. Hence $G = (G_u T_1 S)T_0$. The factor $G_u T_1 S$ is an algebraic subgroup of G, and its intersection with T_0 is finite. Therefore, as is easy to see, there are infinitely many elements of $\mathscr{W}(T_0)$ that leave the elements of $(G_u T_1 S) \cap T_0$ fixed. Clearly, each of these extends to yield an element of $\mathscr{W}(G)$ leaving the elements of $G_u T_1 S$ fixed. By Theorem 3.1, $\mathscr{W}(G)$ can therefore not be an algebraic automorphism group of G, in this case.

Now suppose the dimension of T_0 is not greater than 1. If T_0 were trivial, condition (1) of our theorem would be satisfied. Thus, we are left with the case where T_0 is of dimension 1. Since condition (2) is not satisfied, the toroid T_1 is non-trivial, in this case. Therefore, there are infinitely many morphisms ρ of affine algebraic groups from T_1 to T_0 whose kernels contain the finite group $(G_u T_0 S) \cap T_1$. For each such ρ, we have an element ρ^* of $\mathscr{W}(G)$ such that ρ^* leaves the elements of $G_u T_0 S$ fixed, while $\rho^*(t) = t\rho(t)$ for every element t of T_1. Again by Theorem 3.1, $\mathscr{W}(G)$ is therefore not an algebraic automorphism group of G. $\square$

Notes

1. Evidently, Theorem 4.1 implies that if R and S are irreducible algebraic automorphism groups of G then so is the group generated by R and S in $\mathscr{W}(G)$. The following example shows that this fails in non-zero characteristic.

Let F be an algebraically closed field of non-zero characteristic p, and let G be the 2-dimensional algebraic vector F-group, so that $\mathscr{P}(G)$ is the polynomial algebra $F[x, y]$, where x and y are the usual coordinate functions on G. For every element a of F, define the automorphisms ρ_a and σ_a of G by

$$\rho_a(u, v) = (u, v + au^p), \qquad \sigma_a(u, v) = (u + av^p, v).$$

Clearly, the ρ_a's constitute an algebraic automorphism group R of G, and the σ_a's constitute an algebraic automorphism group S of G, each of R and S being isomorphic, as algebraic group, with the additive group of F. Let γ denote the automorphism $\rho_1 \circ \sigma_1$ of G. Then we have

$$x \circ \gamma = x + y^p, \qquad y \circ \gamma = y + y^{p^2} + x^p,$$

which shows that $\mathscr{P}(G)$ is not locally finite as a module for the group generated by R and S.

2. In the situation of Theorem 4.1, if R and S are algebraic automorphism groups of G, and if R is irreducible, it follows from Theorem 3.1 that $\mathscr{P}(G)$ is locally finite as a right module for the subgroup of $\mathscr{W}(G)$ that is generated by R and S. The following example shows that this can fail if neither R nor S is irreducible.

Let G be the 2-dimensional F-toroid. Then $\mathscr{W}(G)$ is isomorphic with the multiplicative group of the 2 by 2 integer matrices of determinant 1 or -1, as is seen from the proof of Theorem 3.1. Let σ be the automorphism corresponding to the matrix $\begin{pmatrix} -2 & -3 \\ 1 & 1 \end{pmatrix}$, and let τ be the automorphism corresponding to the transpose of this matrix. Then each of σ and τ is of order 3, while $\tau \circ \sigma$ has infinite order.

3. The automorphism groups of affine algebraic groups over fields of characteristic 0 are analyzed by A. Borel and J-P. Serre in [2].

Chapter XVI

The Universal Enveloping Algebra

The universal enveloping algebra of a Lie algebra is the analogue of the usual group algebra of a group. It has the analogous function of exhibiting the category of Lie algebra modules as a category of modules for an associative algebra. This becomes more than an analogy when the universal enveloping algebra is viewed with its full Hopf algebra structure. By dualization, one obtains a commutative Hopf algebra which, in the case where the Lie algebra is that of an irreducible algebraic group over a field of characteristic 0, contains the algebra of polynomial functions of that group as a sub Hopf algebra in a natural fashion. This theme is developed in Section 3.

Section 1 is devoted to the Poincaré–Birkhoff–Witt Theorem, which is needed in Section 2 for establishing the Campbell–Hausdorff formula. This formula concerns the formal exponential map and is decisive in many applications of Lie algebra theory to algebraic groups or Lie groups. Section 4 is devoted to our principal application of the Campbell–Hausdorff formula in perfecting the theory of unipotent algebraic groups over fields of characteristic 0 by reducing it completely to that of nilpotent Lie algebras.

1. Let L be a Lie algebra over a field F. If A is an associative F-algebra, we may consider the Lie algebra $\mathscr{L}(A)$, whose underlying F-space is A and whose Lie composition is defined by $[u, v] = uv - vu$. By a *Lie homomorphism* from L to A, we shall mean a homomorphism of Lie algebras from L to $\mathscr{L}(A)$.

Let $\otimes(L)$ denote the tensor F-algebra built over L, and let $J(L)$ denote the two-sided ideal of $\otimes(L)$ that is generated by the elements of the form $a \otimes b - b \otimes a - [a, b]$, with a and b ranging over L. The associative F-algebra $\otimes(L)/J(L)$ is called the *universal enveloping algebra* of L, and we denote it by $\mathscr{U}(L)$. The canonical injection from L to $\otimes(L)$ yields the canonical map $\gamma: L \to \mathscr{U}(L)$, which is a Lie homomorphism by virtue of the definition of $J(L)$. It is easy to see that the pair $(\mathscr{U}(L), \gamma)$ is characterized up to isomorphisms by the following universal mapping property. *For every Lie homomorphism α from L to an associative algebra A, there is one and only one homomorphism α^* of associative algebras from $\mathscr{U}(L)$ to A such that $\alpha^* \circ \gamma = \alpha$.*

The following *Poincaré–Birkhoff–Witt Theorem* says that the canonical map from $\otimes(L)$ to $\mathscr{U}(L)$ is as non-degenerate as one could wish.

Theorem 1.1. *Let L be a Lie algebra over a field F, and let X be a totally ordered F-basis of L. Let $S(X)$ denote the set of all finite non-decreasing sequences of elements of X. For $(x_1, \ldots, x_n)$ in $S(X)$, put*

$$\gamma(x_1, \ldots, x_n) = \gamma(x_1) \cdots \gamma(x_n)$$

and $\gamma(\varnothing) = 1$. Then γ is a bijection from $S(X)$ to an F-basis of $\mathscr{U}(L)$.

PROOF. An evident "straightening" procedure, based on the relations

$$\gamma(x)\gamma(y) = \gamma(y)\gamma(x) + \gamma([x, y])$$

for all elements x and y of L, shows inductively that the set $\gamma(S(X))$ spans $\mathscr{U}(L)$ over F. Therefore, it suffices to show that γ maps $S(X)$ injectively onto a linearly independent subset of $\mathscr{U}(L)$. This will be clear once we have proved the following fact. *There is an F-linear map σ from $\otimes(L)$ to the symmetric algebra $\mathscr{S}(L)$ built over L satisfying the following conditions.*

(1) $\sigma(1) = 1$;
(2) *if $(x_1, \ldots, x_n)$ belongs to $S(X)$ then*

$$\sigma(x_1 \otimes \cdots \otimes x_n) = x_1 \cdots x_n;$$

(3) *σ annihilates $J(L)$.*

Let $B_0(X) = (1)$ and, for $n > 0$, let $B_n(X)$ be the subset of $\otimes(L)$ consisting of the elements $x_1 \otimes \cdots \otimes x_n$, with each x_i in X. Clearly, $B_n(X)$ is an F-basis of $\otimes^n(L)$. Let $B(X)$ denote the union of the family of $B_n(X)$'s. For each element $u = x_1 \otimes \cdots \otimes x_n$ of $B_n(X)$, define the *disorder* $D(u)$ as the number of index pairs (i, j) such that $i < j$ and $x_i > x_j$. Let $\otimes^n(L)_q$ denote the subspace of $\otimes^n(L)$ that is spanned by the elements u of $B_n(X)$ with $D(u) \leq q$.

Now observe that $J(L)$ is spanned over F by the elements of the form

$$u \otimes x \otimes y \otimes v - u \otimes y \otimes x \otimes v - u \otimes [x, y] \otimes v,$$

where u and v are elements of $B(X)$, and x and y are elements of L. Therefore, it suffices to define σ as an F-linear map satisfying the above conditions (1) and (2) and annihilating each of these elements. Condition (1) serves to define σ on $\otimes^0(L) = F$. Proceeding inductively, suppose that σ has already been defined on $\sum_{n=0}^{r} \otimes^n(L)$, for some $r \geq 0$, so that (1) and (2), for $n \leq r$, are satisfied, and such that σ annihilates $J(L) \cap (\sum_{n=0}^{r} \otimes{}^{n}(L))$. Using (2) for $n = r + 1$, we extend the definition of σ to the domain

$$\otimes^{r+1}(L)_0 + \sum_{n=0}^{r} \otimes^{n}(L).$$

It is easy to see that this does not create a violation of (3).

Now suppose that σ has already been defined on $\otimes^{r+1}(L)_s + \sum_{n=0}^{r} \otimes^{n}(L)$, with some $s \geq 0$, in such a way that σ annihilates the intersection, $J(L)_{r,s}$ say, of $J(L)$ with this space and satisfies (2) for all $n \leq r + 1$. Let w be an element of $B_{r+1}(X)$ such that $D(w) = s + 1$. We may write

$$w = u \otimes x \otimes y \otimes v,$$

where x and y are elements of X such that $x > y$, and u and v are elements of $B(X)$. Evidently, we have $D(u \otimes y \otimes x \otimes v) = D(w) - 1$, so that $u \otimes y \otimes x \otimes v$ belongs $\otimes^{r+1}(L)_s$. Condition (3) demands the equality

$$\sigma(w) = \sigma(u \otimes y \otimes x \otimes v + u \otimes [x, y] \otimes v).$$

Conversely, if σ has been defined on $\otimes^{r+1}(L)_{s+1} + \sum_{n=0}^{r} \otimes^{n}(L)$ so that all these equalities, for all elements w of $B_{r+1}(L)$ with $D(w) = s + 1$ and all possible choices of (x, y), are satisfied then σ annihilates $J(L)_{r,s+1}$. Clearly, this extension of the domain of definition of σ is possible, provided that, for each w, the right side of the above equality is the same for every possible choice of (x, y). Thus, it suffices to show that if w can also be written $u' \otimes x' \otimes y' \otimes v'$ with $x' > y'$ then σ annihilates the element

$$u \otimes y \otimes x \otimes v + u \otimes [x, y] \otimes v - u' \otimes y' \otimes x' \otimes v' - u' \otimes [x', y'] \otimes v'.$$

If the indicated positions in w of the elements x, y, x', y' are all distinct from each other, we have (possibly only after exchanging the primed and the unprimed labels)

$$w = u \otimes x \otimes y \otimes a \otimes x' \otimes y' \otimes b.$$

Then the above element is

$$u \otimes y \otimes x \otimes a \otimes x' \otimes y' \otimes b + u \otimes [x, y] \otimes a \otimes x' \otimes y' \otimes b$$

$$- u \otimes x \otimes y \otimes a \otimes y' \otimes x' \otimes b - u \otimes x \otimes y \otimes a \otimes [x', y'] \otimes b.$$

Addition of suitable elements of $J(L)_{r,s}$ has the effect of replacing each factor $x \otimes y$ with $y \otimes x + [x, y]$, and each factor $x' \otimes y'$ with

$$y' \otimes x' + [x', y'].$$

The resulting sum is 0. This shows that the above element belongs to $J(L)_{r,s}$, so that it is annihilated by σ.

We are left with the case where x' coincides, *literally*, with y. Writing z for y', we have

$$w = u \otimes x \otimes y \otimes z \otimes b.$$

and $x > y > z$. We must show that the element

$$u \otimes y \otimes x \otimes z \otimes b + u \otimes [x, y] \otimes z \otimes b$$

$$-u \otimes x \otimes z \otimes y \otimes b - u \otimes x \otimes [y, z] \otimes b$$

is annihilated by σ. Addition of suitable elements of $J(L)_{r,s}$ has the effect of replacing each factor $x \otimes z$ with $z \otimes x + [x, z]$. Similarly, the resulting factors $y \otimes z$ and $x \otimes y$ may be replaced with

$$z \otimes y + [y, z], \quad \text{and} \quad y \otimes x + [x, y],$$

respectively. Finally, again by adding elements of $J(L)_{r.s}$, we can replace the factors $[y, z] \otimes x$, $y \otimes [x, z]$ and $[x, y] \otimes z$ with

$$x \otimes [y, z] + [[y, z], x] \text{ etc.}$$

After evident cancellations, there remains a sum of three terms, which is 0 by virtue of the Jacobi identity.

This completes the inductive step for extending the domain of definition of σ to $\sum_{n=0}^{r+1} \otimes^n(L)$ by induction on the disorder s. The result is the inductive step for the degree r of the main induction. It is clear that the existence of σ so established implies the theorem. $\qquad\square$

Usually, one identifies L with its canonical image $\gamma(L)$ in $\mathscr{U}(L)$, by means of γ. Now, there is a Lie homomorphism from L to $\mathscr{U}(L) \otimes \mathscr{U}(L)$ sending each element x of L onto the element $x \otimes 1 + 1 \otimes x$ of $\mathscr{U}(L) \otimes \mathscr{U}(L)$. By the universal mapping property, this defines a morphism of F-algebras, δ say, from $\mathscr{U}(L)$ to $\mathscr{U}(L) \otimes \mathscr{U}(L)$. Next, note that $\mathscr{U}(L)$ is the direct F-space sum $F + L\mathscr{U}(L)$. Let ε denote the projection $\mathscr{U}(L) \to F$ with kernel $L\mathscr{U}(L)$. Making an induction on the formal degree with respect to L of expressions for the elements of $\mathscr{U}(L)$, one verifies easily that δ and ε make $\mathscr{U}(L)$ into a coalgebra. Together with the multiplication, μ say, from $\mathscr{U}(L) \otimes \mathscr{U}(L)$ to $\mathscr{U}(L)$, this makes $\mathscr{U}(L)$ into a bialgebra. Finally, it is easy to see that this bialgebra has an antipode, η say, which is characterized as an anti endo-morphism of the F-algebra $\mathscr{U}(L)$ by $\eta(x) = -x$ for every element x of L. Thus, $\mathscr{U}(L)$ *has the structure of a Hopf algebra*. The identities relating to η, μ and δ are established inductively.

2. Let V be a vector space over a field F, and write T for the tensor algebra $\otimes(V)$ built over V. Actually, T has the structure of a Hopf algebra, as follows. The counit ε is the projection $\otimes(V) \to \otimes^0(V) = F$ of the graded structure of $\otimes(V)$. The comultiplication δ is the unique F-algebra homomorphism from T to $T \otimes T$ sending each element t of T onto $t \otimes 1 + 1 \otimes t$. The antipode η is characterized as an anti-endomorphism of the F-algebra T by $\eta(t) = -t$ for every element t of T.

Let L denote the sub Lie algebra of $\mathscr{L}(T)$ that is generated by V. Now suppose that ρ is a linear map from V to an F-Lie algebra M. Let α denote the canonical map from M to $\mathscr{U}(M)$. By the universal mapping property of T in the category of F-algebras, there is one and only one morphism of F-algebras ρ^* from T to $\mathcal{U}(M)$ whose restriction to V is $\alpha \circ \rho$. The restriction of ρ^* to L is clearly a morphism of F-Lie algebras from L to $\alpha(M)$. By Theorem 1.1, the map α is injective. If we compose the restriction of ρ^* to L with the inverse $\alpha(M) \to M$ of α, we obtain a morphism σ of Lie algebras from L to M that extends ρ. Since V generates L as a Lie algebra, there can be at most one such σ. The property of L with respect to V we have thus established means that L is a model for the *free Lie algebra based on* V: for every linear map ρ from V to a Lie algebra M, there is one and only one morphism of Lie algebras from L to M that extends ρ.

Let τ be a Lie homomorphism from L to an associative F-algebra A. There is one and only one morphism τ^* of F-algebras from T to A that extends the restriction of τ to V. Evidently, the restriction of τ^* to L coincides with τ. This shows that T has the same universal mapping property with respect to L as $\mathcal{U}(L)$. Therefore, T *is naturally isomorphic, as a Hopf algebra, with* $\mathcal{U}(L)$. We shall make use of this fact after some general preparation.

Theorem 2.1. *Let L be a Lie algebra over a field F of characteristic 0. Then the space of primitive elements of $\mathcal{U}(L)$ coincides with L.*

PROOF. Choose a totally ordered F-basis (x_α) for L. By Theorem 1.1, the element 1 of F and the ordered monomials $x_{\alpha_1}^{e_1} \cdots x_{\alpha_n}^{e_n}$, where the e_i's are strictly positive integers, and $x_{\alpha_1} < \cdots < x_{\alpha_n}$, constitute an F-basis of $\mathcal{U}(L)$. We have

$$\delta(x_{\alpha_1}^{e_1} \cdots x_{\alpha_n}^{e_n}) = (x_{\alpha_1}^{e_1} \cdots x_{\alpha_n}^{e_n}) \otimes 1 + 1 \otimes (x_{\alpha_1}^{e_1} \cdots x_{\alpha_n}^{e_n}) + \sum,$$

where $\sum$ is the sum of the terms in $(L\mathcal{U}(L)) \otimes (L\mathcal{U}(L))$ resulting from the expansion of $\delta(x_{\alpha_1})^{e_1} \cdots \delta(x_{\alpha_n})^{e_n}$. These terms are

$$C_{e_1, f_1} \cdots C_{e_n, f_n} x_{\alpha_1}^{e_1 - f_1} \cdots x_{\alpha_n}^{e_n - f_n} \otimes x_{\alpha_1}^{f_1} \cdots x_{\alpha_n}^{f_n},$$

where the C_{e_i, f_i}'s are the binomial coefficients, and the summation goes over all n-tuples $(f_1, \ldots, f_n)$ such that $0 \le f_i \le e_i$ and $0 < \sum_{i=1}^{n} f_i < \sum_{i=1}^{n} e_i$. Since F is of characteristic 0, all these binomial coefficients are different from 0. Hence, it follows from the linear independence of the ordered monomials in the x_α's that, if u is a linear combination of such ordered monomials and $\delta(u) = u \otimes 1 + 1 \otimes u$, then u must actually be a linear combination of the x_α's. This means that every primitive element of $\mathcal{U}(L)$ belongs to L. $\qquad\square$

Now let us return to the free Lie algebra L based on the F-space V. If F is of characteristic 0, we know from Theorem 2.1 that L is precisely the space of primitive elements of the Hopf algebra $T = \otimes(V)$, because T may be identified with $\mathcal{U}(L)$ as a Hopf algebra.

Proposition 2.2. *Let F, V, T, L be as above, and assume that F is of characteristic 0. For every element t of T, let D_t be the derivation effected by t in T, so that*

$D_t(u) = tu - ut$. *There is an F-linear projection* π *of* T *onto* L *such that* $\pi(1) = 0$ *and*

$$\pi(t_1 \cdots t_q) = q^{-1}(D_{t_1} \cdots D_{t_{q-1}})(t_q)$$

for every q-tuple $(t_1, \ldots, t_q)$ *of elements of* V, *where* $q > 1$.

PROOF. Clearly, there is one and only one linear endomorphism π of T satisfying the equalities of the proposition and the condition $\pi(v) = v$ for every element v of V. Moreover, it is evident from these properties that $\pi(T)$ is contained in L. It remains only to prove that $\pi(x) = x$ for every element x of L.

Let ρ denote the F-algebra homomorphism from T to $\mathrm{End}_F(T)$ that is determined by the condition $\rho(v) = D_v$ for every element v of V. Making an induction on the degree relative to V, one shows that $\rho(x) = D_x$ for every element x of L. Define the linear endomorphism π' of T such that π' coincides with $n\pi$ on each $\otimes^n(V)$. Then we have

$$\pi'(ut) = \rho(u)(\pi'(t)),$$

for all elements u of T and t of VT. If x and y are elements of L, we have

$$\pi'([x, y]) = \pi'(xy - yx) = \rho(x)(\pi'(y)) - \rho(y)(\pi'(x))$$

$$= D_x(\pi'(y)) - D_y(\pi'(x)) = [x, \pi'(y)] + [\pi'(x), y].$$

Thus, the restriction of π' to L is a derivation of L. Using this, one shows easily by induction on the degree that $\pi'(x) = nx$ for every element x of $L \cap \otimes^n(V)$, whence we have $\pi(y) = y$ for every element y of L. $\qquad\square$

Let A be any F-algebra that is graded by the non-negative integers, and let A_+ denote the ideal of A that consists of the elements whose components of degree 0 are equal to 0. We regard A as a topological algebra by making the powers of A_+ a fundamental system of neighborhoods of 0. As such, A has a completion, A' say, which is a complete topological F-algebra containing A as a dense subalgebra. Here, completeness means that every Cauchy sequence is convergent, and the elements of A' are the usual equivalence classes of Cauchy sequences in A, i.e., of the sequences (a_n) with the property that, for every m, there is an M such that $a_{n+1} - a_n$ belongs to $(A_+)^m$ for every $n \geq M$. The elements of A' are most conveniently viewed as formal infinite sums $\sum_{n \geq 0} t_n$, with each t_n in the homogeneous component A_n of A. We write A'_+ for the ideal $A_+ A'$ of A', and we note that the powers of A'_+ constitute a fundamental system of neighborhoods of 0 for the topology of A'.

Let F be a field of characteristic 0, and suppose that B is any complete topological F-algebra, the topology being defined by the powers of some ideal J. Then we can define the exponential map $\mathrm{Exp}\colon J \to 1 + J$, where $\mathrm{Exp}(x)$ is the limit in B of the Cauchy sequence whose nth term is $\sum_{q=0}^{n}(x^q/q\,!)$. Similarly, we define the map $\mathrm{Log}\colon 1 + J \to J$, where $\mathrm{Log}(1 - x)$ is the limit in B of the Cauchy sequence whose nth term is $-\sum_{q=1}^{n}(x^q/q)$. Note that these maps are continuous and mutually inverse.

In particular, we consider the completions T' and $(T \otimes T)'$, where T is our above tensor algebra $\otimes(V)$. The comultiplication δ from T to $T \otimes T$ is evidently continuous, so that it extends to a morphism of F-algebras from T' to $(T \otimes T)'$, which we shall still denote by δ. We regard $T' \otimes T'$ as a sub F-algebra of $(T \otimes T)'$, in the evident way, and we denote the closure of L in T' by L'. We know that L consists precisely of the primitive elements of T. By an evident continuity argument, this implies that L' consists precisely of all elements x of T' for which $\delta(x) = x \otimes 1 + 1 \otimes x$.

Let X denote the set of all elements x of $1 + T'_+$ for which $\delta(x) = x \otimes x$. Clearly, $XX \subset X$. Moreover, every element x of X is a unit of T', with $x^{-1} = \sum_{n \geq 0}(1 - x)^n$. It follows that $x \otimes x$ is a unit of $(T \otimes T)'$, with $(x \otimes x)^{-1} = x^{-1} \otimes x^{-1}$. Now $\delta(x^{-1}) = \delta(x)^{-1} = x^{-1} \otimes x^{-1}$, showing that x^{-1} belongs to X. Thus, X is a subgroup of the group of units of T'.

Let y be an element of L'. Noting that we have $\mathrm{Exp}(u + v) = \mathrm{Exp}(u)\mathrm{Exp}(v)$ whenever $uv = vu$, we obtain

$$\delta(\mathrm{Exp}(y)) = \mathrm{Exp}(\delta(y)) = \mathrm{Exp}(y \otimes 1 + 1 \otimes y) = \mathrm{Exp}(y) \otimes \mathrm{Exp}(y).$$

Thus, $\mathrm{Exp}(L') \subset X$. Now let x be an element of X. Then we have

$$\delta(\mathrm{Log}(x)) = \delta(\mathrm{Log}(1 - (1 - x))) = \mathrm{Log}(1 \otimes 1 - \delta(1 - x))$$

$$= \mathrm{Log}(\delta(x)) = \mathrm{Log}(x \otimes x).$$

Applying Exp, we see that $\mathrm{Log}(uv) = \mathrm{Log}(u) + \mathrm{Log}(v)$ whenever $uv = vu$. Hence, the above gives

$$\delta(\mathrm{Log}(x)) = \mathrm{Log}(x \otimes 1) + \mathrm{Log}(1 \otimes x)$$

$$= \mathrm{Log}(x) \otimes 1 + 1 \otimes \mathrm{Log}(x).$$

Our conclusion is that $\mathrm{Log}(X) \subset L'$. Thus, Exp *maps L' bijectively onto X, and the inverse map is the restriction of* Log *to* X.

Using Exp and Log, we can transport the group structure of X to a group structure on L'. We examine the "product" of two elements x and y of V. This is expressible as follows.

$$\mathrm{Log}(\mathrm{Exp}(x)\mathrm{Exp}(y)) = \mathrm{Log}(1 - (1 - \mathrm{Exp}(x)\mathrm{Exp}(y)))$$

$$= \sum_{n>0}(-1)^{n+1}n^{-1}(\mathrm{Exp}(x)\mathrm{Exp}(y) - 1)^n.$$

Expansion of the terms of the last sum yields the double sum

$$\sum_{n>0}(-1)^{n+1}n^{-1}\left(\sum_{p_i+q_i>0}(p_1! \cdots p_n!q_1! \cdots q_n!)^{-1}x^{p_1}y^{q_1} \cdots x^{p_n}y^{q_n}\right).$$

Since this sum is an element of L', each of its homogeneous partial sums must lie in L. By Proposition 2.2, we may therefore replace each term with its image under the map π of that Proposition without changing the sum of the whole series. Let us write $\eta(x, y)$ for the resulting series of elements of L. The *Campbell–Hausdorff Formula* is the identity

$$\mathrm{Exp}(x)\mathrm{Exp}(y) = \mathrm{Exp}(\eta(x, y)),$$

in the freely non-commuting variables x and y or, in more rigorous terms, the equality of these series as elements of T', with $V = Fx + Fy$.

We shall write $\eta_k(x, y)$ for the sum of the terms of degree $\leq k$ of the series $\eta(x, y)$. Then, if M is any Lie algebra over F, $\eta_k(x, y)$ defines a map η_k from $M \times M$ to M, where $\eta_k(a, b)$ is the result of substituting a for x and b for y in the expression of $\eta_k(x, y)$ in terms of iterated commutators. For example, we have

$$\eta_1(a, b) = a + b, \qquad \eta_2(a, b) = a + b + \tfrac{1}{2}[a, b],$$

and

$$\eta_3(a, b) = a + b + \tfrac{1}{2}[a, b] + \tfrac{1}{12}[a, [a, b]] + \tfrac{1}{12}[b, [b, a]].$$

3. Let F be a field, (C, δ, ε) an F-coalgebra and (A, μ, u) an F-algebra. We know that $\mathrm{Hom}_F(C, A)$ inherits the structure of an F-algebra in a natural fashion. The dual question, concerning a natural coalgebra structure on $\mathrm{Hom}_F(A, C)$ is somewhat more subtle. Let us regard A as a topological algebra, not necessarily Hausdorff, by making the two-sided ideals of finite codimension in A a fundamental system of neighborhoods of 0. On the other hand, we endow C with the discrete topology. Now let $\mathrm{Hom}'_F(A, C)$ be the F-space of all *continuous* F-linear maps from A to C. We use the topological language only for brevity; an element of $\mathrm{Hom}_F(A, C)$ belongs to $\mathrm{Hom}'_F(A, C)$ if and only if it annihilates some two-sided ideal of finite codimension in A. We shall see that $\mathrm{Hom}'_F(A, C)$ is an F-coalgebra in a natural way.

Writing simply $\mathrm{Hom}(A, C)$ for $\mathrm{Hom}_F(A, C)$, etc., we consider the natural F-linear map

$$\sigma: \mathrm{Hom}(A, C) \otimes \mathrm{Hom}(A, C) \to \mathrm{Hom}(A \otimes A, C \otimes C),$$

where $\sigma(f \otimes g)(u \otimes v) = f(u) \otimes g(v)$. It is easy to see that σ is injective. If f and g belong to $\mathrm{Hom}'(A, C)$, let I and J be two-sided ideals of finite codimension in A such that f annihilates I and g annihilates J. Then $\sigma(f \otimes g)$ annihilates $I \otimes A + A \otimes J$, which is a two-sided ideal of finite codimension in $A \otimes A$. Thus, σ maps $\mathrm{Hom}'(A, C) \otimes \mathrm{Hom}'(A, C)$ injectively into $\mathrm{Hom}'(A \otimes A, C \otimes C)$. We claim that this restriction of σ is also surjective.

In order to see this, consider an element h of $\mathrm{Hom}'(A \otimes A, C \otimes C)$. There is a two-sided ideal K of finite codimension in $A \otimes A$ that is annihilated by h. Let I and J be the intersections of K with the first and second tensor factors A, respectively. Then I and J are two-sided ideals of finite codimension in A. In order to conclude that h belongs to the image of

$\mathrm{Hom}'(A, C) \otimes \mathrm{Hom}'(A, C)$, it is evidently sufficient to prove that the natural map (defined as was σ)

$$\mathrm{Hom}(A/J, C) \otimes \mathrm{Hom}(A/J, C) \to \mathrm{Hom}((A/I) \otimes (A/J), C \otimes C)$$

is surjective. Since A/I and A/J are finite-dimensional, this amounts to showing this in the case where C is finite-dimensional. Since the map is injective, and since then the space on the right has the same dimension as the space on the left, the surjectiveness follows.

Now we know that the natural map (obtained by restricting σ)

$$\sigma': \mathrm{Hom}'(A, C) \otimes \mathrm{Hom}'(A, C) \to \mathrm{Hom}'(A \otimes A, C \otimes C)$$

is a linear isomorphism. Let τ denote the inverse of σ', and define the map γ from $\mathrm{Hom}'(A, C)$ to its tensor square as the following composite

$$\mathrm{Hom}'(A, C) \xrightarrow[\mathrm{Hom}'(\mu, C)]{} \mathrm{Hom}'(A \otimes A, C) \xrightarrow[\mathrm{Hom}'(A \otimes A, \delta)]{} \mathrm{Hom}'(A \otimes A, C \otimes C)$$

$$\xrightarrow[\tau]{} \mathrm{Hom}'(A, C) \otimes \mathrm{Hom}'(A, C).$$

This means that, for f in $\mathrm{Hom}'(A, C)$, we define

$$\gamma(f) = (\sigma')^{-1}(\delta \circ f \circ \mu).$$

One checks in the straightforward fashion that γ is a comultiplication making $\mathrm{Hom}'(A, C)$ into an F-coalgebra. The counit sends each f onto $\varepsilon(f(1))$.

We shall be concerned with the special case where A is actually a Hopf algebra, and where C is the trivial coalgebra F. In this case, let us write A' for $\mathrm{Hom}'(A, F)$. Then A' has the above structure of an F-coalgebra. The map σ is now just the usual injection $A^\circ \otimes A^\circ \to (A \otimes A)^\circ$, where the superscript $^\circ$ denotes the full dual. If I and J are two-sided ideals of finite co-dimension in A then the inverse image in A of $I \otimes A + A \otimes J$ with respect to the comultiplication of A is evidently a two-sided ideal of finite codimension. Using this, we see immediately that A' is a subalgebra of A°. Moreover, one verifies directly from the definitions that the coalgebra structure of A', together with the algebra structure inherited from A°, makes A' into a Hopf algebra. We call this the *Hopf algebra dual to A*.

We are especially interested in the case where A is the universal enveloping algebra $\mathcal{U}(L)$ of a Lie algebra L. Since the comultiplication of $\mathcal{U}(L)$ is commutative, the multiplication of the dual algebra $\mathcal{U}(L)^\circ$ is commutative. Moreover, *if F is of characteristic* 0 *then* $\mathcal{U}(L)^\circ$, *and therefore also* $\mathcal{U}(L)'$, *is an integral domain.*

In order to see this, let (x_α) be a totally ordered F-basis of L, so that, by Theorem 1.1, the elements of $\mathcal{U}(L)$ may be identified with the finite F-linear combinations of the ordered monomials in the x_α's. For each α, let f_α be the element of $\mathcal{U}(L)^\circ$ that takes the value 1 at x_α and the value 0 at every other ordered monomial in the x_α's. It is easy to verify that each monomial $f_{\alpha_1}^{e_1} \cdots f_{\alpha_n}^{e_n}$ takes the value $e_1! \cdots e_n!$ at $x_{\alpha_1}^{e_1} \cdots x_{\alpha_n}^{e_n}$, and the value 0 at every other ordered monomial in the x_α's. Let $\mathscr{U}(L)_d$ denote the sub F-space of $\mathscr{U}(L)$ that is spanned by the ordered monomials in the x_α's of total degree $\leq d$. Then every element of $\mathscr{U}(L)^\circ$ coincides on $\mathscr{U}(L)_d$ with one and only one linear combination of monomials in the f_α's of total degree $\leq d$ (the empty monomial is the projection from $\mathscr{U}(L)$ to F with kernel $L\mathscr{U}(L)$). The result follows easily from this by considering the terms of lowest degree in a product.

Now let G be an affine algebraic F-group, L the Lie algebra of G, and U the universal enveloping algebra $\mathscr{U}(L)$. Note that L is a sub Lie algebra of the Lie algebra $\mathscr{L}(\mathscr{P}(G)^\circ)$ defined from the F-algebra structure of $\mathscr{P}(G)^\circ$. By the universal mapping property of U, it follows that there is one and only one morphism of F-algebras from U to $\mathscr{P}(G)^\circ$ that extends the injection map from L to $\mathscr{P}(G)^\circ$. It is easy to verify from the definitions that the transpose of this morphism is a morphism of F-Hopf algebras from $\mathscr{P}(G)$ to U'. We shall denote this last morphism by π.

If we regard $\mathscr{P}(G)$ as an L-module, via the differential of the left translation representation of G on $\mathscr{P}(G)$, we see that $\mathscr{P}(G)$ is naturally also a U-module (again by the universal mapping property of U). In terms of this U-module structure, the map π is given by $\pi(f)(u) = (u \cdot f)(1_G)$ for every element f of $\mathscr{P}(G)$ and every element u of U.

Now let us suppose that F is of characteristic 0 and that G is irreducible. Then we see immediately from Theorem IV.3.1 that π is injective. We record these facts for reference.

Theorem 3.1. *Let G be an affine algebraic group over a field F. There is a natural morphism of Hopf algebras*

$$\pi \colon \mathscr{P}(G) \to \mathscr{U}(\mathscr{L}(G))',$$

which is injective whenever G is irreducible and F is of characteristic 0.

4. Lemma 4.1. *Let L be a finite-dimensional nilpotent Lie algebra. Then the intersection of the family of powers of the ideal $L\mathscr{U}(L)$ of $\mathscr{U}(L)$ is (0).*

PROOF. There is a basis $(x_1, \ldots, x_n)$ of L such that each $[x_i, x_j]$ belongs to the subspace of L that is spanned by the x_k's with $k > \max(i, j)$. By Theorem 1.1, every element of $\mathscr{U}(L)$ has one and only one expression as a linear combination of ordered monomials $x_1^{e_1} \cdots x_n^{e_n}$. We define a weight function w on $\mathscr{U}(L)$ as follows: (1) $w(1) = 0$; (2) $w(x_i) = 2^i$; (3) the weight of an ordered monomial is the sum of the weights of its factors; (4) the weight of a linear combination of ordered monomials is the minimum of the weights of the monomials occurring with non-zero coefficients. Now one shows by an evident induction on the degree that, if u is an ordered monomial, one has $w(x_i u) \geq w(x_i) + w(u)$. Evidently, this implies that the same holds for every non-zero element u of $\mathscr{U}(L)$. It follows that, given any natural number

p, there is a natural number q such that every non-zero element of $(L\mathcal{U}(L))^q$ has weight at least p. $\square$

Retaining the notation of Lemma 4.1, let $\mathcal{B}(L)$ denote the sub Hopf algebra of $\mathcal{U}(L)'$ consisting of the elements annihilating some power of $L\mathcal{U}(L)$. We call $\mathcal{B}(L)$ the *algebra of nilpotent representative functions on* $\mathcal{U}(L)$. Indeed, the elements of $\mathcal{B}(L)$ are precisely the functions on $\mathcal{U}(L)$ that are associated with nilpotent representations of L, in the evident way.

Now let us assume that our base field F is of characteristic 0. Let $f_1,\ldots,f_n$ be the elements of $\mathcal{U}(L)^\circ$ that we used in Section 3 for showing that $\mathcal{U}(L)^\circ$ is an integral domain, so that $f_i(x_i) = 1$, while f_i takes the value 0 at every other ordered monomial in $x_1,\ldots,x_n$. Evidently, each f_i belongs to $\mathcal{B}(L)$. Moreover, it is clear from what we have seen in Section 3 that Lemma 4.1 implies that every element of $\mathcal{B}(L)$ is a polynomial in $f_1,\ldots,f_n$. We know from Section 3 that the functions f_i are algebraically independent over F. Thus, as an F-algebra, $\mathcal{B}(L)$ is the ordinary polynomial algebra $F[f_1,\ldots,f_n]$. Finally, it is clear from Section 3 that the natural map from $\mathcal{U}(L)$ to $F[f_1,\ldots,f_n]^\circ$ is injective. Evidently, this map sends each element of L to a differentiation of $\mathcal{B}(L)$. Since L is of dimension n, which is equal to the dimension of the space of all differentiations of the polynomial algebra $\mathcal{B}(L)$ in n variables, we conclude that the natural map from $\mathcal{U}(L)$ to $\mathcal{B}(L)^\circ$ identifies L with the Lie algebra of all differentiations of $\mathcal{B}(L)$.

Since $\mathcal{B}(L)$ is an ordinary polynomial algebra, the F-algebra homomorphisms from $\mathcal{B}(L)$ to F separate the elements of $\mathcal{B}(L)$. Therefore, we may regard $\mathcal{B}(L)$ as the algebra of polynomial functions of the affine algebraic F-group $\mathcal{G}(\mathcal{B}(L))$, and it is clear from the definition of $\mathcal{B}(L)$ that the representation of L by derivations of $\mathcal{B}(L)$ is locally nilpotent. By Theorem IV.3.1, this implies that $\mathcal{G}(\mathcal{B}(L))$ is unipotent.

Conversely, let G be a unipotent algebraic F-group. By Theorem VIII.1.1, G is irreducible, so that we may appeal to Theorem 3.1 and identify $\mathcal{P}(G)$ with its canonical image in $\mathcal{U}(\mathcal{L}(G))'$. Since G is unipotent, the representation of $\mathcal{L}(G)$ by derivations of $\mathcal{P}(G)$ is locally nilpotent. With our identification, this means that $\mathcal{P}(G)$ is contained in $\mathcal{B}(\mathcal{L}(G))$. Let us write L for $\mathcal{L}(G)$, and let us consider the restriction morphism from $\mathcal{G}(\mathcal{B}(L))$ to G. Since the Lie algebra of $\mathcal{G}(\mathcal{B}(L))$ is the Lie algebra L of G, the differential of the restriction morphism is an isomorphism from the Lie algebra of $\mathcal{G}(\mathcal{B}(L))$ to the Lie algebra of G. By Theorem VIII.1.1 and the property $\rho \circ \exp = \exp \circ \rho$ of the exponential map, this implies that the restriction morphism is surjective. Moreover, since the differential of this morphism is injective, the kernel is a finite subgroup of $\mathcal{G}(\mathcal{B}(L))$. Since F is of characteristic 0 and $\mathcal{G}(\mathcal{B}(L))$ is unipotent, it follows that the restriction morphism is also injective.

In the case where F is algebraically closed, it follows from this and Proposition III.2.4 that $\mathcal{P}(G) = \mathcal{B}(L)$. In the general case, let F' be an algebraic closure of F. The extended group $G' = \mathcal{G}(\mathcal{P}(G) \otimes F')$ has $L \otimes F'$ as its Lie algebra, and what we have just said shows that $\mathscr{P}(G) \otimes F'$ coincides with $\mathscr{B}(L \otimes F')$. It is not difficult to see that $\mathscr{B}(L \otimes F') = \mathscr{B}(L) \otimes F'$. From $\mathscr{P}(G) \otimes F' = \mathscr{B}(L) \otimes F'$, we obtain $\mathscr{P}(G) = \mathscr{B}(G)$ immediately. We summarize our results in the following theorem.

Theorem 4.2. *Let F be a field of characteristic 0. For every finite-dimensional nilpotent F-Lie algebra L, the affine algebraic F-group $\mathscr{G}(\mathscr{B}(L))$ is unipotent, its algebra of polynomial functions coincides with $\mathscr{B}(L)$ and its Lie algebra coincides with L. This construction yields a functor $\mathscr{G} \circ \mathscr{B}$ from the category of finite-dimensional nilpotent F-Lie algebras to the category of unipotent affine algebraic F-groups, inverse, up to natural isomorphisms, to the Lie algebra functor $\mathscr{L}$, so that these two categories are naturally equivalent.*

Notes

1. In dealing with unipotent groups over fields of characteristic 0, the Campbell–Hausdorff formula can serve to remove the assumption that F be algebraically closed from basic structural theorems. We sketch an example of such an application. Let F be a field of characteristic 0, let G be an affine algebraic F-group, and let U be a unipotent normal algebraic subgroup of G. Let $[G, U]$ denote the subgroup of G that is generated by the elements of the form $xux^{-1}u^{-1}$, with x ranging over G and u ranging over U. The objective is to show that $[G, U]$ is an *algebraic* subgroup of G. Let α denote the adjoint representation of G on $\mathscr{L}(U)$, and let S be the sub F-space of $\mathscr{L}(U)$ that is spanned by $[\mathscr{L}(U), \mathscr{L}(U)]$ and the elements of the form $\alpha(x)(\sigma) - \sigma$ with x in G and σ in $\mathscr{L}(U)$. It is easy to see that S is actually an ideal of $\mathscr{L}(G)$. Using the formal properties of the exponential map and the Campbell–Hausdorff formula, one can show that $[G, U]$ coincides with $\exp(S)$. In particular, it is therefore an algebraic subgroup of G.

2. Using the result of 1, above, one can strengthen the conjugacy part of Theorem VIII.4.3 by showing that the element t of that theorem may be chosen from G_u^{∞}. In fact, the result of 1, above, enables one to proceed inductively, much as in the proof of Theorem VI.3.2.

Chapter XVII

Semisimple Lie Algebras

This chapter is devoted entirely to the classical representation theory of semisimple Lie algebras. Our principal goal is the basic result that, if L is a finite-dimensional semisimple Lie algebra over a field of characteristic 0, then the continuous dual $\mathcal{U}(L)'$ of the universal enveloping algebra is finitely generated as an algebra. This will be used in the final chapter for constructing the "simply connected" affine algebraic group with Lie algebra L. The required finite generation of $\mathcal{U}(L)'$ is obtained from the classification of the finite-dimensional L-modules by the theory of weights.

This theory is based on the technique of analyzing the structure and the representations of a semisimple Lie algebra L by considering the action of a certain *abelian* subalgebra of L, called a Cartan subalgebra, on L and its modules. The Cartan subalgebras are introduced in Section 1, and the elementary facts concerning their actions on L and on L-modules are dealt with in Sections 2 and 3.

Section 4 develops the principal results concerning the structure of L with regard to a Cartan subalgebra. The main representation theoretical results are obtained in Sections 5 and 6.

1. Let L be a finite-dimensional Lie algebra over a field F. A *Cartan subalgebra* of L is a nilpotent sub Lie algebra H that coincides with its stabilizer in L, i.e., has the property that the only elements x of L for which $[x, H]$ is contained in H are the elements of H. It follows from this definition that *a Cartan subalgebra H of L is a maximal nilpotent subalgebra.* In order to see this, suppose that N is a nilpotent sub Lie algebra of L containing H. Consider the representation of H on N/H that is induced from the adjoint representation of N. Since this is nilpotent the assumption $N \neq H$ would yield an element x in $N \setminus H$ such that $[H, x] \subset H$, and this contradicts the definition of H.

Proposition 1.1. *For an element x of L, let L^x denote the subspace of L consisting of all elements that are annihilated by some power of D_x. Then L^x is a sub Lie algebra of L and coincides with its stabilizer in L.*

PROOF. The formula expressing $D_x^n([u, v])$ as a sum of terms $[D_x^p(u), D_x^q(v)]$, where $p + q = n$, shows that $[L^x, L^x] \subset L^x$.

Next, if y is an element of L such that $D_y(L^x) \subset L^x$, we have $[x, y] \in L^x$, because x belongs to L^x. Evidently, this implies that y belongs to L^x. $\qquad\square$

Proposition 1.2. *In the notation of Proposition 1.1, L^x has an F-space complement L_x in L such that $[L^x, L_x] \subset L_x$.*

PROOF. We let L_x be the subspace of L obtained from Fitting's Lemma, so that L_x is the largest subspace of L on which D_x induces a linear automorphism. We have $L_x = D_x^p(L)$ for some positive exponent p, and it remains only to be shown that $[L^x, L_x]$ is contained in L_x.

We shall prove by induction on m that if $D_x^m(y) = 0$ then $D_y(L_x) \subset L_x$. For $m = 0$, the assumption means that $y = 0$, so that the implication holds in the case $m = 0$. If $m = 1$ we have $[x, y] = 0$, whence $D_x D_y = D_y D_x$. Since $L_x = D_x^p(L)$, this shows that $D_y(L_x) \subset L_x$.

Now suppose that $m > 1$ and that the implication has been established in the lower cases. Let y be an element of L such that $D_x^m(y) = 0$, and let u be an element of L_x. Write $u = D_x^p(v)$, with v in L_x. Then we have

$$[y, u] = [y, D_x^p(v)] = D_x^p([y, v]) - \sum_{q=1}^{p} \binom{p}{q} [D_x^q(y), D_x^{p-q}(v)].$$

The first term on the right belongs to L_x, because $L_x = D_x^p(L)$. The terms of the sum that follows belong to L_x by the inductive hypothesis, applied to each $D_x^q(y)$ in the place of y. $\qquad\square$

Theorem 1.3. *Let L be a finite-dimensional Lie algebra over an infinite field F, and let x be an element of L such that L^x is of the smallest possible dimension. Then L^x is a Cartan subalgebra of L.*

PROOF. By Proposition 1.1, it suffices to prove that L^x is a *nilpotent* Lie algebra. By Theorem VII.1.5, it suffices to show that, for every element y of L^x, the restriction of D_y to L^x is nilpotent. Let f_y denote the characteristic polynomial of D_y. By Propositions 1.1 and 1.2, both L^x and L_x are stable under D_y. Therefore, we have $f_y = g_y h_y$, where g_y is the characteristic polynomial of the restriction of D_y to L^x, and h_y is the characteristic polynomial of the restriction of D_y to L_x (we assume $L_x \neq (0)$, as we may).

Let $t, t_1, \ldots, t_m$ be auxiliary variables over F, and let t^n be the highest power of t that divides $g_y(t)$ for every element y of L^x. Let $(y_1, \ldots, y_m)$ be a basis of L^x. There exists a polynomial $u(t_1, \ldots, t_m)$ with coefficients in F such that the coefficient of t^n in $g_y(t)$ is $u(c_1, \ldots, c_m)$, where the c_i's are the coefficients of the y_i's in y. By the definition of n, the polynomial u is not the zero polynomial. Also, there is a polynomial $v(t_1, \ldots, t_m)$ such that

$$h_y(0) = v(c_1, \ldots, c_m).$$

Since no non-zero element of L_x is annihilated by D_x, we must have $h_x(0) \neq 0$.

Therefore, v is not the zero polynomial. Since F is infinite, it contains elements $c_1, \ldots, c_m$ such that $u(c_1, \ldots, c_m)v(c_1, \ldots, c_m) \neq 0$. Now, if

$$y = c_1 y_1 + \cdots + c_m y_m,$$

then this is the coefficient of t^n in $f_y(t)$, so that $f_y(t)$ is not divisible by t^{n+1}.

Since the dimension of L^x is minimal, we have $\dim(L^y) \geq m$. On the other hand, $f_y(t)$ is divisible by $t^{\dim(L^y)}$, and we have seen that $f_y(t)$ is not divisible by t^{n+1}. Hence, we must have $\dim(L^y) \leq n$, so that $m \leq n$. Now let y be an arbitrary element of L^x. By what we have just proved, $g_y(t)$ is divisible by t^m. Since the degree of g_y is equal to m, we must therefore have $g_y(t) = t^m$, which shows that the restriction of D_y to L^x is nilpotent. $\qquad\square$

2. From now on, we assume that our base field F is algebraically closed and of characteristic 0. Let L be an F-Lie algebra, V an L-module, γ an element of L°. Let V_γ denote the subspace of V consisting of all elements that are annihilated by some power of $\rho(x) - \gamma(x)i_V$ for every element x of L, where ρ is the given representation of L on V. If $V_\gamma \neq (0)$ then γ is called a *weight* of ρ, and V_γ is called the corresponding *weight space*.

Theorem 2.1. *Let L be a finite-dimensional nilpotent Lie algebra over the algebraically closed field F of characteristic 0, and let ρ be a representation of L on a finite-dimensional F-space V. Each weight space V_γ is a sub L-module of V, and the set of endomorphisms $\rho(x) - \gamma(x)i_V$ is nilpotent on V_γ. The sum in V of the V_γ's is direct and coincides with V.*

PROOF. It is easy to verify by induction on m that $(\rho(x) - \gamma(x)i_V)^m \rho(y)$ is a sum of endomorphisms of the form $\rho(D_x^p(y))(\rho(x) - \gamma(x)i_V)^q$, with $p + q = m$. Using the nilpotency of L, we see from this that V_γ is a sub L-module of V.

Let M denote the Lie algebra of linear endomorphisms of V_γ that is generated by the restrictions to V_γ of the endomorphisms $\rho(x) - \gamma(x)i_V$. Then $[M, M]$ is contained in the restriction image of $\rho(L)$, which shows that M is solvable. By Theorem VII.1.3, it follows that $[M, M]$ is nilpotent on V_γ. Since each $\rho(x) - \gamma(x)i_V$ is nilpotent on V_γ, it follows from Lemma VII.1.4 that every element of M is nilpotent on V_γ. By Theorem VII.1.5, this implies that M is nilpotent on V_γ.

Let $\gamma_1, \ldots, \gamma_k$ be a finite set of distinct weights. Since F is infinite, there is an element x in L such that the values $\gamma_i(x)$ are all distinct. The endomorphism $\rho(x) - \gamma_i(x)i_V$ is nilpotent on V_{γ_i}. On the other hand, for each j

other than i, the restriction of $\rho(x) - \gamma_i(x)i_V$ to V_{γ_j} is invertible, because it is the sum of the non-zero scalar multiplication by $\gamma_j(x) - \gamma_i(x)$ and the nilpotent endomorphism induced by $\rho(x) - \gamma_j(x)i_V$. Evidently, we may conclude from this that the sum of the family of V_γ''s is direct, whence the set of weights is finite.

It remains only to show that the sum of the family of V_γ's coincides with V. Let $\gamma_1,\ldots,\gamma_k$ be all the weights, and choose x as just above. For every element c of F, let V_c denote the subspace of V consisting of the elements that are annihilated by some power of $\rho(x) - ci_V$. The argument of the very beginning of this proof shows that V_c is a sub L-module of V. By Theorem VII.1.3, every simple sub L-module of V_c is annihilated by $[L, L]$. Since F is algebraically closed, it follows from this and Schur's Lemma that the non-zero simple sub L-modules of V_c are one-dimensional. Therefore, if $V_c \neq (0)$, there is an index i such that $V_c \cap V_{\gamma_i} \neq (0)$. Clearly, we must have $\gamma_i(x) = c$. By the elementary theory of a single linear endomorphism of a finite-dimensional space over an algebraically closed field, V is the sum of the family of V_c's. Therefore, it suffices to show that $V_{\gamma_i(x)}$ coincides with V_{γ_i}.

Let y be an element of L, and let $V_{i,c}$ denote the subspace of $V_{\gamma_i(x)}$ consisting of the elements that are annihilated by some power of $\rho(y) - ci_V$. As above, if $V_{i,c} \neq (0)$, there is an index j such that $V_{i,c} \cap V_{\gamma_j} \neq (0)$. We must have $\gamma_j(x) = \gamma_i(x)$, so that $j = i$. Also, we must have $\gamma_j(y) = c$, and so $\gamma_i(y) = c$. Since $V_{\gamma_i(x)}$ is the sum of the $V_{i,c}$'s, this shows that $V_{\gamma_i(x)} = V_{i,\gamma_i(y)}$. Since this holds for every element y of L, we conclude that $V_{\gamma_i(x)} = V_{\gamma_i}$. $\qquad\square$

Now let L be an arbitrary finite-dimensional Lie algebra over F. Using Theorem 1.3, let us choose a Cartan subalgebra H of L. A weight of the representation of H on L coming from the adjoint representation of L is called a *root* of L with respect to H. The corresponding weight spaces L_γ are called the *root spaces* of L with respect to H.

Theorem 2.2. *Let F be an algebraically closed field of characteristic 0, and let L be a finite-dimensional F-Lie algebra. Let H be a Cartan subalgebra of L. For all elements α and β of H°, one has $[L_\alpha, L_\beta] \subset L_{\alpha+\beta}$, and $L_0 = H$.*

PROOF. The first statement follows immediately from the formula

$$(D_{z,\alpha+\beta})^n([x, y]) = \sum_{k=0}^{n} \binom{n}{k} [(D_{z,\alpha})^k(x), (D_{z,\beta})^{n-k}(y)],$$

where $D_{z,\gamma} = D_z - \gamma(z)i_L$, which is easily established by induction on n.

Since H is nilpotent on H, we have $H \subset L_0$. Since H is nilpotent on L_0, the assumption $L_0 \neq H$ would imply that there is an element v in $L_0 \setminus H$ such that $[H, v] \subset H$, which contradicts the fact that H coincides with its stabilizer in L. Therefore, we must have $L_0 = H$. $\qquad\square$

Theorem 2.3. *Let F, L and H be as in Theorem 2.2, and let B denote the trace form of the adjoint representation of L. For γ in H°, put $d_\gamma = \dim(L_\gamma)$. Then, for all elements x and y of H, one has*

$$B(x, y) = \sum_\gamma d_\gamma \gamma(x)\gamma(y),$$

where the summation goes over the set of all roots γ.

PROOF. By Theorem 2.1, L is the direct sum of the family of sub H-modules L_γ. Hence, if T stands for trace and $D_{x/\gamma}$ for the restriction of D_x to L_γ, we have

$$B(x, x) = T(D_x^2) = \sum_\gamma T(D_{x/\gamma}^2),$$

for every element x of H. Next, we have

$$D_{x/\gamma}^2 - \gamma(x)^2 i_{L_\gamma} = (D_{x/\gamma} - \gamma(x)i_{L_\gamma})(D_{x/\gamma} + \gamma(x)i_{L_\gamma}).$$

The first factor on the right is nilpotent and commutes with the second factor. Therefore, $D_{x/\gamma}^2 - \gamma(x)^2 i_{L_\gamma}$ is nilpotent, whence we have

$$T(D_{x/\gamma}^2) = T(\gamma(x)^2 i_{L_\gamma}) = d_\gamma \gamma(x)^2.$$

This is the required result in the case where $x = y$. Applying this with $x + y$ in the place of x, we obtain the general result. $\qquad\square$

Theorem 2.4. *In the notation of Theorem 2.3, if α and β are elements of H° such that $\alpha + \beta \neq 0$ then $B(L_\alpha, L_\beta) = (0)$.*

PROOF. By Theorem 2.2, if x belongs to L_α and y to L_β, then $D_x D_y$ maps each L_γ into $L_{\alpha+\beta+\gamma}$. Now, it γ is a root then $\alpha + \beta + \gamma$ is either no root or a root distinct from γ. In either case, $D_x D_y$ sends L_γ into the sum of the L_δ's with $\delta \neq \gamma$, which is an H-module complement of L_γ in L. Clearly, this implies that $T(D_x D_y) = 0$. $\qquad\square$

3. Let L be a finite-dimensional semisimple Lie algebra over the algebraically closed field F of characteristic 0. Let H be a Cartan subalgebra of L, and let B denote the trace form of the adjoint representation of L. Let α and β be roots of L with respect to H, with $\beta \neq 0$. We denote by α^β the non-negative integer defined by the condition that $\alpha + m\beta$ be a root for all integers m with $0 \leq m \leq \alpha^\beta$, while this is no longer the case for $m = \alpha^\beta + 1$. Similarly, we define α_β by the condition that $\alpha - m\beta$ be a root for all integers m with $0 \leq m \leq \alpha_\beta$, while this is no longer the case for $m = \alpha_\beta + 1$.

Let α be a root, and let x be a non-zero element of L_α. We know from Theorem VII.2.1 that B is non-degenerate. Therefore, there is an element y in L such that $B(x, y) \neq 0$. By Theorems 2.1 and 2.4, it follows that $-\alpha$ is also a root and that we may choose y from $L_{-\alpha}$. This yields the following conclusions. *If α is a root, so is $-\alpha$, and B induces a non-degenerate pairing from $L_\alpha \times L_{-\alpha}$ to F. In particular, B is non-degenerate on $H \times H$.*

From the last statement, we have that, *for every element α of H°, there is one and only one element h_α in H such that $B(h, h_\alpha) = \alpha(h)$ for every element h of H.*

Suppose that α is a non-zero root. Since the set of endomorphisms $D_h - \alpha(h)i_L$ with h in H is nilpotent on L_α, there is a non-zero element x_α in L_α such that $[h, x_\alpha] = \alpha(h)x_\alpha$ for element h of H. Let y be any element of $L_{-\alpha}$. By Theorem 2.2, we have $[x_\alpha, y] \in H$. Recall from the beginning of Section VII.2 that B satisfies the identity $B([u, v], w) = B(u, [v, w])$. Using this, we find that, for every element h of H, we have

$$B(h, [x_\alpha, y]) = B([h, x_\alpha], y) = \alpha(h)B(x_\alpha, y) = B(h, B(x_\alpha, y)h_\alpha).$$

Since B is non-degenerate on $H \times H$, it follows that

$$[x_\alpha, y] = B(x_\alpha, y)h_\alpha,$$

for every element y of $L_{-\alpha}$. We shall use this fact in proving the next theorem.

Theorem 3.1. *In the notation introduced above, let α and β be roots, with $\alpha \neq 0$. Then the following statements hold:*

(1) *$\alpha(h_\alpha) \sum_\gamma (\gamma_\alpha - \gamma^\alpha)^2 = 4$, the summation going over the set of all roots γ;*
(2) *$2\beta(h_\alpha) = (\beta_\alpha - \beta^\alpha)\alpha(h_\alpha)$;*
(3) *the only F-multiples of α that are roots are $0, \alpha, -\alpha$;*
(4) *$\dim(L_\alpha) = 1$.*

PROOF. Let $U = \sum_{m=-\beta_\alpha}^{\beta^\alpha} L_{\beta+m\alpha}$. It is clear from Theorem 2.2 that U is stable under the endomorphisms D_z with z in $L_\alpha + L_{-\alpha}$. Let x_α be an element of L_α as discussed just above the statement of our theorem. We know that there is an element y in $L_{-\alpha}$ such that $B(x_\alpha, y) \neq 0$. We have shown above that $[x_\alpha, y] = B(x_\alpha, y)h_\alpha$. Therefore, we may choose y so that $[x_\alpha, y] = h_\alpha$. Now $D_{h_\alpha} = [D_{x_\alpha}, D_y]$, whence it is clear that the restriction of D_{h_α} to U is of trace 0. Since $D_{h_\alpha} - (\beta + m\alpha)(h_\alpha)i_L$ is nilpotent on $L_{\beta+m\alpha}$, the trace of the restriction of D_{h_α} to $L_{\beta+m\alpha}$ is equal to $(\beta + m\alpha)(h_\alpha)\dim(L_{\beta+m\alpha})$. Therefore the sum of these elements of F from $m = -\beta_\alpha$ to $m = \beta^\alpha$ is equal to 0. We deduce from this fact that $\alpha(h_\alpha) \neq 0$. Indeed, otherwise we obtain $\beta(h_\alpha) = 0$, and since this then holds for every root β we see from Theorem 2.3 that $B(h, h_\alpha) = 0$ for every element h of H, which contradicts our assumption that $\alpha \neq 0$. Thus, we have $\alpha(h_\alpha) \neq 0$.

Now let $V = H + Fx_\alpha + \sum_{m<0} L_{m\alpha}$. Evidently, V is stable under D_{x_α} and D_y. Hence, we see as above with U that the restriction of D_{h_α} to V is of trace 0. Thus, we have

$$\alpha(h_\alpha)\left(1 + \sum_{m<0} m\,\dim(L_{m\alpha})\right) = 0.$$

Since $\alpha(h_\alpha) \neq 0$, this implies that $L_{m\alpha} = (0)$ for $m < -1$, and $\dim(L_{-\alpha}) = 1$. This gives (4), by replacing α with $-\alpha$. Moreover, this shows that the only *integral* multiples of α that are roots are $0, \alpha, -\alpha$. Hence, we have $0_\alpha = 1 = 0^\alpha$, $\alpha^\alpha = 0$ and $\alpha_\alpha = 2$. In particular, this shows that (2) holds in the case where β is an integral multiple of α.

Now suppose that β is not an integral multiple of α. Then we have from (4) that $\dim(L_{\beta + m\alpha}) = 1$ for every integer m with $-\beta_\alpha \le m \le \beta^\alpha$. We have seen above that the sum of the expressions $(\beta + m\alpha)(h_\alpha)\dim(L_{\beta + m\alpha})$ for these values of m is equal to 0. This yields (2) upon replacing each $\dim(L_{\beta + m\alpha})$ with 1.

From (4), (2) and Theorem 2.3, we obtain

$$\alpha(h_\alpha) = B(h_\alpha, h_\alpha) = \sum_\beta d_\beta \beta(h_\alpha)^2 = \sum_\beta \beta(h_\alpha)^2$$

$$= \frac{1}{4}\sum_\beta (\beta_\alpha - \beta^\alpha)^2 \alpha(h_\alpha)^2,$$

whence we have (1).

Finally, suppose that β is a root of the form $c\alpha$, with c in F. Then it follows from (2) that $c = p/2$, where p is an integer. If $c \ne 0$ we may exchange α and β, and so conclude that $1/c = q/2$, where q is an integer. Now we have $pq = 4$, and we have already seen that the only possible *integral* values of c are $0, 1, -1$. This leaves us with the case where p is odd. Then $pq = 4$ allows only $p = 1$ and $p = -1$, which give $\alpha = 2\beta$ and $\alpha = -2\beta$, respectively. Since the only integral multiples of β that are roots are $0, \beta, -\beta$, these cases are ruled out. Thus (3) holds. $\qquad\square$

Theorem 3.2. *In the above notation, H is abelian, and the roots of L with respect to H span H° over F.*

PROOF. Let α be a non-zero root. Since $\dim(L_\alpha) = 1$, D_h acts as the scalar multiplication by $\alpha(h)$ on L_α, for every element h of H. Now let N be the subspace of H consisting of the elements that are annihilated by every root. By what we have just seen, we have $[N, L_\alpha] = (0)$ and $[H, H] \subset N$. Since L is the sum of H and the family of L_α's with α a non-zero root, N is therefore an ideal of L. Since N is nilpotent and L is semisimple, it follows that $N = (0)$. $\qquad\square$

Proposition 3.3. *In the above notation, let $(\beta_1, \ldots, \beta_r)$ be any maximal F-linearly independent set of roots of L with respect to H. Then every root is a rational linear combination of the β_i's.*

PROOF. Let β be any root. Then $\beta = \sum_{i=1}^r b_i \beta_i$, with each b_i in F. Applying this to h_{β_j}, we obtain

$$B(h_\beta, h_{\beta_j}) = \sum_{i=1}^r b_i B(h_{\beta_i}, h_{\beta_j}).$$

By (1) and (2) of Theorem 3.1, all the values of B occurring here are rational numbers. It follows from Theorem 3.2 and the non-degeneracy of B that the

h_{β_i}'s constitute an F-basis of H and that the determinant of the matrix with entries $B(h_{\beta_i}, h_{\beta_j})$ is different from 0. Therefore, the above system of equations, for $j = 1, \ldots, r$, determines the b_i's as *rational* numbers. $\square$

4. Theorem 4.1. *Let L be a finite-dimensional semisimple Lie algebra over an algebraically closed field F of characteristic 0, and let H be a Cartan sub-algebra of L. There is an F-linearly independent set $(\alpha_1, \ldots, \alpha_r)$ of roots of L with respect to H such that, for every root α, either α or $-\alpha$ is a sum of α_i's, with repetitions allowed.*

PROOF. Let $(\beta_1, \ldots, \beta_r)$ be any maximal F-linearly independent set of roots. By Proposition 3.3, every root is a *rational* linear combination of the β_i's. We define an ordering of the set of all roots from the lexicographic order based on their coefficients when expressed in terms of the β_i's. We call a positive root *simple* if it is not the sum of two positive roots. Let $\alpha_1, \ldots, \alpha_s$ be all the simple roots. Evidently, every positive root is a sum of α_j's. Therefore, all that remains to be proved is that the α_j's are F-linearly independent.

We suppose that this is false and derive a contradiction. Let R denote the space over the field of rational numbers that is spanned by the roots. We know already that the α_j's span R over the field of rational numbers. We know also that the β_i's constitute an F-basis of $R \otimes F$. Since the α_j's are not F-linearly independent and span $R \otimes F$ over F, we must have $s > r$. It follows that the α_j's are not linearly independent over the field of rational numbers either. Hence, there is a non-trivial relation

$$\sum_{i \in I} n_i \alpha_i = \sum_{j \in J} n_j \alpha_j,$$

where the n_i's and n_j's are positive integers, and $I \cap J = \varnothing$. Let

$$h = \sum_{i \in I} n_i h_{\alpha_i}.$$

Then we have also $h = \sum_{j \in J} n_j h_{\alpha_j}$, whence

$$B(h, h) = \sum_{i \in I} \sum_{j \in J} n_i n_j \alpha_i(h_{\alpha_j}).$$

Now observe that $\alpha_i - \alpha_j$ cannot be a root. For otherwise $\alpha_j - \alpha_i$ is also a root, and one of these is positive, so that either α_i or α_j is the sum of two positive roots, contradicting their choice as *simple* roots. By (2) of Theorem 3.1, we have

$$\alpha_i(h_{\alpha_j}) = \tfrac{1}{2}\alpha_j(h_{\alpha_j})((\alpha_i)_{\alpha_j} - (\alpha_i)^{\alpha_j}).$$

By what we have just seen, $(\alpha_i)_{\alpha_j} = 0$. By (1) of Theorem 3.1, $\alpha_j(h_{\alpha_j}) > 0$. Hence, we have $\alpha_i(h_{\alpha_j}) \leq 0$, so that the above gives $B(h, h) \leq 0$. By Theorem 2.3, $B(h, h)$ is a sum of terms $\gamma(h)^2$, with γ ranging over the set of all roots. Since each $\gamma(h)$ is a rational number, it follows that each $\gamma(h)$ must be 0. By Theorem 3.2, this implies that $h = 0$. Hence, we have

$$\sum_{i \in I} n_i \alpha_i = 0 = \sum_{j \in J} n_j \alpha_j.$$

Since the n_i's and n_j's are positive integers and the α_i's and α_j's are positive roots, these relations cannot hold, unless both I and J are empty, so that we have the required contradiction. $\qquad\square$

A system $(\alpha_1, \ldots, \alpha_r)$ as in Theorem 4.1 is called a *fundamental system of roots*.

Let V denote the space over the field of rational numbers that is spanned by the h_α's, where α ranges over the set of all roots. We know that the restriction of B to $V \times V$ is positive definite and takes on only rational values. Thus, V carries a rational-valued inner product: $u \cdot v = B(u, v)$. In particular, we have $h_\alpha \cdot v = \alpha(v)$ for every root α and every element v of V. Let P_α be the kernel of α in V, where α is a non-zero root. In geometrical terms, P_α is the hyperplane orthogonal to h_α in V. Let r_α denote the reflection of V in P_α, so that

$$r_\alpha(v) = v - 2\left(\frac{h_\alpha \cdot v}{h_\alpha \cdot h_\alpha}\right)h_\alpha = v - 2\,\frac{\alpha(v)}{\alpha(h_\alpha)}\,h_\alpha.$$

In particular, if β is a root, we see from (2) of Theorem 3.1 that

$$r_\alpha(h_\beta) = h_\beta - (\beta_\alpha - \beta^\alpha)h_\alpha.$$

By the definitions of β_α and β^α, the element $\beta - (\beta_\alpha - \beta^\alpha)\alpha$ of H° is a root. If β is a non-zero root then $r_\alpha(h_\beta) \neq 0$, because r_α is a linear automorphism of V. This shows that $\beta - (\beta_\alpha - \beta^\alpha)\alpha$ is a *non-zero* root if β is a non-zero root. We shall write $\pi_\alpha(\beta)$ for this non-zero root, so that

$$r_\alpha(h_\beta) = h_{\pi_\alpha(\beta)}.$$

The group of linear automorphisms of V that is generated by the reflections r_α is called the *Weyl group* of (L, H), and it is denoted by W. A *Weyl chamber* is a maximal convex subset of $V \setminus \cup_\alpha P_\alpha$. If $(\alpha_1, \ldots, \alpha_r)$ is a fundamental system of roots, then the set $C(\alpha_1, \ldots, \alpha_r)$ of all points v in V such that $\alpha_i(v) > 0$ for each i is evidently a Weyl chamber. We have

$$\pi_\alpha(\beta)(r_\alpha(v)) = r_\alpha(h_\beta) \cdot r_\alpha(v) = h_\beta \cdot v = \beta(v),$$

which shows that $r_\alpha(P_\beta) = P_{\pi_\alpha(\beta)}$. Thus, the elements of the Weyl group permute the P_α's among themselves, and hence permute also the Weyl chambers among themselves. A *wall* of a Weyl chamber is a non-trivial intersection of its "boundary" with one of the P_α's.

Theorem 4.2. *Let F, L, H be as in Theorem 4.1, and let $(\alpha_1, \ldots, \alpha_r)$ be a fundamental system of roots. Then the Weyl group is generated already by the r_{α_i}'s. Furthermore, the Weyl group acts transitively on the set of Weyl chambers, and every non-zero root is the image of an α_i by an element of the root permutation group generated by the π_{α_j}'s.*

PROOF. For every non-zero root α, we denote the linear automorphism of V° dual to r_α by π_α. Note that V° may be identified with the space R spanned over the field of rational numbers by the roots, and that the above π_α is then the restriction to the set of roots of the new π_α. Let ρ stand for half the sum of all the positive roots. From the original definition of π_{α_i} and Theorem 4.1, we see that, if α is a positive root other than α_i, then $\pi_{\alpha_i}(\alpha)$ is a positive root, while $\pi_{\alpha_i}(\alpha_i) = -\alpha_i$. It follows immediately from this that

$$\pi_{\alpha_i}(\rho) = \rho - \alpha_i.$$

Let W' be the subgroup of W that is generated by the r_{α_i}'s, let C be a Weyl chamber, and let v be a point in C. Choose an element t from W' for which $\rho(t(v))$ is as large as possible. Then we have $\rho(t(v)) \geq \rho(r_{\alpha_i}t(v))$ for each i. On the other hand,

$$\rho(r_{\alpha_i}t(v)) = \pi_{\alpha_i}(\rho)(t(v)) = (\rho - \alpha_i)(t(v)) = \rho(t(v)) - \alpha_i(t(v)).$$

Hence, we must have $\alpha_i(t(v)) \geq 0$. Since v is in a Weyl chamber, so is $t(v)$, and we must actually have $\alpha_i(t(v)) > 0$. Thus, $t(v)$ belongs to $C(\alpha_1, \ldots, \alpha_r)$. Since the Weyl chambers are permuted among themselves by the elements of the Weyl group, this proves that $t(C) = C(\alpha_1, \ldots, \alpha_r)$.

Now let α be any non-zero root, and let C be a Weyl chamber one of whose walls lies in P_α. Then, with t as above, one of the walls of $C(\alpha_1, \ldots, \alpha_r)$ lies in $t(P_\alpha)$, i.e., there is an index i such that $t(P_\alpha) = P_{\alpha_i}$. Hence, if π is the linear automorphism of V° dual to t, we must have $\pi(\alpha) = \pm\,\alpha_i$. If $\pi(\alpha) = -\alpha_i$ then $(\pi_{\alpha_i}\pi)(\alpha) = \alpha_i$. Hence, in any case, there is an element s in W' such that the corresponding root permutation sends α onto α_i. Finally, we have $r_\alpha = s^{-1}r_{\alpha_i}s \in W'$. $\qquad\square$

5. Let L be a finite-dimensional semisimple Lie algebra over the algebraically closed field F of characteristic 0, and let H be a Cartan subalgebra of L. Let V be an L-module. If γ is a weight of the representation of H on V then we shall call γ simply a weight of V, or a weight of the representation of L on V. We denote by V^γ the sub F-space of V consisting of all elements v such that $h \cdot v = \gamma(h)v$ for every element h of H. In the cases of interest to us, and in particular whenever V is finite-dimensional, this space V^γ coincides with the weight space V_γ defined in Section 2, as we shall see shortly. Throughout this section, we shall refer to V^γ as the weight space belonging to γ. Observe that if γ is a weight then $V^\gamma \neq (0)$.

We choose a fundamental system $(\alpha_1, \ldots, \alpha_r)$ of roots, and we make an ordering of the set of roots such that α_i is positive for each i. We say that γ is a *dominant weight* of V if the following two conditions are satisfied:

(1) $L_\alpha \cdot V^\gamma = (0)$ for every positive root α;
(2)   there is an element v in V^γ such that the sub L-module of V generated by v coincides with V.

Theorem 5.1. *Let F, L and H be as above, and let V be an L-module having a dominant weight γ. Then V^γ is 1-dimensional, each weight space V^δ is finite-dimensional, and V is the direct F-space sum of the family of V^δ's. Moreover, γ is the only dominant weight of V, and every weight of V is of the form $\gamma - \sum_{i=1}^r m_i \alpha_i$, where the m_i's are non-negative integers. Finally, for every element ρ of H°, there is one and only one isomorphism class of simple L-modules having ρ as a dominant weight.*

PROOF. Choose an element v from V^γ that generates V as an L-module. Consider the subspace $Fv + \sum_\delta V^\delta$ of V, where δ ranges over the set of weights $\gamma - \sum_{i=1}^r m_i \alpha_i$, other than γ, as described in the statement of the theorem. Let $(\beta_1, \ldots, \beta_m)$ be the set of all positive roots. For each non-zero root α, choose an element x_α from L such that $L_\alpha = Fx_\alpha$. Then the elements of $\mathcal{U}(L)$ are the F-linear combinations of the products

$$x_{-\beta_1}^{e_1} \cdots x_{-\beta_m}^{e_m} u x_{\beta_1}^{f_1} \cdots x_{\beta_m}^{f_m},$$

where u belongs to $\mathcal{U}(H)$, and the e_i's and f_i's are non-negative integers. Since the x_{β_i}'s annihilate v, such a product maps v into our subspace $Fv + \sum_\delta V^\delta$, so that this subspace coincides with $\mathcal{U}(L) \cdot v = V$. More precisely, each V^δ is spanned over F by the transforms $x_{-\beta_1}^{e_1} \cdots x_{-\beta_m}^{e_m} \cdot v$ with $\sum_{j=1}^m e_j \beta_j = \gamma - \delta$. This shows that $V^\gamma = Fv$, that every weight of V is of the form $\gamma - \sum_{i=1}^r m_i \alpha_i$, where the m_i's are non-negative integers, and that every weight space is finite-dimensional.

If w is an element of V^δ that is annihilated by L_α for each positive root α then $\mathcal{U}(L) \cdot w$ cannot contain v, unless $\delta = \gamma$, as is evident from the above. Therefore, γ is the only dominant weight of V.

In order to prove the last statement of the theorem, let

$$N_+ = \sum_{\alpha > 0} L_\alpha, \qquad N_- = \sum_{\alpha > 0} L_{-\alpha}.$$

Each of these is clearly a nilpotent sub Lie algebra of L. Put $U_+ = \mathcal{U}(N_+)$, $U_- = \mathcal{U}(N_-)$ and $U_0 = \mathcal{U}(H)$, regarding each as a subalgebra of $\mathcal{U}(L)$. Also, write U for $\mathcal{U}(L)$. Let ρ be an element of H°, and let J_ρ be the left ideal of U that is generated by N_+ and the elements $h - \rho(h)$ with h in H. If V is an L-module having ρ as its dominant weight and J denotes the annihilator of V^ρ in U then, as a U-module, V is isomorphic with U/J. Evidently, $J_\rho \subset J$, so that V is a homomorphic image of U/J_ρ. Therefore, it will suffice to prove that, *for every ρ in H°, there is one and only one maximal left ideal in U that contains J_ρ.*

Let K_ρ be the left ideal of $U_0 U_+$ that is generated by N_+ and the elements $h - \rho(h)$ with h in H, so that $J_\rho = U_- K_\rho$. As an F-space, U is the tensor product $U_- \otimes (U_0 U_+)$, and hence $J_\rho = U_- \otimes K_\rho$. We have

$$(U_0 U_+)/K_\rho \approx U_0/(U_0 \cap K_\rho),$$

and $U_0 \cap K_\rho$ is the ideal of U_0 that is generated by the elements $h - \rho(h)$, so that $U_0 \cap K_\rho \neq U_0$. Therefore, $J_\rho \neq U$.

Now let I be any left ideal of U containing J_ρ, and consider the sub L-module I/J_ρ of the L-module U/J_ρ, with L acting via the multiplication of U. Clearly, ρ is a dominant weight of U/J_ρ, the coset $1 + J_\rho$ being an F-space generator of $(U/J_\rho)^\rho$. By the part of the theorem already proved, we have

$$U/J_\rho = (U/J_\rho)^\rho + \sum_\delta (U/J_\rho)^\delta,$$

where the δ's are as described above. It is easy to see that the submodule I/J_ρ, like every submodule, coincides with the sum of the family of its intersections with the weight subspaces of U/J_ρ. If the intersection of I/J_ρ with $(U/J_\rho)^\rho$ is not (0) then $(U/J_\rho)^\rho$ is contained in I/J_ρ, because it is 1-dimensional, and it follows that $I = U$. If the intersection of I/J_ρ with $(U/J_\rho)^\rho$ is (0), we have $I/J_\rho \subset \sum_\delta (U/J_\rho)^\delta$. Let M be the inverse image of $\sum_\delta (U/J_\rho)^\delta$ in U. Then we have $J_\rho \subset I \subset M$, and $M \neq U$. It is clear from this that there is only one *maximal* left ideal of U containing J_ρ. $\qquad\square$

6. Retaining the notation of Section 5, let us put

$$h_i = \frac{2}{\alpha_i(h_{\alpha_i})}\, h_{\alpha_i}.$$

Generally, if α is any non-zero root, we put

$$H_\alpha = \frac{2}{\alpha(h_\alpha)}\, h_\alpha,$$

so that $H_{\alpha_i} = h_i$.

We shall use the Weyl group in its dual form, as a group of linear automorphisms of the rational subspace R of H° that is spanned by the roots. We know from Theorem 4.2 that it is generated by the π_{α_i}'s. It follows from the definitions and (2) of Theorem 3.1 that, for every non-zero root α and every element ρ of R, we have $\pi_\alpha(\rho) = \rho - \rho(H_\alpha)\alpha$.

Theorem 6.1. *Let V be a finite-dimensional simple L-module. Then V has a dominant weight. If δ is any weight of V and σ is an element of the Weyl group, then $\sigma(\delta)$ is also a weight of V. If α is a root such that $\delta + \alpha$ is a weight of V then $L_\alpha \cdot V^\delta \neq (0)$. For a simple L-module with dominant weight γ to be finite-dimensional, it is sufficient that each $\gamma(h_i)$ be a non-negative integer, and it is necessary that each $\gamma(h_i)$ be non-negative and that $\delta(H_\alpha)$ be an integer for every weight δ and every root α.*

PROOF. The set of weights of V is finite and non-empty. Pick a weight δ and a non-zero element v of V^δ. Since $\mathscr{U}(L) \cdot v = V$, every weight of V is the sum of δ and an integral linear combination of the α_i's. This yields an ordering of the weights compatible with that of the roots. Clearly, the largest weight of V is a dominant weight.

Let δ be a weight of V, and let α be a non-zero root. Define δ_α and δ^α as in the case where δ is a root from the condition that $\delta + k\alpha$ be a weight. Choose a non-zero element u_0 from $V^{\delta - \delta_\alpha \alpha}$. Choose x from L_α and y from $L_{-\alpha}$ so that $[x, y] = H_\alpha$. Put $u_j = x^j \cdot u_0$, and let J be the smallest non-negative integer j for which $u_{j+1} = 0$. Agreeing that $u_{-1} = 0$, we claim that $y \cdot u_j = c_j u_{j-1}$ with some c_j in F for all indices j in the interval $[0, J]$. This holds for $j = 0$, with $c_0 = 0$, because $y \cdot u_0$ belongs to $V^{\delta - (1 + \delta_\alpha)\alpha}$, which is (0) since $\delta - (1 + \delta_\alpha)\alpha$ is no longer a weight of V, by the definition of δ_α. Suppose our claim has already been established for some $j < J$. Then we have

$$y \cdot u_{j+1} = y \cdot (x \cdot u_j) = x \cdot (y \cdot u_j) + [y, x] \cdot u_j = c_j u_j - H_\alpha \cdot u_j$$

$$= (c_j - (\delta + (j - \delta_\alpha)\alpha)(H_\alpha))u_j.$$

Thus, our claim holds for $j + 1$, with

$$c_{j+1} = c_j - (\delta + (j - \delta_\alpha)\alpha)(H_\alpha).$$

This establishes our claim inductively. Since $u_{J+1} = 0$, the above, with $j = J$, gives

$$c_J = (\delta + (J - \delta_\alpha)\alpha)(H_\alpha).$$

Hence we obtain

$$0 = \sum_{j=0}^{J-1} (c_{j+1} - c_j) - c_J = - \sum_{j=0}^{J} (\delta + (j - \delta_\alpha)\alpha)(H_\alpha),$$

which yields

$$(J + 1)\delta(H_\alpha) = (J + 1)\delta_\alpha \alpha(H_\alpha) - \tfrac{1}{2}J(J + 1)\alpha(H_\alpha).$$

Since $\alpha(H_\alpha) = 2$, our result is

$$\delta(H_\alpha) = 2\delta_\alpha - J.$$

In exactly the same way, using $-\alpha$ in the place of α, we find that

$$\delta(H_{-\alpha}) = 2\delta_{-\alpha} - J',$$

i.e.,

$$\delta(H_\alpha) = J' - 2\delta^\alpha,$$

where J' is the analogue of J.

   Since $\delta + (j - \delta_\alpha)\alpha$ is a weight of V for each j in $[0, J]$, we have

$$J - \delta_\alpha \leq \delta^\alpha.$$

Similarly, since $\delta + (j - \delta_{-\alpha})(-\alpha)$, i.e., $\delta - (j - \delta^\alpha)\alpha$ is a weight of V for each j in $[0, J']$, we have $J' - \delta^\alpha \leq \delta_\alpha$. Thus, each of J and J' is at most equal to $\delta_\alpha + \delta^\alpha$. From the above, we know that $2\delta_\alpha - J = J' - 2\delta^\alpha$, i.e., $J + J' = 2(\delta_\alpha + \delta^\alpha)$. It follows that $J = \delta_\alpha + \delta^\alpha = J'$. With the above, this yields $\delta(H_\alpha) = \delta_\alpha - \delta^\alpha$. In particular, this shows that $\delta(H_\alpha)$ is an integer.

We know that $\delta + (j - \delta_\alpha)\alpha$ is a weight of V for every j in

$$[0, J] = [0, \delta_\alpha + \delta^\alpha],$$

and hence, in particular, for $j = \delta^\alpha$. Thus, $\delta + (\delta^\alpha - \delta_\alpha)\alpha$ is a weight of V, i.e., $\pi_\alpha(\delta)$ is a weight of V. Since the π_α's generate the Weyl group, it follows that $\sigma(\delta)$ is a weight of V for every element σ of the Weyl group.

If γ is the dominant weight of V then, since $\pi_\alpha(\gamma)$ is a weight of V, Theorem 5.1 shows that $\gamma \geq \pi_\alpha(\gamma)$. This says that $\gamma \geq \gamma - \gamma(H_\alpha)\alpha$. If $\alpha > 0$, this gives $\gamma(H_\alpha) \geq 0$. In particular, $\gamma(h_i) \geq 0$ for each i.

If $\delta + \alpha$ is a weight of V then $\delta^\alpha > 0$, so that J is strictly greater than δ_α. Therefore, in the notation of the definition of J,

$$0 \neq u_{1+\delta_\alpha} = x \cdot u_{\delta_\alpha} \in L_\alpha \cdot V^\delta,$$

which shows that $L_\alpha \cdot V^\delta \neq (0)$.

Now let V be any simple L-module having a dominant weight γ, and suppose that $\gamma(h_i)$ is a non-negative integer for each i. Choose x_i from L_{α_i} and y_i from $L_{-\alpha_i}$ such that $[x_i, y_i] = h_i$. Let $L_i = Fx_i + Fy_i + Fh_i$. This is a sub Lie algebra of L, and its only ideals are (0) and L_i. In particular, L_i is a semisimple Lie algebra. Let v be a non-zero element of V^γ, and let V_i denote the sub L_i-module of V that is generated by v. For every element ρ of H°, we denote the restriction of ρ to Fh_i by ρ', while we keep i fixed. Then V_i has γ' as its dominant weight. We know from the proof of Theorem 5.1 that V_i is spanned over F by the elements $y_i^k \cdot v$. For $j \neq i$, we have $[x_j, y_i] = 0$, and $x_j \cdot v = 0$ for every j. It follows that each x_j with $j \neq i$ annihilates V_i. By Theorem 5.1, we have

$$V_i = Fv + \sum_\tau V_i^\tau,$$

where τ ranges over the set of elements $\gamma' - k\alpha_i'$ of $(Fh_i)^\circ$, with $k = 1, 2, \ldots$.

We claim that V_i is a simple L_i-module. In order to prove this, consider any sub L_i-module, P say, of V_i other than V_i. We have

$$P = P \cap (Fv) + \sum_\tau P \cap V_i^\tau.$$

Since $P \neq V_i$, we must have $P \cap (Fv) = (0)$, so that $P \subset \sum_\tau V_i^\tau$. Now, if $\tau = \gamma' - k\alpha_i'$, then $V_i^\tau = Fy_i^k \cdot v \subset V^{\gamma - k\alpha_i}$. Thus, we have $P \subset \sum_{k>0} V^{\gamma - k\alpha_i}$, and we know from the above that $x_j \cdot P \subset P$ for each index j.

By the part of the theorem we have already proved, applied to the simple sub L-modules of L, we know that if α, β and $\alpha + \beta$ are roots and $\alpha \neq 0$ then $[L_\alpha, L_\beta] \neq (0)$. If, moreover, α and β are positive it follows that

$$[L_\alpha, L_\beta] = L_{\alpha+\beta},$$

because $L_{\alpha+\beta}$ is 1-dimensional. Therefore, we conclude from the above that $x_\alpha \cdot P \subset P$ for every positive root α. Since P is contained in $\sum_{\delta \neq \gamma} V^\delta$, and since this space is stable under the action of H and the $L_{-\alpha}$'s with α a positive root, it follows that the sub L-module of V that is generated by P

is contained in $\sum_{\delta \neq \gamma} V^\delta$. Since V is a simple L-module, this implies that $P = (0)$. Our conclusion is that V_i is a simple L_i-module with dominant weight γ'.

On the other hand, we can exhibit a *finite-dimensional* simple L_i-module with dominant weight γ'. This module has an F-basis $f_0, \ldots, f_{\gamma(h_i)}$, and the action of L_i is given by the following formulas:

$$x_i \cdot f_k = k f_{k-1},$$

$$y_i \cdot f_k = (\gamma(h_i) - k) f_{k+1},$$

$$h_i \cdot f_k = (\gamma(h_i) - 2k) f_k.$$

By the unicity part of Theorem 5.1, this L_i-module is isomorphic with V_i, so that V_i is finite-dimensional.

Let M be any finite-dimensional sub L_i-module of V. We have

$$L_i \cdot (L \cdot M) \subset L \cdot (L_i \cdot M) + [L_i, L] \cdot M \subset L \cdot M,$$

so that $L \cdot M$ is again a finite-dimensional L_i-module. This shows that the sum of the family of all finite-dimensional sub L_i-modules of V is a sub L-module of V, so that it coincides with V. In other words, V is locally finite as an L_i-module.

Now let δ be any weight of V. The space $\sum_k V^{\delta + k\alpha_i}$, where k ranges over the set of integers, is evidently a sub L_i-module of V. Since V is a locally finite L_i-module, there is a *finite-dimensional* non-zero L_i-module $M \subset \sum_k V^{\delta + k\alpha_i}$. We may apply our above argument involving the indices J and J', operating on M and letting x_i and y_i play the roles of the former x and y. In this way, we obtain the conclusion that $\pi_{\alpha_i}(\delta)$ is also a weight of V. It follows, as in the finite-dimensional case, that $\sigma(\delta)$ is a weight of V for every weight δ of V and every element σ of the Weyl group.

We know from Theorem 5.1 that every V^δ is finite-dimensional. Therefore, in order to prove that V is finite-dimensional, it suffices to show that the set of weights of V is finite. Let δ be a weight of V, and let μ be the largest among the weights $\sigma(\delta)$ with σ in the Weyl group. Then $\mu \geq \pi_\alpha(\mu)$ for every non-zero root α. Let R denote the space spanned by the set of all roots over the field of rational numbers. Since each $\gamma(h_i)$ is an integer, R contains γ, and hence all the weights of V. We have

$$\pi_\alpha(\mu) = \mu - 2 \frac{\mu \cdot \alpha}{\alpha \cdot \alpha} \alpha.$$

Hence, if $\alpha > 0$, we have $\mu \cdot \alpha \geq 0$. Thus, every weight of V is the transform by an element of the Weyl group of a weight μ with the property that $\mu \cdot \alpha \geq 0$ for every positive root α. Since $\mu = \gamma - \beta$, where β is a sum of positive roots, we have

$$\gamma \cdot \gamma = \mu \cdot \mu + 2\mu \cdot \beta + \beta \cdot \beta \geq \mu \cdot \mu,$$

which shows that the set of these weights μ is bounded in R. Since these weights are all of the form $\gamma - \sum_i m_i \alpha_i$, where the m_i's are non-negative integers, they make up a *discrete* subset of R. It follows that the set of μ's is finite. Since the Weyl group is finite, this implies that the set of all weights of V is finite. $\qquad\qquad\qquad\qquad\qquad\qquad\qquad\qquad\qquad\qquad\qquad\qquad$ $\square$

Notes

1. Let G be an irreducible algebraic group over an algebraically closed field of characteristic 0, and suppose that $\mathscr{L}(G)$ is semisimple. Then the Cartan subalgebras of $\mathscr{L}(G)$ are the Lie algebras of the maximal toroids in G and, via this correspondence, the Weyl group of this chapter is isomorphic with that of the end of Chapter XIII.

2. A concise account of the representation theory of semisimple Lie algebras is given by J-P. Serre in [15], part of which has been embodied above.

3. In the notation of Theorem 6.1, let V be a finite-dimensional L-module, δ a weight of V and α a non-zero root. *If k is an integer such that $\delta + k\alpha$ is a weight of V then k must belong to the interval* $[-\delta_\alpha, \delta^\alpha]$. In order to see this, suppose that there are integers $k > \delta^\alpha$ such that $\delta + k\alpha$ is a weight, and let $k_1, \ldots, k_p$ be all of them. By the definition of δ^α, we must actually have $k_i > 1 + \delta^\alpha$ for every i. Put

$$W = \sum_{i=1}^{p} V^{\delta + k_i \alpha}.$$

Clearly, W is stable under $H + L_\alpha + L_{-\alpha}$. As in the proof of Theorem 3.1, one sees that the trace of the endomorphism of W corresponding to H_α is 0. This yields

$$\sum_{i=1}^{p} (\delta(H_\alpha) + 2k_i)\dim(V^{\delta + k_i \alpha}) = 0,$$

whence we see that $\delta(H_\alpha) + 2(1 + \delta^\alpha) < 0$. Since $\delta(H_\alpha) = \delta_\alpha - \delta^\alpha$, this is a contradiction. Similarly, one sees that there can be no integer $k < -\delta_\alpha$ such that $\delta + k\alpha$ is a weight of V.

Chapter XVIII

From Lie Algebras to Groups

In this final chapter, we apply the above results on Lie algebras in order to bring the Lie theory of algebraic groups over fields of characteristic 0 to a satisfactory state of completeness.

Section 1 establishes the adjoint criterion for a Lie algebra to be that of an algebraic group. This is refined in Section 2, where it is shown that the isomorphism classes of Lie algebras satisfying the adjoint criterion are in bijective correspondence with the isomorphism classes of irreducible algebraic groups with unipotent centers.

In Section 3, we deal with the basic facts concerning group coverings and simply connected groups.

Section 4 prepares the ground for Section 5, which provides a construction of the simply connected algebraic group with a given Lie algebra. For this to exist, it is necessary and sufficient that the radical of the Lie algebra be nilpotent. An irreducible affine algebraic group over an algebraically closed field of characteristic 0 has a universal covering in the category of affine algebraic groups if and only if its radical is unipotent.

1. Let G be an affine algebraic group over an algebraically closed field F of characteristic 0, and let α denote the adjoint representation of G. Then $\alpha(G)$ is an algebraic subgroup of $\mathrm{Aut}_F(\mathscr{L}(G))$, and its Lie algebra is the image of $\mathscr{L}(G)$ in the Lie algebra $\mathrm{End}_F(\mathscr{L}(G))$ of $\mathrm{Aut}_F(\mathscr{L}(G))$ under the adjoint representation $\tau \mapsto D_\tau$ of $\mathscr{L}(G)$. Thus, the image of $\mathscr{L}(G)$ under its adjoint representation is an *algebraic* sub Lie algebra of $\mathscr{L}(\mathrm{Aut}_F(\mathscr{L}(G)))$, in the sense of Section VIII.3. We are concerned with a converse of this result, whose crude part is as follows.

Theorem 1.1. *Let F be a field of characteristic 0, and suppose that L is a finite-dimensional F-Lie algebra having the property that its adjoint image in $\mathrm{End}_F(L)$ is the Lie algebra of an irreducible algebraic subgroup H of $\mathrm{Aut}_F(L)$. Then there is an affine algebraic F-group G such that $\mathscr{L}(G)$ is isomorphic with L. In the case where F is algebraically closed, G may be so chosen that its center is unipotent.*

PROOF. Appealing to Theorem VIII.4.3, we make a semidirect product decomposition $H = H_u \rtimes P$, where P is linearly reductive. Let $\alpha\colon L \to \mathscr{L}(H)$ be the surjective Lie algebra homomorphism defined by the adjoint representation of L, and let M denote the maximum nilpotent ideal of L. Now, $\mathscr{L}(H)$ is the semidirect Lie algebra sum $\mathscr{L}(H_u) + \mathscr{L}(P)$. Since the action of $\mathscr{L}(P)$ on L is semisimple and $\mathscr{L}(H_u)$ is an ideal of $\mathscr{L}(H)$ whose action on L is nilpotent, it follows that $\alpha(M) = \mathscr{L}(H_u)$.

Let us write S for the sub Lie algebra $\alpha^{-1}(\mathscr{L}(P))$ of L, so that $L = M + S$. Evidently, S and $M \cap S$ are stable under the action of $\mathscr{L}(P)$ on L. Since this action is semisimple, it follows that there is an $\mathscr{L}(P)$-module complement, T say, for $M \cap S$ in S. The restriction of α to T is a linear isomorphism from T to $\mathscr{L}(P)$, because the kernel of α is contained in M. Moreover, we have

$$[T, T] = \alpha(T)(T) = \mathscr{L}(P)(T) \subset T,$$

showing that T is a sub Lie algebra of L. From $L = M + S$, we see that L is the semidirect Lie algebra sum $M + T$.

Now let us consider the unipotent affine algebraic F-group $\mathscr{G}(\mathscr{B}(M))$ obtained from M by Theorem XVI.4.2. Via the composite functor $\mathscr{G} \circ \mathscr{B}$, the action of P by Lie algebra automorphisms on M determines an action of P by affine algebraic group automorphisms on $\mathscr{G}(\mathscr{B}(M))$. Using this, we define a semidirect product group $\mathscr{G}(\mathscr{B}(M)) \rtimes P$. The algebra of polynomial functions of $\mathscr{G}(\mathscr{B}(M))$ is $\mathscr{B}(M)$. In proving Theorem XV.3.1, we showed that the algebra of polynomial functions of a unipotent algebraic group over a field of characteristic 0 is locally finite under the action of the group of algebraic group automorphisms. In particular, $\mathscr{B}(M)$ is locally finite as a right P-module. It is easy to see from this that $\mathscr{G}(\mathscr{B}(M)) \rtimes P$ is an affine algebraic F-group whose algebra of polynomial functions is $\mathscr{B}(M) \otimes \mathscr{P}(P)$. The Lie algebra of this semidirect product is the semidirect Lie algebra sum $M + \mathscr{L}(P)$, and we have seen above that this is isomorphic with L.

Now let Z denote the center of $\mathscr{G}(\mathscr{B}(M)) \rtimes P$. Then $\mathscr{L}(Z)$ may be identified with the center of L, so that $\mathscr{L}(Z) \subset M$. This shows that Z_1 is contained in $\mathscr{G}(\mathscr{B}(M))$. Since $\mathscr{G}(\mathscr{B}(M))$ is unipotent, so is therefore Z_1, and we have $Z_1 = Z \cap \mathscr{G}(\mathscr{B}(M))$. Now it is clear that $Z = Z_1 \times D$, where D is a finite central subgroup of P.

If F is algebraically closed, then the factor group $(\mathscr{G}(\mathscr{B}(M)) \rtimes P)/D$ is an affine algebraic F-group whose Lie algebra is isomorphic with $M + \mathscr{L}(P)$, via the differential of the canonical morphism. The center of this factor group coincides with the canonical image of Z, because the original group is irreducible and the kernel of the canonical morphism is the *finite* group D. Thus, the center of our factor group is the canonical image of Z_1, and this is unipotent. $\qquad\Box$

2. Let L be a finite-dimensional Lie algebra, and let M be the maximum nilpotent ideal of L. We say that L is *regular* if the centralizer of M in L is contained in M.

Lemma 2.1. *Let L be a finite-dimensional Lie algebra over a field of characteristic 0. There is one and only one direct Lie algebra decomposition*

$$L = L_0 + L_1,$$

where L_0 is semisimple and L_1 is regular.

PROOF. Let K denote the centralizer of M in L, and let T be the radical of K. Since K is an ideal of L, its radical T is contained in the radical of L. An application of Theorem VII.3.2 shows that $[T, T] \subset M$, whence

$$[T, [T, T]] = (0).$$

Thus, T is a nilpotent ideal of L, so that we must have $T \subset M$ and $[T, K] = (0)$. Using Theorem VII.3.1, we write K as a semidirect Lie algebra sum $T + S$, where S is a semisimple sub Lie algebra of K. Since $[T, K] = (0)$, this is actually a direct Lie algebra sum, and $S = [K, K]$.

We put $L_0 = [K, K]$, and we define L_1 as the centralizer of L_0 in L. From the fact that L_0 is a semisimple ideal of L, it follows that L is the direct Lie algebra sum $L_0 + L_1$. In fact, if x is any element of L, the derivation effected by x in L_0 is of the form D_y with y in L_0, by Proposition VII.2.6. Now $x = y + (x - y)$, and $x - y$ belongs to L_1. Evidently, the maximum nilpotent ideal of L_1 is M, and the centralizer of M in L_1 is T, which we know to lie in M. Thus, L_1 is regular.

Now consider any direct Lie algebra decomposition $L = U + V$, where U is semisimple and V is regular. Evidently, the maximum nilpotent ideal of V coincides with M, and we have $K = U + (K \cap M)$. This shows that $U = [K, K] = L_0$. Clearly, V is the centralizer of U in L, so that $V = L_1$. $\quad\Box$

In connection with the relations between isomorphisms of algebraic groups and isomorphisms of their Lie algebras, there is an unavoidable difficulty, as follows. Let H be an irreducible algebraic group, and let $\mathcal{J}(H)$ denote the ideal $\mathcal{L}(H_u) + [\mathcal{L}(H), \mathcal{L}(H)]$ of $\mathcal{L}(H)$. We assume that the base field is algebraically closed and of characteristic 0. Then

$$\mathcal{J}(H) = \mathcal{L}(H_u[H, H]).$$

Let $\mathcal{Z}(H)$ denote centralizer of $\mathcal{L}(H)$ in $\mathcal{L}(H_u)$. Now suppose that s is a linear map from $\mathcal{L}(H)$ to $\mathcal{Z}(H)$ whose kernel contains $\mathcal{J}(H)$. Let σ be the map from $\mathscr{L}(H)$ to $\mathscr{L}(H)$ given by $\sigma(x) = x + s(x)$ for every element x of $\mathscr{L}(H)$. One verifies directly that σ is a Lie algebra automorphism of $\mathscr{L}(H)$. However, *unless $s = 0$, σ is not the differential of an algebraic group automorphism of H.* Indeed, if τ is a morphism of algebraic groups from H to H such that $\tau^{\cdot} = \sigma$, then $\tau(h) = h\rho(h)$, with $\rho(h)$ in the centralizer of H in H_u. Now ρ is a morphism of algebraic groups from H to H_u, and the kernel of ρ contains H_u. Since H/H_u is linearly reductive, it follows that ρ is trivial, so that τ is the identity map.

Let us call Lie algebra automorphisms like the above σ *exceptional.*

Theorem 2.2. *Let F be an algebraically closed field of characteristic 0, and suppose that G and H are irreducible affine algebraic F-groups with unipotent centers. Suppose that ρ is a Lie algebra isomorphism from $\mathscr{L}(G)$ to $\mathscr{L}(H)$. Then there is an algebraic group isomorphism σ from G to H such that $\sigma^{\cdot} \circ \rho^{-1}$ is an exceptional automorphism of $\mathscr{L}(H)$.*

PROOF. Let the subscripts 0 and 1 on Lie algebras indicate the components of the decomposition given by Lemma 2.1. Clearly, a Lie algebra isomorphism from $\mathscr{L}(G)$ to $\mathscr{L}(H)$ restricts to Lie algebra isomorphisms $\mathscr{L}(G)_0 \to \mathscr{L}(H)_0$ and $\mathscr{L}(G)_1 \to \mathscr{L}(H)_1$. It is clear from the definition of these components that they are the Lie algebras of irreducible normal algebraic subgroups G_0 and G_1 of G, and H_0 and H_1 of H. Moreover, the Lie algebra of $G_0 \cap G_1$ is (0), so that $G_0 \cap G_1$ is a finite central subgroup of G. Since the center of G is unipotent, it follows that $G_0 \cap G_1$ is trivial, so that G is the direct product $G_0 \times G_1$. Similarly, H is the direct product $H_0 \times H_1$.

This reduces the theorem to two special cases, where $\mathscr{L}(G)$ and $\mathscr{L}(H)$ are semisimple, and where $\mathscr{L}(G)$ and $\mathscr{L}(H)$ are regular. In the first case, the adjoint representations of G and H are isomorphisms identifying G and H with the irreducible components of the identity in the groups of all Lie algebra automorphisms of $\mathscr{L}(G)$ and $\mathscr{L}(H)$. The proof of this fact consists in combining the result of the end of Section XV.3 concerning the image of the adjoint representation with the remark that the kernel of the adjoint representation is a finite central subgroup and hence trivial. Now, if ρ is a Lie algebra isomorphism from $\mathscr{L}(G)$ to $\mathscr{L}(H)$, the required isomorphism from G to H is the map sending each element x of G, viewed as an automorphism of $\mathscr{L}(G)$, onto the automorphism $\rho \circ x \circ \rho^{-1}$ of $\mathscr{L}(H)$.

We are left with the case where $\mathscr{L}(G)$ and $\mathscr{L}(H)$ are regular. In this case, we make a semidirect product decomposition $G = G_u \rtimes P$, where P is an irreducible linearly reductive algebraic subgroup of G. Then $\mathscr{L}(P)$ is the direct Lie algebra sum of its center, T say, and a semisimple Lie algebra, S say. Moreover, the adjoint representation of T on $\mathscr{L}(G)$ is semisimple. Evidently, the maximum nilpotent ideal M of $\mathscr{L}(G)$ is of the form $\mathscr{L}(G_u) + T_1$, with $T_1 \subset T$. The adjoint representation of T_1 on $\mathscr{L}(G)$ is semisimple, as well as nilpotent, so that T_1 lies in the center of $\mathscr{L}(G)$. Since the center of G is unipotent, we must therefore have $T_1 \subset \mathscr{L}(G_u)$, whence $T_1 = (0)$. Thus,

$M = \mathscr{L}(G_u)$. Since $\mathscr{L}(G)$ is regular, it follows that the centralizer of $\mathscr{L}(G_u)$ in $\mathscr{L}(P)$ is (0), which implies that the centralizer of G_u in P is finite. Since this group centralizer is a normal subgroup of the irreducible group P, it must lie in the center of P, and therefore in the center of G. Since the center of G is unipotent, it follows that the centralizer of G_u in P is trivial. Therefore, the adjoint representation of P on $\mathscr{L}(G_u)$ is an isomorphism of affine algebraic groups $\alpha: P \to \alpha(P)$, where $\alpha(P)$ is an irreducible algebraic subgroup of the group of all Lie algebra automorphisms of $\mathscr{L}(G_u)$, the Lie algebra of $\alpha(P)$ being the adjoint image of $\mathscr{L}(P)$ in the Lie algebra of all derivations of $\mathscr{L}(G_u)$.

The given Lie algebra isomorphism ρ from $\mathscr{L}(G)$ to $\mathscr{L}(H)$ must send the maximum nilpotent ideal $\mathscr{L}(G_u)$ of $\mathscr{L}(G)$ onto the maximum nilpotent ideal $\mathscr{L}(H_u)$ of $\mathscr{L}(H)$. It is clear from Theorem XVI.4.2 that there is an algebraic group isomorphism σ_u from G_u to H_u whose differential is the restriction of ρ to $\mathscr{L}(G_u)$.

From the situation in $\mathscr{L}(G)$, it is clear that $\mathscr{L}(H)$ is the semidirect Lie algebra sum $\mathscr{L}(H_u) + \rho(\mathscr{L}(P))$, that the adjoint representation of $\rho(\mathscr{L}(P))$ on $\mathscr{L}(H)$ is semisimple, and that the adjoint representation of $\rho(\mathscr{L}(P))$ on $\mathscr{L}(H_u)$ is injective. Let Q denote the smallest algebraic subgroup of H whose Lie algebra contains $\rho(\mathscr{L}(P))$. It is clear from the results of Chapter IV that the sub Q-modules of $\mathscr{L}(H)$ coincide with the sub $\rho(\mathscr{L}(P))$-modules. Therefore, the adjoint representation of Q on $\mathscr{L}(H)$ is semisimple. It follows that Q_u is contained in the center of H. Since the center of H is unipotent, it is contained in H_u, whence $Q_u \subset H_u$.

Since $\mathscr{L}(P)$ is the direct sum of its center T and the semisimple ideal S, the group Q is the product in H of the element-wise commuting subgroups $H_{\rho(T)}$ and $H_{\rho(S)}$, where we use the notation of Theorem IV.2.2. The semisimple Lie algebra $\rho(S)$ is an *algebraic* sub Lie algebra of $\mathscr{L}(H)$, whence we have $\mathscr{L}(H_{\rho(S)}) = \rho(S)$. On the other hand, $\mathscr{L}(H_{\rho(T)})$ lies in the center of $\mathscr{L}(Q)$. Consequently, $\mathscr{L}(Q)$ is the direct Lie algebra sum $\mathscr{L}(H_{\rho(T)}) + \rho(S)$. Moreover, it is now clear that $H_{\rho(T)}$ is the direct product $Q_u \times R$, where R is a toroid contained in the center of Q.

For every element x of $\rho(T)$, let us write

$$x = s(x) + (x - s(x))$$

with $s(x)$ in $\mathscr{L}(Q_u)$ and $x - s(x)$ in $\mathscr{L}(R)$. Next, observe that $\mathscr{L}(H)$, as an F-space, is the direct sum of $\mathscr{L}(H_u)$, $\rho(S)$ and $\rho(T)$, and that $[\mathscr{L}(H), \mathscr{L}(H)]$ is contained in $\mathscr{L}(H_u) + \rho(S)$. Therefore, we have

$$\rho(T) \cap (\mathscr{L}(H_u) + [\mathscr{L}(H), \mathscr{L}(H)]) = (0)$$

Since $\mathscr{L}(Q_u)$ is contained in $\mathscr{Z}(H)$, it is clear from this that s can be extended to yield a linear map from $\mathscr{L}(H)$ to $\mathscr{Z}(H)$ whose kernel contains $\mathscr{L}(H_u) + [\mathscr{L}(H), \mathscr{L}(H)]$. If η is the isomorphism from $\mathscr{L}(G)$ to $\mathscr{L}(H)$ that is obtained by composing ρ with the exceptional automorphism of $\mathscr{L}(H)$ defined by $-s$, we have

$$\eta(\mathscr{L}(P)) = \mathscr{L}(R) + \rho(S) = \mathscr{L}(RH_{\rho(S)}).$$

This reduces the theorem to the case where $\rho(\mathscr{L}(P))$ is the Lie algebra of a linearly reductive subgroup Q of H such that H is the semidirect product $H_u \rtimes Q$. We shall show that, in this case, ρ is the differential of an algebraic group isomorphism σ from G to H.

As in the case of G and P treated above, we see that the adjoint representation of Q on $\mathscr{L}(H_u)$ is actually an isomorphism of affine algebraic groups $\beta: Q \to \beta(Q)$, where $\beta(Q)$ is an irreducible algebraic subgroup of the group of all Lie algebra automorphisms of $\mathscr{L}(H_u)$, its Lie algebra being the adjoint image of $\rho(\mathscr{L}(P))$ in the Lie algebra of all derivations of $\mathscr{L}(H_u)$. Clearly, we have

$$\beta(Q) = \rho_u \circ \alpha(P) \circ \rho_u^{-1},$$

where ρ_u is the isomorphism from $\mathscr{L}(G_u)$ to $\mathscr{L}(H_u)$ obtained from ρ by restriction. Hence, we have an isomorphism of algebraic groups $\sigma_r: P \to Q$, where

$$\sigma_r(p) = \beta^{-1}(\rho_u \circ \alpha(p) \circ \rho_u^{-1})$$

for every element p of P. The required isomorphism σ from G to H is given by $\sigma(xp) = \sigma_u(x)\sigma_r(p)$ for every element x of G_u and every element p of P. $\square$

Combining Theorems 1.1 and 2.2, we have the result that *if L is a finite-dimensional Lie algebra over the algebraically closed field F of characteristic 0 whose adjoint image is algebraic, then there is one and only one isomorphism class of irreducible affine algebraic F-groups with unipotent centers whose Lie algebras are isomorphic with L.*

If G is as in Theorem 2.2 then it is easy to see that the group of exceptional automorphisms of $\mathscr{L}(G)$ is normalized by the image in $\mathrm{Aut}(\mathscr{L}(G))$ of the group $\mathscr{W}(G)$ of all algebraic group automorphisms of G. Directly, from the definition, one sees that the group of exceptional automorphisms is an *algebraic* subgroup of the group of all Lie algebra automorphisms. Theorem 2.2 yields the following corollary.

Corollary 2.3. *Let F be an algebraically closed field of characteristic 0, and let G be an irreducible affine algebraic F-group whose center is unipotent. Then the group of all Lie algebra automorphisms of $\mathscr{L}(G)$ is the semidirect product $E \rtimes W$, where E is the group of exceptional automorphisms, and W is the canonical isomorphic image of $\mathscr{W}(G)$.*

PROOF. By Theorem XV.4.3, $\mathscr{W}(G)$ is an algebraic automorphism group of G, and thus is an affine algebraic F-group in the canonical fashion of Chapter XV. By Proposition XV.2.3, the canonical map from $\mathscr{W}(G)$ to $\mathrm{Aut}(\mathscr{L}(G))$ is a morphism of affine algebraic groups. It follows from Corollary IV.3.2 that this morphism is injective. Since F is of characteristic 0, it is therefore an algebraic group isomorphism from $\mathscr{W}(G)$ to W. By Theorem 2.2, EW coincides with the group of all Lie algebra automorphisms of $\mathscr{L}(G)$. Since $E \cap W$ is trivial, it follows that the group of automorphisms of $\mathscr{L}(G)$ is the semidirect product $E \rtimes W$. $\square$

3. Suppose that G and H are irreducible affine algebraic groups over an algebraically closed field F. A surjective morphism of affine algebraic F-groups from G to H is called a *group covering* if its kernel K is finite (and hence central in G) and the induced bijective morphism $G/K \to H$ is an isomorphism of algebraic groups. If F is of characteristic 0 then the last requirement is always satisfied, so that then a group covering is simply a surjective morphism between irreducible groups with finite kernel. We say that an irreducible algebraic group H is *simply connected* if every covering to H is an isomorphism of algebraic groups.

Proposition 3.1. *Let G be a simply connected affine algebraic group over an algebraically closed field, and let N be a normal irreducible algebraic subgroup of G. Then G/N is simply connected.*

PROOF. Let π denote the canonical morphism from G to G/N, and consider a group covering $\eta: H \to G/N$. We must show that the kernel, K say, of η is trivial. Consider the fibered product $P = H \times_{(\eta, \pi)} G$, i.e., the algebraic subgroup of the direct product $H \times G$ consisting of the elements (h, g) such that $\eta(h) = \pi(g)$. The factor group of P by the canonical image of N is isomorphic with H. Since H and N are irreducible, it follows that P is irreducible. Let γ denote projection from P to G. It is easy to see that γ is a group covering whose kernel is the canonical image of K in P. Since G is simply connected, it follows that K is trivial. $\qquad\square$

Let G be any affine algebraic group over an algebraically closed field. The subgroup of G that is generated by the family of all irreducible solvable normal algebraic subgroups of G is still a member of this family. It is called the *radical* of G. Evidently, it contains G_u whenever G_u is irreducible.

Proposition 3.2. *Let G be a simply connected affine algebraic group over an algebraically closed field of characteristic 0. Then the radical of G is unipotent.*

PROOF. Write P for G/G_u, and note that P is an irreducible linearly reductive affine algebraic group. Appealing to Theorems VII.1.2 and VIII.3.3, we see that $\mathscr{L}([P, P])$ is semisimple. By Proposition 3.1, P, and hence also $P/[P, P]$, is simply connected. On the other hand, $P/[P, P]$ is an irreducible abelian linearly reductive algebraic group, and therefore a toroid. Evidently, a non-trivial toroid is not simply connected. Hence, we have $P = [P, P]$, so that $\mathscr{L}(P)$ is semisimple. This implies that P has no non-trivial irreducible solvable normal algebraic subgroup, because the Lie algebra of such a subgroup is a solvable ideal of $\mathscr{L}(P)$. Therefore, G_u coincides with the radical of G. $\qquad\square$

Theorem 3.3. *Let G be an irreducible affine algebraic group over an algebraically closed field of characteristic 0, and let R denote the radical of G. Then G is simply connected if and only if R is unipotent and G/R is simply connected.*

PROOF. The necessity of the conditions has already been established in Proposition 3.1 and 3.2. Now suppose that the conditions are satisfied, and consider a group covering $\eta: H \to G$, with kernel K. Let S denote the irreducible component of the neutral element in $\eta^{-1}(R)$. Then S is an irreducible normal algebraic subgroup of H, and η induces a group covering $\eta': H/S \to G/R$. Since G/R is simply connected, the kernel of η' is trivial, which means that $K \subset S$. Clearly, the restriction of η to S is a group covering of R. We have just seen that the kernel of this group covering coincides with K. Since R is solvable, so is S. Hence, $S = S_u \rtimes T$, where T is a toroid. Since R is unipotent, $\eta(T)$ must be trivial, so that $T \subset K$. Since K is finite, this shows that T is trivial, which means that S is unipotent. Finally, since the base field is of characteristic 0, the finite subgroup K of S must therefore be trivial. $\qquad\square$

Proposition 3.4. *Let G be as in Theorem 3.3. If G is simply connected, so is every irreducible normal algebraic subgroup of G.*

PROOF. Let R denote the radical of G, and let N be any irreducible normal algebraic subgroup of G. Clearly, $N \cap R$ is a normal algebraic subgroup of N, and $N/(N \cap R)$ may be identified with its canonical image in G/R. Since $\mathscr{L}(G/R)$ is semisimple, so is the ideal $\mathscr{L}(N/(N \cap R))$ of $\mathscr{L}(G/R)$, and $\mathscr{L}(G/R)$ is the direct Lie algebra sum of $\mathscr{L}(N/(N \cap R))$ and a complementary semisimple ideal, J say. By Theorem VIII.3.2, J is an algebraic sub Lie algebra of $\mathscr{L}(G/R)$. Thus, there is an irreducible normal algebraic subgroup Q of G/R such that $\mathscr{L}(Q) = J$. It follows that G/R is the image of the direct product $(N/(N \cap R)) \times Q$ by the evident morphism of algebraic groups. Since the intersection of J with the Lie algebra of $N/(N \cap R)$ is (0), the intersection of Q with $N/(N \cap R)$ is finite, so that the kernel of our morphism of algebraic groups is finite. By Proposition 3.1, G/R is simply connected, which implies that this finite kernel is trivial. Hence, $N/(N \cap R)$ is isomorphic with $(G/R)/Q$. Since Q is irreducible, it follows, by virtue of Proposition 3.1, that $N/(N \cap R)$ is simply connected. By Proposition 3.2, R is unipotent, so that $N \cap R$ is unipotent. Now Theorem 3.3 applies and shows that N is simply connected. $\qquad\square$

4. Let L be a finite-dimensional Lie algebra over a field F, and suppose that L is a semidirect Lie algebra sum $K + H$, where K is an ideal and H a complementary sub Lie algebra of L. Then we may write $\mathscr{U}(L)$ as the tensor product F-algebra $\mathscr{U}(H) \otimes \mathscr{U}(K)$. We describe a corresponding tensor product decomposition of the continuous dual $\mathscr{U}(L)'$. Note that L acts on $\mathscr{U}(L)'$ by F-algebra derivations, making $\mathscr{U}(L)'$ an L-module. The transform of an element f of $\mathscr{U}(L)'$ by an element x of L is given by

$$(x \cdot f)(u) = f(ux).$$

Via the canonical morphism $L \to L/K = H$, we obtain an identification of $\mathscr{U}(H)'$ with the K-annihilated sub F-algebra $(\mathscr{U}(L)')^K$ of $\mathscr{U}(L)'$.

On the other hand, we make $\mathscr{U}(K)'$ into an L-module as follows. Let f be an element of $\mathscr{U}(K)'$. If k is an element of K, we define the transform $k \cdot f$ as above by

$$(k \cdot f)(u) = f(uk).$$

If h is an element of H, we define $h \cdot f$ by

noting that, for every element u of $\mathscr{U}(K)$, the element $uh - hu$ of $\mathscr{U}(L)$ actually belongs to $\mathscr{U}(K)$, owing to the fact that K is an ideal of L. One verifies directly that these definitions, extended additively, make $\mathscr{U}(K)'$ into an L-module, the main point being that

$$h \cdot (k \cdot f) - k \cdot (h \cdot f) = [h, k] \cdot f.$$

This shows also that, if f is the restriction to $\mathscr{U}(K)$ of an element of $\mathscr{U}(L)'$, then f generates a *finite-dimensional* sub L-module of $\mathscr{U}(K)'$.

Conversely, suppose that f is an element of $\mathscr{U}(K)'$ belonging to some finite-dimensional sub L-module, T say, of $\mathscr{U}(K)'$. Using the canonically induced $\mathscr{U}(L)$-module structure of T, define the function f^+ on $\mathscr{U}(L)$ by $f^+(u) = (u \cdot f)(1)$ for every element u of $\mathscr{U}(L)$. The finite-dimensionality of T implies that f^+ belongs to $\mathscr{U}(L)'$. It follows directly from the definition that the restriction of f^+ to $\mathscr{U}(K)$ coincides with f, and that f^+ is annihilated by H under the *right* L-module structure of $\mathscr{U}(L)'$, which is given by $(f \cdot u)(v) = f(uv)$.

Our conclusion is that *the restriction image of $\mathscr{U}(L)'$ in $\mathscr{U}(K)'$ coincides with the sum of the family of all finite-dimensional sub L-modules of $\mathscr{U}(K)'$, and the map sending each f onto f^+ pre inverts the restriction map and has the right H-annihilated part $^H(\mathscr{U}(L)')$ of $\mathscr{U}(L)'$ as its image.*

Finally, we show that *the multiplication map*

$$(\mathscr{U}(L)')^K \otimes\, {}^H(\mathscr{U}(L)') \to \mathscr{U}(L)',$$

is an isomorphism of F-algebras. In order to show that this map is surjective, consider an arbitrary element f of $\mathscr{U}(L)'$, and let $(f_1, \ldots, f_n)$ be an F-basis of the space spanned by the L-transforms of f. This yields elements $g_1, \ldots, g_n$ of $\mathscr{U}(L)'$ such that

$$v \cdot f = \sum_{i=1}^{n} g_i(v) f_i,$$

for every element v of $\mathscr{U}(L)$. If u is in $\mathscr{U}(H)$ and v is in $\mathscr{U}(K)$, we obtain

$$f(uv) = \sum_{i=1}^{n} f_i(u) g_i(v).$$

Let g_i' be the element of $^H(\mathscr{U}(L)')$ obtained by first restricting g_i to $\mathscr{U}(K)$ and then applying the above map, indicated by $^+$. Similarly, let f_i' denote the element of $(\mathcal{U}(L)')^K$ corresponding to the restriction of f_i to $\mathcal{U}(H)$. If δ is the comultiplication of $\mathcal{U}(L)$, we have

$$(f_i'g_i')(uv) = (f_i' \otimes g_i')(\delta(u)\delta(v)).$$

Writing this out, and noting that g_i' vanishes on $H\mathcal{U}(L)$ and f_i' vanishes on $\mathcal{U}(L)K$, we see that it reduces to the single term $f_i'(u)g_i'(v) = f_i(u)g_i(v)$. Hence, the above expression for $f(uv)$ yields

$$f(uv) = \left(\sum_{i=1}^{n} f_i'g_i' \right)(uv),$$

whence $f = \sum_{i=1}^{n} f_i'g_i'$, showing that f belongs to the image of

$$(\mathcal{U}(L)')^K \otimes {}^H(\mathcal{U}(L)')$$

We have just seen that $(fg)(uv) = f(u)g(v)$ whenever $f \in (\mathcal{U}(L)')^K$, $g \in {}^H(\mathcal{U}(L)'), u \in \mathcal{U}(H)$ and $v \in \mathcal{U}(K)$. From this, it is easy to see that our multiplication map is also injective.

Now let us suppose that the ideal K of L is nilpotent. Let f be an element of $\mathcal{U}(K)'$ that vanishes on some power of the ideal generated by K. Then it is clear that every L-transform of f vanishes on the *same* power of the ideal generated by K. Therefore, the sub L-module of $\mathcal{U}(K)'$ generated by f is finite-dimensional, so that f belongs to the restriction image of $\mathcal{U}(L)'$. Thus, if Q denotes the sub Hopf algebra of $\mathcal{U}(L)'$ consisting of the elements whose restrictions to $\mathcal{U}(K)$ lie in the algebra $\mathcal{B}(K)$ of nilpotent representative functions, we have $Q = (\mathcal{U}(L)')^K \otimes {}^HQ$, and the restriction map induces an isomorphism from ${}^HQ$ to $\mathcal{B}(K)$.

We apply the above to the following situation. Let L be a finite-dimensional Lie algebra over a field F of characteristic 0, and assume that the radical, R say, of L is nilpotent. We have a semidirect decomposition $L = R + S$, where S is a semisimple sub Lie algebra of L. Extending the notation of Section XVI.4, we define $\mathcal{B}(L)$ as the sub Hopf algebra of $\mathcal{U}(L)'$ consisting of the elements whose restrictions to $\mathcal{U}(R)$ belong to $\mathcal{B}(R)$. From the above, we see that, as an F-algebra, $\mathcal{B}(L)$ is isomorphic with $\mathcal{U}(S)' \otimes \mathcal{B}(R)$.

We claim that $\mathcal{U}(S)'$ *is finitely generated as an F-algebra*. First, consider the case where F is algebraically closed. In this case, we fix a Cartan subalgebra of S and a fundamental system of roots. By Theorem XVII.6.1, there is a set $(\gamma_1, \ldots, \gamma_r)$ of weights such that the finite sums, with repetitions allowed, of the γ_i's are precisely the dominant weights of the finite-dimensional simple S-modules. For each i, choose a finite-dimensional simple S-module M_i having γ_i as its dominant weight. Now observe that, if $p_1, \ldots, p_r$ are non-negative integers, then the tensor product S-module

$$\otimes^{p_1}(M_1) \otimes \cdots \otimes^{p_r}(M_r)$$

has an S-simple component whose dominant weight is $\sum_{i=1}^{r} p_i\gamma_i$. Consequently, every finite-dimensional S-module is isomorphic with a sum of submodules of tensor products of the M_i's. It follows immediately from this that every representative function on $\mathcal{U}(S)$ is a sum of products of representative functions associated with one of the M_i's. In particular, $\mathcal{U}(S)'$ is finitely generated as an F-algebra.

If F is not algebraically closed, let F' be an algebraic closure of F, and consider the Lie algebra $S' = S \otimes F'$. Evidently, $\mathcal{U}(S') = \mathcal{U}(S) \otimes F'$. Let M' be a finite-dimensional S'-module. Choose an F-basis $(s_1, \ldots, s_n)$ of S, and an F'-basis $(u_1, \ldots, u_k)$ of M'. Write each $s_i \cdot u_j$ as an F'-linear combination of the u_t's, and let K be the field generated over F by the coefficients in F' that appear. Then K is of finite F-dimension, so that the K-space spanned by the u_j's is a sub S-module, M say, of M' that is of finite dimension as an F-space. We have $M' = M \otimes_K F'$, which is an S'-module homomorphic image of $M \otimes_F F'$. This shows that $\mathcal{U}(S')'$ coincides with $\mathcal{U}(S)' \otimes F'$. Moreover, it shows that every finite-dimensional S'-module is semisimple. From $S = [S, S]$, we have $S' = [S', S']$. Therefore, Theorem VII.1.2 shows that S' is semisimple, and we have from the above that $\mathcal{U}(S)' \otimes F'$ is finitely generated as an F'-algebra. Clearly, this implies that $\mathcal{U}(S)'$ is finitely generated as an F-algebra.

We know from Section XVI.4 that $\mathcal{B}(R)$ is finitely generated as an F-algebra. Therefore, we may summarize as follows.

Theorem 4.1. *Let L be a finite-dimensional Lie algebra with nilpotent radical R over a field of characteristic 0. Then the Hopf algebra $\mathcal{B}(L)$ of the R-nilpotent representative functions on $\mathcal{U}(L)$ is finitely generated as an F-algebra, and F-algebra isomorphic with $\mathcal{U}(L/R)' \otimes \mathcal{B}(R)$.*

5. Let us suppose that our base field F is algebraically closed and of characteristic 0. Then the Hopf algebra $\mathcal{B}(L)$ of Theorem 4.1 may be regarded as the algebra of polynomial functions of the affine algebraic F-group $\mathcal{G}(\mathcal{B}(L))$. We wish to show that the Lie algebra of this group may be identified with L. The evaluations at the elements of R annihilate the factor $\mathcal{U}(L/R)'$, and we know from Theorem XVI.4.2 that they make up the Lie algebra R of $\mathcal{G}(\mathcal{B}(R))$ on the factor $\mathcal{B}(R)$. On the other hand, the evaluations at the elements of $S = L/R$ annihilate the factor $\mathcal{B}(R)$. Therefore, we shall have the desired conclusion once we have established it in the case where L is semisimple.

In that case, let G be the irreducible component of the identity in the group of all Lie algebra automorphisms of L. We know that $\mathcal{L}(G)$ may be identified with L, via the adjoint representation of L. By Theorem XVI.3.1, $\mathcal{P}(G)$ may therefore be identified with a sub Hopf algebra of $\mathcal{U}(L)'$. It follows that the map associating with each element of L the evaluation of $\mathcal{U}(L)'$ at that element is an injective Lie algebra homomorphism from L to the Lie algebra of $\mathcal{G}(\mathcal{U}(L)')$.

Now write K for $\mathcal{G}(\mathcal{U}(L)')$, and let M be a finite-dimensional sub K-module of $\mathcal{U}(L)'$ that generates $\mathcal{U}(L)'$ as an F-algebra. We may identify K with its image in $\mathrm{Aut}_F(M)$. The image of L in $\mathrm{End}_F(M)$ is a sub Lie algebra of $\mathcal{L}(K)$. Since $L = [L, L]$, the image of L in $\mathrm{End}_F(M)$ is an *algebraic* sub Lie algebra of $\mathcal{L}(K)$. Let K_L be the corresponding irreducible algebraic subgroup of K, so that $\mathcal{L}(K_L)$ is the image of L in $\mathrm{End}_F(M)$. Now we drop M from our situation, merely retaining the conclusion that there is an irreducible algebraic subgroup K_L of K whose Lie algebra is the canonical image of L in $\mathcal{L}(K)$. Let Q be the annihilator of K_L in $\mathcal{U}(L)'$. Then Q is stable under the action of K_L, and hence stable also under the action of $\mathcal{U}(L)$. Therefore, if f is an element of Q and u an element of $\mathcal{U}(L)$, we have

$$(u \cdot f)(1_K) = 0,$$

i.e., $f(u) = 0$. Thus, $Q = (0)$, whence $K_L = K$ and $L = \mathcal{L}(K)$.

We shall write $\mathcal{G}(L)$ for the affine algebraic F-group $\mathcal{G}(\mathcal{B}(L))$.

Theorem 5.1. *Let F be an algebraically closed field of characteristic* 0, *and let L be a finite-dimensional F-Lie algebra with nilpotent radical R. Then $\mathcal{G}(L)$ is simply connected, its Lie algebra may be identified with L, and its algebra of polynomial functions is $\mathcal{B}(L)$. If G is an irreducible affine algebraic F-group with unipotent radical, and if σ is a surjective homomorphism from L to $\mathcal{L}(G)$, then there is a surjective morphism of affine algebraic groups $\sigma^+ : \mathcal{G}(L) \to G$ whose differential coincides with σ.*

PROOF. What remains to be proved is the last statement concerning σ, and the fact that $\mathcal{G}(L)$ is simply connected. However, if the last statement is applied to the inverse of the differential of a group covering of $\mathcal{G}(L)$, it shows that the group covering is an isomorphism. Thus, it suffices to prove the last statement of the theorem.

The surjective Lie algebra homomorphism σ defines a surjective morphism of Hopf algebras from $\mathcal{U}(L)$ to $\mathcal{U}(\mathcal{L}(G))$ in the canonical fashion. This dualizes into an injective morphism of Hopf algebras from $\mathcal{U}(\mathcal{L}(G))'$ to $\mathcal{U}(L)'$. Composing this with the morphism π of Theorem XVI.3.1, we obtain an injective morphism of Hopf algebras σ^* from $\mathcal{P}(G)$ to $\mathcal{U}(L)'$.

The restriction of σ to R is a surjective Lie algebra homomorphism from R to the radical of $\mathcal{L}(G)$. Since the radical of G is unipotent, $\mathcal{P}(G)$ is locally nilpotent as a module for the radical of $\mathcal{L}(G)$. This shows that σ^* actually sends $\mathcal{P}(G)$ into the sub Hopf algebra $\mathcal{B}(L)$ of $\mathcal{U}(L)'$. The transpose of σ^* is the required morphism σ^+. $\qquad\square$

Theorem 5.2. *Let F be an algebraically closed field of characteristic* 0, *and let $\eta : H \to G$ be a covering of irreducible affine algebraic F-groups. Suppose that T is a simply connected affine algebraic F-group, and τ is a morphism of affine algebraic F-groups from T to G. Then there is one and only one morphism of affine algebraic F-groups $\tau^\eta : T \to H$ such that $\eta \circ \tau^\eta = \tau$.*

PROOF. By Theorem 5.1, we may identify T with $\mathcal{G}(\mathcal{L}(T))$. Now $\tau(T)$ is an irreducible algebraic subgroup of G, and there is evidently an irreducible algebraic subgroup K of H such that the restriction of η to K is a group covering of $\tau(T)$. Therefore, we now assume without loss of generality that τ is surjective. Then the radical of G is the image under τ of the radical of T, whence the radical of G is unipotent. The radical of H is the irreducible component of the neutral element in the inverse image of the radical of G, with respect to η. Since unipotent affine algebraic F-groups are simply connected, it follows that the restriction of η to the radical of H is an isomorphism onto the radical of G. Therefore, the radical of H is unipotent. Now the differential of η is a Lie algebra isomorphism from $\mathscr{L}(H)$ to $\mathscr{L}(G)$, and $(\eta')^{-1} \circ \tau'$ is a surjective Lie algebra homomorphism from $\mathscr{L}(T)$ to $\mathscr{L}(H)$. Applying Theorem 5.1, we conclude that there is a morphism of affine algebraic groups $\tau'' : \mathscr{G}(\mathscr{L}(T)) \to H$ whose differential is $(\eta')^{-1} \circ \tau'$. We know that, over a field of characteristic 0, a morphism of algebraic groups is determined by its differential. Hence, $\eta \circ \tau'' = \tau$ and there is only one such τ''. $\qquad\square$

Notes

1. *If L is a finite-dimensional Lie algebra over a field F of characteristic 0, then $\mathscr{U}(L)'$ is finitely generated as an F-algebra if and only if $L = [L, L]$.*

In order to see this, first suppose that the condition is satisfied, and let R denote the radical of L. One sees readily that R is nilpotent, and that the restriction image of $\mathscr{U}(L)'$ in $\mathscr{U}(R)'$ coincides with $\mathscr{B}(R)$. Hence, $\mathscr{U}(L)'$ coincides with the finitely generated F-algebra $\mathscr{B}(L)$. If $L \neq [L, L]$, choose an ideal M of codimension 1 in L. Then $\mathscr{U}(L/M)'$ may be identified with its canonical image in $\mathscr{U}(L)'$. It is not hard to determine $\mathscr{U}(L/M)'$ explicitly. As an F-algebra, it is isomorphic with a polynomial algebra $T[x]$, where T is the group algebra over F of the additive group of F. The field of fractions of T has infinite transcendence degree over F, so that $\mathscr{U}(L)'$ cannot be finitely generated.

2. The following examples show that Theorem 2.2 cannot be improved so as to admit homomorphisms other than isomorphisms. In each example, there are no non-trivial exceptional automorphisms.

Let F be an algebraically closed field of characteristic 0. Let H be the additive group of F, and let G be the semidirect product of H with the multiplicative group of F, where

$$(a, u)(b, v) = (a + ub, uv).$$

Then $\mathscr{L}(G)$ has an F-basis (x, y), where $[x, y] = y$, $\mathscr{L}(H) = Fy$ and Fx is the Lie algebra of the multiplicative group of F. Let ρ be the Lie algebra homomorphism from $\mathscr{L}(G)$ to $\mathscr{L}(H)$ that sends x onto y and annihilates y. Then ρ is surjective. However, ρ is not the differential of an algebraic group homomorphism σ from G to H, because then σ would map the multiplicative group of F onto its additive group, which is impossible.

Let Q denote the group of all 2 by 2 matrices of determinant 1 with entries in F. The center of Q is of order 2, and we let G be the factor group of Q modulo its center. Let V denote the canonical 2-dimensional representation space for Q. Accordingly, construct the semidirect product $H = V \rtimes Q$. The group covering $Q \to G$ induces a Lie algebra isomorphism $\mathscr{L}(Q) \to \mathscr{L}(G)$. The inverse of this isomorphism yields an injective Lie algebra homomorphism $\rho: \mathscr{L}(G) \to \mathscr{L}(H)$. Both G and H have trivial center, and $\mathscr{L}(G)$ is semisimple. Here again, ρ cannot be the differential of a homomorphism of algebraic groups $\sigma: G \to H$, because then σ would yield an inverse of the group covering $Q \to G$, which is impossible.

3. Theorem 1.1 is due to M. Goto.

References

[1] A. Borel, *Linear Algebraic Groups* (lecture notes taken by H. Bass), Benjamin, New York, 1969.

[2] A. Borel and J-P. Serre, Théorèmes de Finitude en Cohomologie Galoisienne, Comment. Math. Helv. **39**, 111–163 (1964).

[3] N. Bourbaki, *Groupes et Algèbres de Lie*, Ch. I, Act. Sc. et Ind. 1285, Hermann, Paris, 1960.

[4] C. Chevalley, *Théorie des Groupes de Lie*, Tome II, Hermann, Paris, 1951.

[5] C. Chevalley, *Théorie des Groupes de Lie*, Tome III, Hermann, Paris, 1955.

[6] C. Chevalley, *Classification des Groupes de Lie Algébriques* (mimeographed lecture notes), Ecole Norm. Sup., Paris, 1956–1958.

[7] J. E. Humphreys, *Linear Algebraic Groups*, GTM 21, Springer-Verlag, New York, 1975.

[8] N. Jacobson, *Lie Algebras*, Interscience, New York, 1962.

[9] J. W. Milnor and J. C. Moore, On the Structure of Hopf Algebras, Ann. Math. **81**, 211–264 (1965).

[10] G. D. Mostow, Fully Reducible Subgroups of Algebraic Groups, Am. J. Math. **78**, 200–221 (1956).

[11] D. Mumford, *Introduction to Algebraic Geometry* (preliminary version of the first three chapters) Math. Dept., Harvard Univ., Cambridge, MA.

[12] M. Rosenlicht, Some Basic Theorems on Algebraic Groups, Am. J. Math. **78**, 401–443 (1956).

[13] M. Rosenlicht, Extensions of Vector Groups by Abelian Varieties, Am. J. Math. **80**, 685–714 (1958).

[14] J-P. Serre, *Groupes Algébriques et Corps de Classes*, Act. Sc. et Ind. 1264, Hermann, Paris, 1959.

[15] J-P. Serre, *Algèbres de Lie Semi-simples Complexes*, Benjamin, New York, 1966.

[16] R. Steinberg, *Lectures on Chevalley Groups* (notes by J. Faulkner and R. Wilson), Math. Dept., Yale Univ., New Haven, CT, 1967.

[17] M. E. Sweedler, *Hopf Algebras*, Benjamin, New York, 1969.

[18] H. Weyl, *The Classical Groups*, Princeton Univ. Press, Princeton, NJ, 1946.

Index

Graduate Texts in Mathematics

Soft and hard cover editions are available for each volume up to Vol. 14, hard cover only from Vol. 15.